Bose–Einstein Condensation in Dilute Gases

In 1925 Einstein predicted that at low temperatures particles in a gas could all reside in the same quantum state. This peculiar gaseous state, a Bose–Einstein condensate, was produced in the laboratory for the first time in 1995 using the powerful laser-cooling methods developed in recent years. These condensates exhibit quantum phenomena on a large scale, and investigating them has become one of the most active areas of research in contemporary physics.

The study of Bose–Einstein condensates in dilute gases encompasses a number of different subfields of physics, including atomic, condensed matter, and nuclear physics. The authors of this textbook explain this exciting new subject in terms of basic physical principles, without assuming detailed knowledge of any of these subfields. This pedagogical approach therefore makes the book useful for anyone with a general background in physics, from undergraduates to researchers in the field.

Chapters cover the statistical physics of trapped gases, atomic properties, the cooling and trapping of atoms, interatomic interactions, structure of trapped condensates, collective modes, rotating condensates, superfluidity, interference phenomena and trapped Fermi gases. Problem sets are also included in each chapter.

CHRISTOPHER PETHICK graduated with a D.Phil. in 1965 from the University of Oxford, and he had a research fellowship there until 1970. During the years 1966–69 he was a postdoctoral fellow at the University of Illinois at Urbana–Champaign, where he joined the faculty in 1970, becoming Professor of Physics in 1973. Following periods spent at the Landau Institute for Theoretical Physics, Moscow and at Nordita (Nordic Institute for Theoretical Physics), Copenhagen, as a visiting scientist, he accepted a permanent position at Nordita in 1975, and divided his time for many years between Nordita and the University of Illinois. Apart from the subject of the present book, Professor Pethick's main research interests are condensed matter physics (quantum liquids, especially ^{3}He, ^{4}He and superconductors) and astrophysics (particularly the properties of dense matter and the interiors of neutron stars). He is also the co-author of *Landau Fermi-Liquid Theory: Concepts and Applications* (1991).

HENRIK SMITH obtained his mag. scient. degree in 1966 from the University of Copenhagen and spent the next few years as a postdoctoral fellow at Cornell University and as a visiting scientist at the Institute for Theoretical Physics, Helsinki. In 1972 he joined the faculty of the University

of Copenhagen where he became dr. phil. in 1977 and Professor of Physics in 1978. He has also worked as a guest scientist at the Bell Laboratories, New Jersey. Professor Smith's research field is condensed matter physics and low-temperature physics including quantum liquids and the properties of superfluid ^{3}He, transport properties of normal and superconducting metals, and two-dimensional electron systems. His other books include *Transport Phenomena* (1989) and *Introduction to Quantum Mechanics* (1991).

The two authors have worked together on problems in low-temperature physics, in particular on the superfluid phases of liquid ^{3}He, superconductors and dilute quantum gases. This book derives from graduate-level lectures given by the authors at the University of Copenhagen.

Bose–Einstein Condensation in Dilute Gases

C. J. Pethick
Nordita

H. Smith
University of Copenhagen

PUBLISHED BY THE PRESS SYNDICATE OF THE UNIVERSITY OF CAMBRIDGE
The Pitt Building, Trumpington Street, Cambridge, United Kingdom

CAMBRIDGE UNIVERSITY PRESS
The Edinburgh Building, Cambridge CB2 2RU, UK
40 West 20th Street, New York, NY 10011-4211, USA
477 Williamstown Road, Port Melbourne,VIC 3207, Australia
Ruiz de Alarcón 13, 28014, Madrid, Spain
Dock House, The Waterfront, Cape Town 8001, South Africa

http://www.cambridge.org

© C. J. Pethick, H. Smith 2002

This book is in copyright. Subject to statutory exception
and to the provisions of relevant collective licensing agreements,
no reproduction of any part may take place without
the written permission of Cambridge University Press.

First published 2002

Printed in the United Kingdom at the University Press, Cambridge

Typeface Computer Modern 11/14pt.   *System* LATEX 2_ε [DBD]

A catalogue record of this book is available from the British Library

Library of Congress Cataloguing in Publication Data

Pethick, Christopher.
Bose–Einstein condensation in dilute gases / C. J. Pethick, H. Smith.
p.   cm.
Includes bibliographical references and index.
ISBN 0 521 66194 3 – ISBN 0 521 66580 9 (pb.)
1. Bose–Einstein condensation. I. Smith, H. 1939– II. Title.
QC175.47.B65 P48 2001
530.4'2–dc21   2001025622

ISBN 0 521 66194 3   hardback
ISBN 0 521 66580 9   paperback

Contents

Preface		*page* xi
1	**Introduction**	**1**
1.1	Bose–Einstein condensation in atomic clouds	4
1.2	Superfluid ^{4}He	6
1.3	Other condensates	8
1.4	Overview	10
	Problems	13
	References	14
2	**The non-interacting Bose gas**	**16**
2.1	The Bose distribution	16
	2.1.1 Density of states	18
2.2	Transition temperature and condensate fraction	21
	2.2.1 Condensate fraction	23
2.3	Density profile and velocity distribution	24
	2.3.1 The semi-classical distribution	27
2.4	Thermodynamic quantities	29
	2.4.1 Condensed phase	30
	2.4.2 Normal phase	32
	2.4.3 Specific heat close to T_c	32
2.5	Effect of finite particle number	35
2.6	Lower-dimensional systems	36
	Problems	37
	References	38
3	**Atomic properties**	**40**
3.1	Atomic structure	40
3.2	The Zeeman effect	44

3.3	Response to an electric field	49
3.4	Energy scales	55
	Problems	57
	References	57
4	**Trapping and cooling of atoms**	**58**
4.1	Magnetic traps	59
	4.1.1 The quadrupole trap	60
	4.1.2 The TOP trap	62
	4.1.3 Magnetic bottles and the Ioffe–Pritchard trap	64
4.2	Influence of laser light on an atom	67
	4.2.1 Forces on an atom in a laser field	71
	4.2.2 Optical traps	73
4.3	Laser cooling: the Doppler process	74
4.4	The magneto-optical trap	78
4.5	Sisyphus cooling	81
4.6	Evaporative cooling	90
4.7	Spin-polarized hydrogen	96
	Problems	99
	References	100
5	**Interactions between atoms**	**102**
5.1	Interatomic potentials and the van der Waals interaction	103
5.2	Basic scattering theory	107
	5.2.1 Effective interactions and the scattering length	111
5.3	Scattering length for a model potential	114
5.4	Scattering between different internal states	120
	5.4.1 Inelastic processes	125
	5.4.2 Elastic scattering and Feshbach resonances	131
5.5	Determination of scattering lengths	139
	5.5.1 Scattering lengths for alkali atoms and hydrogen	142
	Problems	144
	References	144
6	**Theory of the condensed state**	**146**
6.1	The Gross–Pitaevskii equation	146
6.2	The ground state for trapped bosons	149
	6.2.1 A variational calculation	151
	6.2.2 The Thomas–Fermi approximation	154
6.3	Surface structure of clouds	158
6.4	Healing of the condensate wave function	161

	Problems	163
	References	163
7	**Dynamics of the condensate**	**165**
7.1	General formulation	165
	7.1.1 The hydrodynamic equations	167
7.2	Elementary excitations	171
7.3	Collective modes in traps	178
	7.3.1 Traps with spherical symmetry	179
	7.3.2 Anisotropic traps	182
	7.3.3 Collective coordinates and the variational method	186
7.4	Surface modes	193
7.5	Free expansion of the condensate	195
7.6	Solitons	196
	Problems	201
	References	202
8	**Microscopic theory of the Bose gas**	**204**
8.1	Excitations in a uniform gas	205
	8.1.1 The Bogoliubov transformation	207
	8.1.2 Elementary excitations	209
8.2	Excitations in a trapped gas	214
	8.2.1 Weak coupling	216
8.3	Non-zero temperature	218
	8.3.1 The Hartree–Fock approximation	219
	8.3.2 The Popov approximation	225
	8.3.3 Excitations in non-uniform gases	226
	8.3.4 The semi-classical approximation	228
8.4	Collisional shifts of spectral lines	230
	Problems	236
	References	237
9	**Rotating condensates**	**238**
9.1	Potential flow and quantized circulation	238
9.2	Structure of a single vortex	240
	9.2.1 A vortex in a uniform medium	240
	9.2.2 A vortex in a trapped cloud	245
	9.2.3 Off-axis vortices	247
9.3	Equilibrium of rotating condensates	249
	9.3.1 Traps with an axis of symmetry	249
	9.3.2 Rotating traps	251

9.4	Vortex motion	254
	9.4.1 Force on a vortex line	255
9.5	The weakly-interacting Bose gas under rotation	257
	Problems	261
	References	262

10 Superfluidity — 264

10.1	The Landau criterion	265
10.2	The two-component picture	267
	10.2.1 Momentum carried by excitations	267
	10.2.2 Normal fluid density	268
10.3	Dynamical processes	270
10.4	First and second sound	273
10.5	Interactions between excitations	280
	10.5.1 Landau damping	281
	Problems	287
	References	288

11 Trapped clouds at non-zero temperature — 289

11.1	Equilibrium properties	290
	11.1.1 Energy scales	290
	11.1.2 Transition temperature	292
	11.1.3 Thermodynamic properties	294
11.2	Collective modes	298
	11.2.1 Hydrodynamic modes above T_c	301
11.3	Collisional relaxation above T_c	306
	11.3.1 Relaxation of temperature anisotropies	310
	11.3.2 Damping of oscillations	315
	Problems	318
	References	319

12 Mixtures and spinor condensates — 320

12.1	Mixtures	321
	12.1.1 Equilibrium properties	322
	12.1.2 Collective modes	326
12.2	Spinor condensates	328
	12.2.1 Mean-field description	330
	12.2.2 Beyond the mean-field approximation	333
	Problems	335
	References	336

13	**Interference and correlations**	**338**
13.1	Interference of two condensates	338
	13.1.1 Phase-locked sources	339
	13.1.2 Clouds with definite particle number	343
13.2	Density correlations in Bose gases	348
13.3	Coherent matter wave optics	350
13.4	The atom laser	354
13.5	The criterion for Bose–Einstein condensation	355
	13.5.1 Fragmented condensates	357
	Problems	359
	References	359
14	**Fermions**	**361**
14.1	Equilibrium properties	362
14.2	Effects of interactions	366
14.3	Superfluidity	370
	14.3.1 Transition temperature	371
	14.3.2 Induced interactions	376
	14.3.3 The condensed phase	378
14.4	Boson–fermion mixtures	385
	14.4.1 Induced interactions in mixtures	386
14.5	Collective modes of Fermi superfluids	388
	Problems	391
	References	392
Appendix. Fundamental constants and conversion factors		394
Index		397

Preface

The experimental discovery of Bose–Einstein condensation in trapped atomic clouds opened up the exploration of quantum phenomena in a qualitatively new regime. Our aim in the present work is to provide an introduction to this rapidly developing field.

The study of Bose–Einstein condensation in dilute gases draws on many different subfields of physics. Atomic physics provides the basic methods for creating and manipulating these systems, and the physical data required to characterize them. Because interactions between atoms play a key role in the behaviour of ultracold atomic clouds, concepts and methods from condensed matter physics are used extensively. Investigations of spatial and temporal correlations of particles provide links to quantum optics, where related studies have been made for photons. Trapped atomic clouds have some similarities to atomic nuclei, and insights from nuclear physics have been helpful in understanding their properties.

In presenting this diverse range of topics we have attempted to explain physical phenomena in terms of basic principles. In order to make the presentation self-contained, while keeping the length of the book within reasonable bounds, we have been forced to select some subjects and omit others. For similar reasons and because there now exist review articles with extensive bibliographies, the lists of references following each chapter are far from exhaustive. A valuable source for publications in the field is the archive at Georgia Southern University: http://amo.phy.gasou.edu/bec.html

This book originated in a set of lecture notes written for a graduate-level one-semester course on Bose–Einstein condensation at the University of Copenhagen. We have received much inspiration from contacts with our colleagues in both experiment and theory. In particular we thank Gordon Baym and George Kavoulakis for many stimulating and helpful discussions over the past few years. Wolfgang Ketterle kindly provided us with the cover illustration and Fig. 13.1. The illustrations in the text have been prepared by Janus Schmidt, whom we thank for a pleasant collaboration. It is a pleasure to acknowledge the continuing support of Simon Capelin and Susan Francis at the Cambridge University Press, and the careful copy-editing of the manuscript by Brian Watts.

Copenhagen   Christopher Pethick   Henrik Smith

1
Introduction

Bose–Einstein condensates in dilute atomic gases, which were first realized experimentally in 1995 for rubidium [1], sodium [2], and lithium [3], provide unique opportunities for exploring quantum phenomena on a macroscopic scale.[1] These systems differ from ordinary gases, liquids, and solids in a number of respects, as we shall now illustrate by giving typical values of some physical quantities.

The particle density at the centre of a Bose–Einstein condensed atomic cloud is typically 10^{13}–10^{15} cm^{-3}. By contrast, the density of molecules in air at room temperature and atmospheric pressure is about 10^{19} cm^{-3}. In liquids and solids the density of atoms is of order 10^{22} cm^{-3}, while the density of nucleons in atomic nuclei is about 10^{38} cm^{-3}.

To observe quantum phenomena in such low-density systems, the temperature must be of order 10^{-5} K or less. This may be contrasted with the temperatures at which quantum phenomena occur in solids and liquids. In solids, quantum effects become strong for electrons in metals below the Fermi temperature, which is typically 10^4–10^5 K, and for phonons below the Debye temperature, which is typically of order 10^2 K. For the helium liquids, the temperatures required for observing quantum phenomena are of order 1 K. Due to the much higher particle density in atomic nuclei, the corresponding degeneracy temperature is about 10^{11} K.

The path that led in 1995 to the first realization of Bose–Einstein condensation in dilute gases exploited the powerful methods developed over the past quarter of a century for cooling alkali metal atoms by using lasers. Since laser cooling alone cannot produce sufficiently high densities and low temperatures for condensation, it is followed by an evaporative cooling stage, in

---

[1] Numbers in square brackets are references, to be found at the end of each chapter.

which the more energetic atoms are removed from the trap, thereby cooling the remaining atoms.

Cold gas clouds have many advantages for investigations of quantum phenomena. A major one is that in the Bose–Einstein condensate, essentially all atoms occupy the same quantum state, and the condensate may be described very well in terms of a mean-field theory similar to the Hartree–Fock theory for atoms. This is in marked contrast to liquid ^{4}He, for which a mean-field approach is inapplicable due to the strong correlations induced by the interaction between the atoms. Although the gases are dilute, interactions play an important role because temperatures are so low, and they give rise to collective phenomena related to those observed in solids, quantum liquids, and nuclei. Experimentally the systems are attractive ones to work with, since they may be manipulated by the use of lasers and magnetic fields. In addition, interactions between atoms may be varied either by using different atomic species, or, for species that have a Feshbach resonance, by changing the strength of an applied magnetic or electric field. A further advantage is that, because of the low density, 'microscopic' length scales are so large that the structure of the condensate wave function may be investigated directly by optical means. Finally, real collision processes play little role, and therefore these systems are ideal for studies of interference phenomena and atom optics.

The theoretical prediction of Bose–Einstein condensation dates back more than 75 years. Following the work of Bose on the statistics of photons [4], Einstein considered a gas of non-interacting, massive bosons, and concluded that, below a certain temperature, a finite fraction of the total number of particles would occupy the lowest-energy single-particle state [5]. In 1938 Fritz London suggested the connection between the superfluidity of liquid ^{4}He and Bose–Einstein condensation [6]. Superfluid liquid ^{4}He is the prototype Bose–Einstein condensate, and it has played a unique role in the development of physical concepts. However, the interaction between helium atoms is strong, and this reduces the number of atoms in the zero-momentum state even at absolute zero. Consequently it is difficult to measure directly the occupancy of the zero-momentum state. It has been investigated experimentally by neutron scattering measurements of the structure factor at large momentum transfers [7], and the measurements are consistent with a relative occupation of the zero-momentum state of about 0.1 at saturated vapour pressure and about 0.05 near the melting curve [8].

The fact that interactions in liquid helium reduce dramatically the occupancy of the lowest single-particle state led to the search for weakly-interacting Bose gases with a higher condensate fraction. The difficulty with most substances is that at low temperatures they do not remain gaseous, but form solids, or, in the case of the helium isotopes, liquids, and the effects of interaction thus become large. In other examples atoms first combine to form molecules, which subsequently solidify. As long ago as in 1959 Hecht [9] argued that spin-polarized hydrogen would be a good candidate for a weakly-interacting Bose gas. The attractive interaction between two hydrogen atoms with their electronic spins aligned was then estimated to be so weak that there would be no bound state. Thus a gas of hydrogen atoms in a magnetic field would be stable against formation of molecules and, moreover, would not form a liquid, but remain a gas to arbitrarily low temperatures.

Hecht's paper was before its time and received little attention, but his conclusions were confirmed by Stwalley and Nosanow [10] in 1976, when improved information about interactions between spin-aligned hydrogen atoms was available. These authors also argued that because of interatomic interactions the system would be a superfluid as well as being Bose–Einstein condensed. This latter paper stimulated the quest to realize Bose–Einstein condensation in atomic hydrogen. Initial experimental attempts used a high magnetic field gradient to force hydrogen atoms against a cryogenically cooled surface. In the lowest-energy spin state of the hydrogen atom, the electron spin is aligned opposite the direction of the magnetic field (H↓), since then the magnetic moment is in the same direction as the field. Spin-polarized hydrogen was first stabilized by Silvera and Walraven [11]. Interactions of hydrogen with the surface limited the densities achieved in the early experiments, and this prompted the Massachusetts Institute of Technology (MIT) group led by Greytak and Kleppner to develop methods for trapping atoms purely magnetically. In a current-free region, it is impossible to create a local maximum in the magnitude of the magnetic field. To trap atoms by the Zeeman effect it is therefore necessary to work with a state of hydrogen in which the electronic spin is polarized parallel to the magnetic field (H↑). Among the techniques developed by this group is that of evaporative cooling of magnetically trapped gases, which has been used as the final stage in all experiments to date to produce a gaseous Bose–Einstein condensate. Since laser cooling is not feasible for hydrogen, the gas is precooled cryogenically. After more than two decades of heroic experimental work, Bose–Einstein condensation of atomic hydrogen was achieved in 1998 [12].

As a consequence of the dramatic advances made in laser cooling of alkali atoms, such atoms became attractive candidates for Bose–Einstein condensation, and they were used in the first successful experiments to produce a gaseous Bose–Einstein condensate. Other atomic species, among them noble gas atoms in excited states, are also under active investigation, and in 2001 two groups produced condensates of metastable ^{4}He atoms in the lowest spin-triplet state [13, 14].

The properties of interacting Bose fluids are treated in many texts. The reader will find an illuminating discussion in the volume by Nozières and Pines [15]. A collection of articles on Bose–Einstein condensation in various systems, prior to its discovery in atomic vapours, is given in [16], while more recent theoretical developments have been reviewed in [17]. The 1998 Varenna lectures describe progress in both experiment and theory on Bose–Einstein condensation in atomic gases, and contain in addition historical accounts of the development of the field [18]. For a tutorial review of some concepts basic to an understanding of Bose–Einstein condensation in dilute gases see Ref. [19].

1.1 Bose–Einstein condensation in atomic clouds

Bosons are particles with integer spin. The wave function for a system of identical bosons is symmetric under interchange of any two particles. Unlike fermions, which have half-odd-integer spin and antisymmetric wave functions, bosons may occupy the same single-particle state. An order-of-magnitude estimate of the transition temperature to the Bose–Einstein condensed state may be made from dimensional arguments. For a uniform gas of free particles, the relevant quantities are the particle mass m, the number density n, and the Planck constant $h = 2\pi\hbar$. The only energy that can be formed from $\hbar$, n, and m is $\hbar^2 n^{2/3}/m$. By dividing this energy by the Boltzmann constant k we obtain an estimate of the condensation temperature T_c,

$$T_c = C \frac{\hbar^2 n^{2/3}}{mk}. \tag{1.1}$$

Here C is a numerical factor which we shall show in the next chapter to be equal to approximately 3.3. When (1.1) is evaluated for the mass and density appropriate to liquid ^{4}He at saturated vapour pressure one obtains a transition temperature of approximately 3.13 K, which is close to the temperature below which superfluid phenomena are observed, the so-called lambda point[2] (T_λ= 2.17 K at saturated vapour pressure).

An equivalent way of relating the transition temperature to the particle density is to compare the thermal de Broglie wavelength λ_T with the

---

[2] The name lambda point derives from the measured shape of the specific heat as a function of temperature, which near the transition resembles the Greek letter λ.

mean interparticle spacing, which is of order $n^{-1/3}$. The thermal de Broglie wavelength is conventionally defined by

$$\lambda_T = \left(\frac{2\pi\hbar^2}{mkT}\right)^{1/2}. \tag{1.2}$$

At high temperatures, it is small and the gas behaves classically. Bose–Einstein condensation in an ideal gas sets in when the temperature is so low that λ_T is comparable to $n^{-1/3}$. For alkali atoms, the densities achieved range from 10^{13} cm^{-3} in early experiments to 10^{14}–10^{15} cm^{-3} in more recent ones, with transition temperatures in the range from 100 nK to a few μK. For hydrogen, the mass is lower and the transition temperatures are correspondingly higher.

In experiments, gases are non-uniform, since they are contained in a trap, which typically provides a harmonic-oscillator potential. If the number of particles is N, the density of gas in the cloud is of order N/R^3, where the size R of a thermal gas cloud is of order $(kT/m\omega_0^2)^{1/2}$, ω_0 being the angular frequency of single-particle motion in the harmonic-oscillator potential. Substituting the value of the density $n \sim N/R^3$ at $T = T_c$ into Eq. (1.1), one sees that the transition temperature is given by

$$kT_c = C_1 \hbar \omega_0 N^{1/3}, \tag{1.3}$$

where C_1 is a numerical constant which we shall later show to be approximately 0.94. The frequencies for traps used in experiments are typically of order 10^2 Hz, corresponding to $\omega_0 \sim 10^3$ s^{-1}, and therefore, for particle numbers in the range from 10^4 to 10^7, the transition temperatures lie in the range quoted above. Estimates of the transition temperature based on results for a uniform Bose gas are therefore consistent with those for a trapped gas.

In the original experiment [1] the starting point was a room-temperature gas of rubidium atoms, which were trapped and cooled to about 10 μK by bombarding them with photons from laser beams in six directions – front and back, left and right, up and down. Subsequently the lasers were turned off and the atoms trapped magnetically by the Zeeman interaction of the electron spin with an inhomogeneous magnetic field. If we neglect complications caused by the nuclear spin, an atom with its electron spin parallel to the magnetic field is attracted to the minimum of the magnetic field, while one with its electron spin antiparallel to the magnetic field is repelled. The trapping potential was provided by a quadrupole magnetic field, upon which a small oscillating bias field was imposed to prevent loss of particles at the centre of the trap. Some more recent experiments have employed other magnetic field configurations.

In the magnetic trap the cloud of atoms was cooled further by evaporation. The rate of evaporation was enhanced by applying a radio-frequency magnetic field which flipped the electronic spin of the most energetic atoms from up to down. Since the latter atoms are repelled by the trap, they escape, and the average energy of the remaining atoms falls. It is remarkable that no cryogenic apparatus was involved in achieving the record-low temperatures in the experiment [1]. Everything was held at room temperature except the atomic cloud, which was cooled to temperatures of the order of 100 nK.

So far, Bose–Einstein condensation has been realized experimentally in dilute gases of rubidium, sodium, lithium, hydrogen, and metastable helium atoms. Due to the difference in the properties of these atoms and their mutual interaction, the experimental study of the condensates has revealed a range of fascinating phenomena which will be discussed in later chapters. The presence of the nuclear and electronic spin degrees of freedom adds further richness to these systems when compared with liquid ^{4}He, and it gives the possibility of studying multi-component condensates. From a theoretical point of view, much of the appeal of Bose–Einstein condensed atomic clouds stems from the fact that they are dilute in the sense that the scattering length is much less than the interparticle spacing. This makes it possible to calculate the properties of the system with high precision. For a *uniform* dilute gas the relevant theoretical framework was developed in the 1950s and 60s, but the presence of a confining potential – essential to the observation of Bose–Einstein condensation in atomic clouds – gives rise to new features that are absent for uniform systems.

1.2 Superfluid ^{4}He

Many of the concepts used to describe properties of quantum gases were developed in the context of liquid ^{4}He. The helium liquids are exceptions to the rule that liquids solidify when cooled to sufficiently low temperatures, because the low mass of the helium atom makes the zero-point energy large enough to overcome the tendency to crystallization. At the lowest temperatures the helium liquids solidify only under a pressure in excess of 25 bar (2.5 MPa) for ^{4}He and 34 bar for the lighter isotope ^{3}He.

Below the lambda point, liquid ^{4}He becomes a superfluid with many remarkable properties. One of the most striking is the ability to flow through narrow channels without friction. Another is the existence of quantized vorticity, the quantum of circulation being given by h/m ($= 2\pi\hbar/m$). The occurrence of frictionless flow led Landau and Tisza to introduce a two-fluid description of the hydrodynamics. The two fluids – the normal and the superfluid components – are interpenetrating, and their densities depend on temperature. At very low temperatures the density of the normal component vanishes, while the density of the superfluid component approaches the total density of the liquid. The superfluid density is therefore generally quite different from the density of particles in the condensate, which for liquid ^{4}He is only about 10 % or less of the total, as mentioned above. Near the transition temperature to the normal state the situation is reversed: here the superfluid density tends towards zero as the temperature approaches the lambda point, while the normal density approaches the density of the liquid.

The properties of the normal component may be related to the elementary excitations of the superfluid. The concept of an elementary excitation plays a central role in the description of quantum systems. In an ideal gas an elementary excitation corresponds to the addition of a single particle in a momentum eigenstate. Interactions modify this picture, but for low excitation energies there still exist excitations with well-defined energies. For small momenta the excitations in liquid ^{4}He are sound waves or *phonons*. Their dispersion relation is linear, the energy ϵ being proportional to the magnitude of the momentum p,

$$\epsilon = sp, \tag{1.4}$$

where the constant s is the velocity of sound. For larger values of p, the dispersion relation shows a slight upward curvature for pressures less than 18 bar, and a downward one for higher pressures. At still larger momenta, $\epsilon(p)$ exhibits first a local maximum and subsequently a local minimum. Near this minimum the dispersion relation may be approximated by

$$\epsilon(p) = \Delta + \frac{(p - p_0)^2}{2m^*}, \tag{1.5}$$

where m^* is a constant with the dimension of mass and p_0 is the momentum at the minimum. Excitations with momenta close to p_0 are referred to as *rotons*. The name was coined to suggest the existence of vorticity associated with these excitations, but they should really be considered as short-wavelength phonon-like excitations. Experimentally, one finds at zero pressure that m^* is 0.16 times the mass of a ^{4}He atom, while the constant Δ, the energy gap, is given by $\Delta/k = 8.7$ K. The roton minimum occurs at a wave number $p_0/\hbar$ equal to 1.9×10^8 cm^{-1} (see Fig. 1.1). For excitation

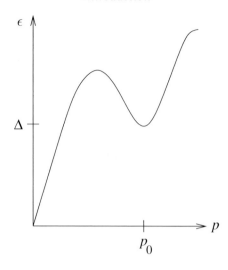

Fig. 1.1. The spectrum of elementary excitations in superfluid ^{4}He. The minimum roton energy is Δ, corresponding to the momentum p_0.

energies greater than 2Δ the excitations become less well-defined since they can decay into two rotons.

The elementary excitations obey Bose statistics, and therefore in thermal equilibrium the distribution function f^0 for the excitations is given by

$$f^0 = \frac{1}{e^{\epsilon(p)/kT} - 1}. \tag{1.6}$$

The absence of a chemical potential in this distribution function is due to the fact that the number of excitations is not a conserved quantity: the energy of an excitation equals the difference between the energy of an excited state and the energy of the ground state for a system containing the same number of particles. The number of excitations therefore depends on the temperature, just as the number of phonons in a solid does. This distribution function Eq. (1.6) may be used to evaluate thermodynamic properties.

1.3 Other condensates

The concept of Bose–Einstein condensation finds applications in many systems other than liquid ^{4}He and the clouds of spin-polarized boson alkali atoms, atomic hydrogen, and metastable helium atoms discussed above. Historically, the first of these were superconducting metals, where the bosons are pairs of electrons with opposite spin. Many aspects of the behaviour of superconductors may be understood qualitatively on the basis of the idea that pairs of electrons form a Bose–Einstein condensate, but the properties of superconductors are quantitatively very different from those of a weakly-interacting gas of pairs. The important physical point is that the binding energy of a pair is small compared with typical atomic energies, and at the temperature where the condensate disappears the pairs themselves break up. This situation is to be contrasted with that for the atomic systems, where the energy required to break up an atom is the ionization energy, which is of order electron volts. This corresponds to temperatures of tens of thousands of degrees, which are much higher than the temperatures for Bose–Einstein condensation.

Many properties of high-temperature superconductors may be understood in terms of Bose–Einstein condensation of pairs, in this case of holes rather than electrons, in states having predominantly d-like symmetry in contrast to the s-like symmetry of pairs in conventional metallic superconductors. The rich variety of magnetic and other behaviour of the superfluid phases of liquid ^{3}He is again due to condensation of pairs of fermions, in this case ^{3}He atoms in triplet spin states with p-wave symmetry. Considerable experimental effort has been directed towards creating Bose–Einstein condensates of excitons, which are bound states of an electron and a hole [20], and of biexcitons, molecules made up of two excitons [21].

Bose–Einstein condensation of pairs of fermions is also observed experimentally in atomic nuclei, where the effects of neutron–neutron, proton–proton, and neutron–proton pairing may be seen in the excitation spectrum as well as in reduced moments of inertia. A significant difference between nuclei and superconductors is that the size of a pair in bulk nuclear matter is large compared with the nuclear size, and consequently the manifestations of Bose–Einstein condensation in nuclei are less dramatic than they are in bulk systems. Theoretically, Bose–Einstein condensation of nucleon pairs is expected to play an important role in the interiors of neutron stars, and observations of glitches in the spin-down rate of pulsars have been interpreted in terms of neutron superfluidity. The possibility of mesons, either pions or kaons, forming a Bose–Einstein condensate in the cores of neutron stars has been widely discussed, since this would have far-reaching consequences for theories of supernovae and the evolution of neutron stars [22].

In the field of nuclear and particle physics the ideas of Bose–Einstein condensation also find application in the understanding of the vacuum as a condensate of quark–antiquark ($u\bar{u}$, $d\bar{d}$ and $s\bar{s}$) pairs, the so-called chiral condensate. This condensate gives rise to particle masses in much the same way as the condensate of electron pairs in a superconductor gives rise to the gap in the electronic excitation spectrum.

This brief account of the rich variety of contexts in which the physics of Bose–Einstein condensation plays a role, shows that an understanding of the phenomenon is of importance in many branches of physics.

1.4 Overview

To assist the reader, we give here a 'road map' of the material we cover. We begin, in Chapter 2, by discussing Bose–Einstein condensation for non-interacting gases in a confining potential. This is useful for developing understanding of the phenomenon of Bose–Einstein condensation and for application to experiment, since in dilute gases many quantities, such as the transition temperature and the condensate fraction, are close to those predicted for a non-interacting gas. We also discuss the density profile and the velocity distribution of an atomic cloud at zero temperature. When the thermal energy kT exceeds the spacing between the energy levels of an atom in the confining potential, the gas may be described semi-classically in terms of a particle distribution function that depends on both position and momentum. We employ the semi-classical approach to calculate thermodynamic quantities. The effect of finite particle number on the transition temperature is estimated, and Bose–Einstein condensation in lower-dimensional systems is discussed.

In experiments to create a Bose–Einstein condensate in a dilute gas the particles used have been primarily alkali atoms and hydrogen, whose spins are non-zero. The new methods to trap and cool atoms that have been developed in recent years make use of the basic atomic structure of these atoms, which is the subject of Chapter 3. There we also study the energy levels of an atom in a static magnetic field, which is a key element in the physics of trapping, and discuss the atomic polarizability in an oscillating electric field.

A major experimental breakthrough that opened up this field was the development of laser cooling techniques. In contrast to so many other proposals which in practice work less well than predicted theoretically, these turned out to be far more effective than originally estimated. Chapter 4 describes magnetic traps, the use of lasers in trapping and cooling, and evaporative cooling, which is the key final stage in experiments to make Bose–Einstein condensates.

In Chapter 5 we consider atomic interactions, which play a crucial role in evaporative cooling and also determine many properties of the condensed state. At low energies, interactions between particles are characterized by the scattering length a, in terms of which the total scattering cross section at low energies is given by $8\pi a^2$ for identical bosons. At first sight, one might expect that, since atomic sizes are typically of order the Bohr radius, scattering lengths would also be of this order. In fact they are one or two orders of magnitude larger for alkali atoms, and we shall show how this may be understood in terms of the long-range part of the interatomic force, which is due to the van der Waals interaction. We also show that the sign of the effective interaction at low energies depends on the details of the short-range part of the interaction. Following that we extend the theory to take into account transitions between channels corresponding to the different hyperfine states for the two atoms. We then estimate rates of inelastic processes, which are a mechanism for loss of atoms from traps, and present the theory of Feshbach resonances, which may be used to tune atomic interactions by varying the magnetic field. Finally we list values of the scattering lengths for the alkali atoms currently under investigation.

The ground-state energy of clouds in a confining potential is the subject of Chapter 6. While the scattering lengths for alkali atoms are large compared with atomic dimensions, they are small compared with atomic separations in gas clouds. As a consequence, the effects of atomic interactions in the ground state may be calculated very reliably by using a pseudopotential proportional to the scattering length. This provides the basis for a mean-field description of the condensate, which leads to the Gross–Pitaevskii equation. From this we calculate the energy using both variational methods and the Thomas–Fermi approximation. When the atom–atom interaction is attractive, the system becomes unstable if the number of particles exceeds a critical value, which we calculate in terms of the trap parameters and the scattering length. We also consider the structure of the condensate at the surface of a cloud, and the characteristic length for healing of the condensate wave function.

In Chapter 7 we discuss the dynamics of the condensate at zero temperature, treating the wave function of the condensate as a classical field. We derive the coupled equations of motion for the condensate density and velocity, and use them to determine the elementary excitations in a uniform gas and in a trapped cloud. We describe methods for calculating collective properties of clouds in traps. These include the Thomas–Fermi approximation and a variational approach based on the idea of collective coordinates. The methods are applied to treat oscillations in both spherically-symmetric and anisotropic traps, and the free expansion of the condensate. We show that, as a result of the combined influence of non-linearity and dispersion, there exist soliton solutions to the equations of motion for a Bose–Einstein condensate.

The microscopic, quantum-mechanical theory of the Bose gas is treated in Chapter 8. We discuss the Bogoliubov approximation and show that it gives the same excitation spectrum as that obtained from classical equations of motion in Chapter 7. At higher temperatures thermal excitations deplete the condensate, and to treat these situations we discuss the Hartree–Fock and Popov approximations. Finally we analyse collisional shifts of spectral lines, such as the 1S–2S two-photon absorption line in spin-polarized hydrogen, which is used experimentally to probe the density of the gas, and lines used as atomic clocks.

One of the characteristic features of a superfluid is its response to rotation, in particular the occurrence of quantized vortices. We discuss in Chapter 9 properties of vortices in atomic clouds and determine the critical angular velocity for a vortex state to be energetically favourable. We also calculate the force on a moving vortex line from general hydrodynamic considerations. The nature of the lowest-energy state for a given angular momentum is considered, and we discuss the weak-coupling limit, in which the interaction energy is small compared with the energy quantum of the harmonic-oscillator potential.

In Chapter 10 we treat some basic aspects of superfluidity. The Landau criterion for the onset of dissipation is discussed, and we introduce the two-fluid picture, in which the condensate and the excitations may be regarded as forming two interpenetrating fluids, each with temperature-dependent densities. We calculate the damping of collective modes in a homogeneous gas at low temperatures, where the dominant process is Landau damping. As an application of the two-fluid picture we derive the dispersion relation for the coupled sound-like modes, which are referred to as first and second sound.

Chapter 11 deals with particles in traps at non-zero temperature. The effects of interactions on the transition temperature and thermodynamic properties are considered. We also discuss the coupled motion of the condensate and the excitations at temperatures below T_c. We then present calculations for modes above T_c, both in the hydrodynamic regime, when collisions are frequent, and in the collisionless regime, where we obtain the mode attenuation from the kinetic equation for the particle distribution function.

Chapter 12 discusses properties of mixtures of bosons, either different bosonic isotopes, or different internal states of the same isotope. In the former case, the theory may be developed along lines similar to those for a single-component system. For mixtures of two different internal states of the same isotope, which may be described by a spinor wave function, new possibilities arise because the number of atoms in each state is not conserved. We derive results for the static and dynamic properties of such mixtures. An interesting result is that for an antiferromagnetic interaction between atomic spins, the simple Gross–Pitaevskii treatment fails, and the ground state may be regarded as a Bose–Einstein condensate of *pairs* of atoms, rather than of single atoms.

In Chapter 13 we take up a number of topics related to interference and correlations in Bose–Einstein condensates and applications to matter wave optics. First we describe interference between two Bose–Einstein condensed clouds, and explore the reasons for the appearance of an interference pattern even though the phase difference between the wave functions of particles in the two clouds is not fixed initially. We then demonstrate the suppression of density fluctuations in a Bose–Einstein condensed gas. Following that we consider how properties of coherent matter waves may be investigated by manipulating condensates with lasers. The final section considers the question of how to characterize Bose–Einstein condensation microscopically.

Trapped Fermi gases are considered in Chapter 14. We first show that interactions generally have less effect on static and dynamic properties of fermions than they do for bosons, and we then calculate equilibrium properties of a free Fermi gas in a trap. The interaction can be important if it is attractive, since at sufficiently low temperatures the fermions are then expected to undergo a transition to a superfluid state similar to that for electrons in a metallic superconductor. We derive expressions for the transition temperature and the gap in the excitation spectrum at zero temperature, and we demonstrate that they are suppressed due to the modification of the interaction between two atoms by the presence of other atoms. We also consider how the interaction between fermions is altered by the addition of bosons and show that this can enhance the transition temperature. Finally we briefly describe properties of sound modes in a superfluid Fermi gas, since measurement of collective modes has been proposed as a probe of the transition to a superfluid state.

Problems

PROBLEM 1.1 Consider an ideal gas of ^{87}Rb atoms at zero temperature, confined by the harmonic-oscillator potential

$$V(r) = \frac{1}{2}m\omega_0^2 r^2,$$

where m is the mass of a ^{87}Rb atom. Take the oscillator frequency ω_0 to be given by $\omega_0/2\pi = 150$ Hz, which is a typical value for traps in current use. Determine the ground-state density profile and estimate its width. Find the root-mean-square momentum and velocity of a particle. What is the density at the centre of the trap if there are 10^4 atoms?

PROBLEM 1.2 Determine the density profile for the gas discussed in Problem 1.1 in the classical limit, when the temperature T is much higher than the condensation temperature. Show that the central density may be written as $N/R_{\rm th}^3$ and determine $R_{\rm th}$. At what temperature does the mean distance between particles at the centre of the trap become equal to the thermal de Broglie wavelength λ_T? Compare the result with the transition temperature (1.3).

PROBLEM 1.3 Estimate the number of rotons contained in 1 cm^3 of liquid ^{4}He at temperatures $T = 1$ K and $T = 100$ mK at saturated vapour pressure.

References

[1] M. H. Anderson, J. R. Ensher, M. R. Matthews, C. E. Wieman, and E. A. Cornell, *Science* **269**, 198 (1995).

[2] K. B. Davis, M.-O. Mewes, M. R. Andrews, N. J. van Druten, D. S. Durfee, D. M. Kurn, and W. Ketterle, *Phys. Rev. Lett.* **75**, 3969 (1995).

[3] C. C. Bradley, C. A. Sackett, J. J. Tollett, and R. G. Hulet, *Phys. Rev. Lett.* **75**, 1687 (1995); C. C. Bradley, C. A. Sackett, and R. G. Hulet, *Phys. Rev. Lett.* **78**, 985 (1997).

[4] S. N. Bose, *Z. Phys.* **26**, 178 (1924). Bose's paper dealt with the statistics of photons, for which the total number is not a fixed quantity. He sent his paper to Einstein asking for comments. Recognizing its importance, Einstein translated the paper and submitted it for publication. Subsequently, Einstein extended Bose's treatment to massive particles, whose total number is fixed.

[5] A. Einstein, *Sitzungsberichte der Preussischen Akademie der Wissenschaften, Physikalisch-mathematische Klasse* (1924) p. 261; (1925) p. 3.

[6] F. London, *Nature* **141**, 643 (1938); *Phys. Rev.* **54**, 947 (1938).

[7] E. C. Svensson and V. F. Sears, in *Progress in Low Temperature Physics*, Vol. XI, ed. D. F. Brewer, (North-Holland, Amsterdam, 1987), p. 189.

[8] P. E. Sokol, in Ref. [16], p. 51.

[9] C. E. Hecht, *Physica* **25**, 1159 (1959).

[10] W. C. Stwalley and L. H. Nosanow, *Phys. Rev. Lett.* **36**, 910 (1976).

[11] I. F. Silvera and J. T. M. Walraven, *Phys. Rev. Lett.* **44**, 164 (1980).

[12] D. G. Fried, T. C. Killian, L. Willmann, D. Landhuis, S. C. Moss, D. Kleppner, and T. J. Greytak, *Phys. Rev. Lett.* **81**, 3811 (1998).

[13] A. Robert, O. Sirjean, A. Browaeys, J. Poupard, S. Nowak, D. Boiron, C. Westbrook, and A. Aspect, *Science* **292**, 461 (2001).

[14] F. Pereira Dos Santos, J. Léonard, J. Wang, C. J. Barrelet, F. Perales, E. Rasel, C. S. Unnikrishnan, M. Leduc, and C. Cohen-Tannoudji, *Phys. Rev. Lett.* **86**, 3459 (2001).

[15] P. Nozières and D. Pines, *The Theory of Quantum Liquids, Vol. II*, (Addison-Wesley, Reading, Mass., 1990).

[16] *Bose–Einstein Condensation*, ed. A. Griffin, D. W. Snoke, and S. Stringari, (Cambridge Univ. Press, Cambridge, 1995).

[17] F. Dalfovo, S. Giorgini, L. P. Pitaevskii, and S. Stringari, *Rev. Mod. Phys.* **71**, 463 (1999).

[18] *Bose–Einstein Condensation in Atomic Gases*, Proceedings of the Enrico Fermi International School of Physics, Vol. CXL, ed. M. Inguscio, S. Stringari, and C. E. Wieman, (IOS Press, Amsterdam, 1999).

[19] A. J. Leggett, *Rev. Mod. Phys.* **73**, 307 (2001).

[20] K. E. O'Hara, L. Ó Súilleabháin, and J. P. Wolfe, *Phys. Rev.* B **60**, 10 565 (1999).

[21] A. Mysyrowicz, in Ref. [16], p. 330.

[22] G. E. Brown, in Ref. [16], p. 438.

2
The non-interacting Bose gas

The topic of Bose–Einstein condensation in a uniform, non-interacting gas of bosons is treated in most textbooks on statistical mechanics [1]. In the present chapter we discuss the properties of a non-interacting Bose gas in a trap. We shall calculate equilibrium properties of systems in a semi-classical approximation, in which the energy spectrum is treated as a continuum. For this approach to be valid the temperature must be large compared with $\Delta\epsilon/k$, where $\Delta\epsilon$ denotes the separation between neighbouring energy levels. As is well known, at temperatures below the Bose–Einstein condensation temperature, the lowest energy state is not properly accounted for if one simply replaces sums by integrals, and it must be included explicitly.

The statistical distribution function is discussed in Sec. 2.1, as is the single-particle density of states, which is a key ingredient in the calculations of thermodynamic properties. Calculations of the transition temperature and the fraction of particles in the condensate are described in Sec. 2.2. In Sec. 2.3 the semi-classical distribution function is introduced, and from this we determine the density profile and the velocity distribution of particles. Thermodynamic properties of Bose gases are calculated as functions of the temperature in Sec. 2.4. The final two sections are devoted to effects not captured by the simplest version of the semi-classical approximation: corrections to the transition temperature due to a finite particle number (Sec. 2.5), and thermodynamic properties of gases in lower dimensions (Sec. 2.6).

2.1 The Bose distribution

For non-interacting bosons in thermodynamic equilibrium, the mean occupation number of the single-particle state ν is given by the Bose distribution function,

$$f^0(\epsilon_\nu) = \frac{1}{e^{(\epsilon_\nu - \mu)/kT} - 1}, \tag{2.1}$$

where ϵ_ν denotes the energy of the single-particle state for the particular trapping potential under consideration. Since the number of particles is conserved, unlike the number of elementary excitations in liquid ^{4}He, the chemical potential μ enters the distribution function (2.1). The chemical potential is determined as a function of N and T by the condition that the total number of particles be equal to the sum of the occupancies of the individual levels. It is sometimes convenient to work in terms of the quantity $\zeta = \exp(\mu/kT)$, which is known as the *fugacity*. If we take the zero of energy to be that of the lowest single-particle state, the fugacity is less than unity above the transition temperature and equal to unity (to within terms of order $1/N$, which we shall generally neglect) in the condensed state. In Fig. 2.1 the distribution function (2.1) is shown as a function of energy for various values of the fugacity.

At high temperatures, the effects of quantum statistics become negligible, and the distribution function (2.1) is given approximately by the Boltzmann distribution

$$f^0(\epsilon_\nu) \simeq e^{-(\epsilon_\nu - \mu)/kT}. \tag{2.2}$$

For particles in a box of volume V the index ν labels the allowed wave vectors $\mathbf{q}$ for plane-wave states $V^{-1/2} \exp(i\mathbf{q} \cdot \mathbf{r})$, and the particle energy is $\epsilon = \hbar^2 q^2/2m$. The distribution (2.2) is thus a Maxwellian one for the velocity $v = \hbar q/m$.

At high temperatures the chemical potential is much less than $\epsilon_{\min}$, the energy of the lowest single-particle state, since the mean occupation number of any state is much less than unity, and therefore, in particular, $\exp[(\mu - \epsilon_{\min})/kT] \ll 1$. As the temperature is lowered, the chemical potential rises and the mean occupation numbers increase. However, the chemical potential cannot exceed $\epsilon_{\min}$, otherwise the Bose distribution function (2.2) evaluated for the lowest single-particle state would be negative, and hence unphysical. Consequently the mean occupation number of any excited single-particle state cannot exceed the value $1/\{\exp[(\epsilon_\nu - \epsilon_{\min})/kT] - 1\}$. If the total number of particles in excited states is less than N, the remaining particles must be accommodated in the single-particle ground state, whose occupation number can be arbitrarily large: the system has a Bose–Einstein condensate. The highest temperature at which the condensate exists is referred to as the Bose–Einstein transition temperature and we shall denote it by T_c. As we

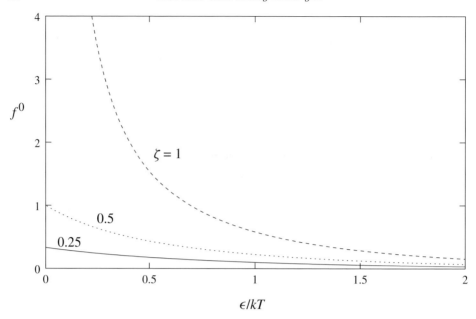

Fig. 2.1. The Bose distribution function f^0 as a function of energy for different values of the fugacity ζ. The value $\zeta = 1$ corresponds to temperatures below the transition temperature, while $\zeta = 0.5$ and $\zeta = 0.25$ correspond to $\mu = -0.69kT$ and $\mu = -1.39kT$, respectively.

shall see in more detail in Sec. 2.2, the energy dependence of the single-particle density of states at low energies determines whether or not Bose–Einstein condensation will occur for a particular system. In the condensed state, at temperatures below T_c, the chemical potential remains equal to $\epsilon_{\min}$, to within terms of order kT/N, which is small for large N, and the occupancy of the single-particle ground state is macroscopic in the sense that a finite fraction of the particles are in this state. The number of particles N_0 in the single-particle ground state equals the total number of particles N minus the number of particles N_{ex} occupying higher-energy (excited) states.

2.1.1 Density of states

When calculating thermodynamic properties of gases it is common to replace sums over states by integrals, and to use a density of states in which details of the level structure are smoothed out. This procedure fails for a Bose–Einstein condensed system, since the contribution from the lowest state is not properly accounted for. However, it does give a good approximation to the contribution from excited states, and we shall now calculate these smoothed densities of states for a number of different situations.

Throughout most of this book we shall assume that all particles are in one particular internal (spin) state, and therefore we generally suppress the part of the wave function referring to the internal state. In Chapters 12–14 we discuss a number of topics where internal degrees of freedom come into play.

In three dimensions, for a free particle in a particular internal state, there is on average one quantum state per volume $(2\pi\hbar)^3$ of phase space. The region of momentum space for which the magnitude of the momentum is less than p has a volume $4\pi p^3/3$ equal to that of a sphere of radius p and, since the energy of a particle of momentum $\mathbf{p}$ is given by $\epsilon_\mathbf{p} = p^2/2m$, the total number of states $G(\epsilon)$ with energy less than ϵ is given by

$$G(\epsilon) = V\frac{4\pi}{3}\frac{(2m\epsilon)^{3/2}}{(2\pi\hbar)^3} = V\frac{2^{1/2}}{3\pi^2}\frac{(m\epsilon)^{3/2}}{\hbar^3}, \tag{2.3}$$

where V is the volume of the system. Quite generally, the number of states with energy between ϵ and $\epsilon + d\epsilon$ is given by $g(\epsilon)d\epsilon$, where $g(\epsilon)$ is the density of states. Therefore

$$g(\epsilon) = \frac{dG(\epsilon)}{d\epsilon}, \tag{2.4}$$

which, from Eq. (2.3), is thus given by

$$g(\epsilon) = \frac{Vm^{3/2}}{2^{1/2}\pi^2\hbar^3}\epsilon^{1/2}. \tag{2.5}$$

For free particles in d dimensions the corresponding result is $g(\epsilon) \propto \epsilon^{(d/2-1)}$, and therefore the density of states is independent of energy for a free particle in two dimensions.

Let us now consider a particle in the anisotropic harmonic-oscillator potential

$$V(\mathbf{r}) = \frac{1}{2}(K_1 x^2 + K_2 y^2 + K_3 z^2), \tag{2.6}$$

which we will refer to as a harmonic trap. Here the quantities K_i denote the three force constants, which are generally unequal. The corresponding classical oscillation frequencies ω_i are given by $\omega_i^2 = K_i/m$, and we shall therefore write the potential as

$$V(\mathbf{r}) = \frac{1}{2}m(\omega_1^2 x^2 + \omega_2^2 y^2 + \omega_3^2 z^2). \tag{2.7}$$

The energy levels, $\epsilon(n_1, n_2, n_3)$, are then

$$\epsilon(n_1, n_2, n_3) = (n_1 + \frac{1}{2})\hbar\omega_1 + (n_2 + \frac{1}{2})\hbar\omega_2 + (n_3 + \frac{1}{2})\hbar\omega_3, \quad (2.8)$$

where the numbers n_i assume all integer values greater than or equal to zero.

We now determine the number of states $G(\epsilon)$ with energy less than a given value ϵ. For energies large compared with $\hbar\omega_i$, we may treat the n_i as continuous variables and neglect the zero-point motion. We therefore introduce a coordinate system defined by the three variables $\epsilon_i = \hbar\omega_i n_i$, in terms of which a surface of constant energy (2.8) is the plane $\epsilon = \epsilon_1 + \epsilon_2 + \epsilon_3$. Then $G(\epsilon)$ is proportional to the volume in the first octant bounded by the plane,

$$G(\epsilon) = \frac{1}{\hbar^3 \omega_1 \omega_2 \omega_3} \int_0^{\epsilon} d\epsilon_1 \int_0^{\epsilon - \epsilon_1} d\epsilon_2 \int_0^{\epsilon - \epsilon_1 - \epsilon_2} d\epsilon_3 = \frac{\epsilon^3}{6\hbar^3 \omega_1 \omega_2 \omega_3}. \quad (2.9)$$

Since $g(\epsilon) = dG/d\epsilon$, we obtain a density of states given by

$$g(\epsilon) = \frac{\epsilon^2}{2\hbar^3 \omega_1 \omega_2 \omega_3}. \quad (2.10)$$

For a d-dimensional harmonic-oscillator potential, the analogous result is

$$g(\epsilon) = \frac{\epsilon^{d-1}}{(d-1)! \prod_{i=1}^{d} \hbar\omega_i}. \quad (2.11)$$

We thus see that in many contexts the density of states varies as a power of the energy, and we shall now calculate thermodynamic properties for systems with a density of states of the form

$$g(\epsilon) = C_\alpha \epsilon^{\alpha - 1}, \quad (2.12)$$

where C_α is a constant. In three dimensions, for a gas confined by rigid walls, α is equal to $3/2$. The corresponding coefficient may be read off from Eq. (2.5), and it is

$$C_{3/2} = \frac{V m^{3/2}}{2^{1/2} \pi^2 \hbar^3}. \quad (2.13)$$

The coefficient for a three-dimensional harmonic-oscillator potential ($\alpha = 3$), which may be obtained from Eq. (2.10), is

$$C_3 = \frac{1}{2\hbar^3 \omega_1 \omega_2 \omega_3}. \quad (2.14)$$

For particles in a box or in a harmonic-oscillator potential, α is equal to half the number of classical degrees of freedom per particle.

2.2 Transition temperature and condensate fraction

The transition temperature T_c is defined as the highest temperature at which the macroscopic occupation of the lowest-energy state appears. When the number of particles, N, is sufficiently large, we may neglect the zero-point energy in (2.8) and thus equate the lowest energy $\epsilon_{\min}$ to zero, the minimum of the potential (2.6). Corrections to the transition temperature arising from the zero-point energy will be considered in Sec. 2.5. The number of particles in excited states is given by

$$N_{\text{ex}} = \int_0^\infty d\epsilon g(\epsilon) f^0(\epsilon). \tag{2.15}$$

This achieves its greatest value for $\mu = 0$, and the transition temperature T_c is determined by the condition that the total number of particles can be accommodated in excited states, that is

$$N = N_{\text{ex}}(T_c, \mu = 0) = \int_0^\infty d\epsilon g(\epsilon) \frac{1}{e^{\epsilon/kT_c} - 1}. \tag{2.16}$$

When (2.16) is written in terms of the dimensionless variable $x = \epsilon/kT_c$, it becomes

$$N = C_\alpha (kT_c)^\alpha \int_0^\infty dx \frac{x^{\alpha-1}}{e^x - 1} = C_\alpha \Gamma(\alpha) \zeta(\alpha) (kT_c)^\alpha, \tag{2.17}$$

where $\Gamma(\alpha)$ is the gamma function and $\zeta(\alpha) = \sum_{n=1}^\infty n^{-\alpha}$ is the Riemann zeta function. In evaluating the integral in (2.17) we expand the Bose function in powers of e^{-x}, and use the fact that $\int_0^\infty dx x^{\alpha-1} e^{-x} = \Gamma(\alpha)$. The result is

$$\int_0^\infty dx \frac{x^{\alpha-1}}{e^x - 1} = \Gamma(\alpha) \zeta(\alpha). \tag{2.18}$$

Table 2.1 lists $\Gamma(\alpha)$ and $\zeta(\alpha)$ for selected values of α.

From (2.17) we now find

$$kT_c = \frac{N^{1/\alpha}}{[C_\alpha \Gamma(\alpha) \zeta(\alpha)]^{1/\alpha}}. \tag{2.19}$$

For a three-dimensional harmonic-oscillator potential, α is 3 and C_3 is given by Eq. (2.14). From (2.19) we then obtain a transition temperature given by

$$kT_c = \frac{\hbar \bar{\omega} N^{1/3}}{[\zeta(3)]^{1/3}} \approx 0.94 \hbar \bar{\omega} N^{1/3}, \tag{2.20}$$

where

$$\bar{\omega} = (\omega_1 \omega_2 \omega_3)^{1/3} \tag{2.21}$$

Table 2.1. *The gamma function Γ and the Riemann zeta function ζ for selected values of α.*

α	$\Gamma(\alpha)$	$\zeta(\alpha)$
1	1	∞
1.5	$\sqrt{\pi}/2 = 0.886$	2.612
2	1	$\pi^2/6 = 1.645$
2.5	$3\sqrt{\pi}/4 = 1.329$	1.341
3	2	1.202
3.5	$15\sqrt{\pi}/8 = 3.323$	1.127
4	6	$\pi^4/90 = 1.082$

is the geometric mean of the three oscillator frequencies. The result (2.20) may be written in the useful form

$$T_c \approx 4.5 \left(\frac{\bar{f}}{100 \text{ Hz}}\right) N^{1/3} \text{ nK}, \tag{2.22}$$

where $\bar{f} = \bar{\omega}/2\pi$.

For a uniform Bose gas in a three-dimensional box of volume V, corresponding to $\alpha = 3/2$, the constant $C_{3/2}$ is given by Eq. (2.13) and thus the transition temperature is given by

$$kT_c = \frac{2\pi}{[\zeta(3/2)]^{2/3}} \frac{\hbar^2 n^{2/3}}{m} \approx 3.31 \frac{\hbar^2 n^{2/3}}{m}, \tag{2.23}$$

where $n = N/V$ is the number density. For a uniform gas in two dimensions, α is equal to 1, and the integral in (2.17) diverges. Thus Bose–Einstein condensation in a two-dimensional box can occur only at zero temperature. However, a two-dimensional Bose gas can condense at non-zero temperature if the particles are confined by a harmonic-oscillator potential. In that case $\alpha = 2$ and the integral in (2.17) is finite. We shall return to gases in lower dimensions in Sec. 2.6.

It is useful to introduce the *phase-space density*, which we denote by ϖ. This is defined as the number of particles contained within a volume equal to the cube of the thermal de Broglie wavelength, $\lambda_T^3 = (2\pi\hbar^2/mkT)^{3/2}$,

$$\varpi = n \left(\frac{2\pi\hbar^2}{mkT}\right)^{3/2}. \tag{2.24}$$

If the gas is classical, this is a measure of the typical occupancy of single-particle states. The majority of occupied states have energies of order kT

or less, and therefore the number of states per unit volume that are occupied significantly is of order the total number of states per unit volume with energies less than kT, which is approximately $(mkT/\hbar^2)^{3/2}$ according to (2.3). The phase-space density is thus the ratio between the particle density and the number of significantly occupied states per unit volume. The Bose–Einstein phase transition occurs when $\varpi = \zeta(3/2) \approx 2.612$, according to (2.23). The criterion that ϖ should be comparable with unity indicates that low temperatures and/or high particle densities are necessary for condensation.

The existence of a well-defined phase transition for particles in a harmonic-oscillator potential is a consequence of our assumption that the separation of single-particle energy levels is much less than kT. For an isotropic harmonic oscillator, with $\omega_1 = \omega_2 = \omega_3 = \omega_0$, this implies that the energy quantum $\hbar\omega_0$ should be much less than kT_c. Since T_c is given by Eq. (2.20), the condition is $N^{1/3} \gg 1$. If the finiteness of the particle number is taken into account, the transition becomes smooth.

2.2.1 Condensate fraction

Below the transition temperature the number N_{ex} of particles in excited states is given by Eq. (2.15) with $\mu = 0$,

$$N_{\text{ex}}(T) = C_\alpha \int_0^\infty d\epsilon \, \epsilon^{\alpha-1} \frac{1}{e^{\epsilon/kT} - 1}. \tag{2.25}$$

Provided the integral converges, that is $\alpha > 1$, we may use Eq. (2.18) to write this result as

$$N_{\text{ex}} = C_\alpha \Gamma(\alpha) \zeta(\alpha) (kT)^\alpha. \tag{2.26}$$

Note that this result does not depend on the total number of particles. However, if one makes use of the expression (2.19) for T_c, it may be rewritten in the form

$$N_{\text{ex}} = N \left(\frac{T}{T_c}\right)^\alpha. \tag{2.27}$$

The number of particles in the condensate is thus given by

$$N_0(T) = N - N_{\text{ex}}(T) \tag{2.28}$$

or

$$N_0 = N \left[1 - \left(\frac{T}{T_c}\right)^\alpha\right]. \tag{2.29}$$

For particles in a box in three dimensions, α is $3/2$, and the number of excited particles n_{ex} per unit volume may be obtained from Eqs. (2.26) and (2.13). It is

$$n_{\text{ex}} = \frac{N_{\text{ex}}}{V} = \zeta(3/2)\left(\frac{mkT}{2\pi\hbar^2}\right)^{3/2}. \quad (2.30)$$

The occupancy of the condensate is therefore given by the well-known result $N_0 = N[1 - (T/T_c)^{3/2}]$.

For a three-dimensional harmonic-oscillator potential ($\alpha = 3$), the number of particles in the condensate is

$$N_0 = N\left[1 - \left(\frac{T}{T_c}\right)^3\right]. \quad (2.31)$$

In all cases the transition temperatures T_c are given by (2.19) for the appropriate value of α.

2.3 Density profile and velocity distribution

The cold clouds of atoms which are investigated at microkelvin temperatures typically contain of order 10^4–10^7 atoms. It is not feasible to apply the usual techniques of low-temperature physics to these systems for a number of reasons. First, there are rather few atoms, second, the systems are metastable, so one cannot allow them to come into equilibrium with another body, and third, the systems have a lifetime which is of order seconds to minutes.

Among the quantities that can be measured is the density profile. One way to do this is by absorptive imaging. Light at a resonant frequency for the atom will be absorbed on passing through an atomic cloud. Thus by measuring the absorption profile one can obtain information about the density distribution. The spatial resolution can be improved by allowing the cloud to expand before measuring the absorptive image. A drawback of this method is that it is destructive, since absorption of light changes the internal states of atoms and heats the cloud significantly. To study time-dependent phenomena it is therefore necessary to prepare a new cloud for each time point. An alternative technique is to use phase-contrast imaging [2, 3]. This exploits the fact that the refractive index of the gas depends on its density, and therefore the optical path length is changed by the medium. By allowing a light beam that has passed through the cloud to interfere with a reference beam that has been phase shifted, changes in optical path length may be converted into intensity variations, just as in phase-contrast microscopy.

The advantage of this method is that it is almost non-destructive, and it is therefore possible to study time-dependent phenomena using a single cloud.

The distribution of particles after a cloud is allowed to expand depends not only on the initial density distribution, but also on the initial velocity distribution. Consequently it is important to consider both density and velocity distributions.

In the ground state of the system, all atoms are condensed in the lowest single-particle quantum state and the density distribution $n(\mathbf{r})$ reflects the shape of the ground-state wave function $\phi_0(\mathbf{r})$ for a particle in the trap since, for non-interacting particles, the density is given by

$$n(\mathbf{r}) = N|\phi_0(\mathbf{r})|^2, \tag{2.32}$$

where N is the number of particles. For an anisotropic harmonic oscillator the ground-state wave function is

$$\phi_0(\mathbf{r}) = \frac{1}{\pi^{3/4}(a_1 a_2 a_3)^{1/2}} e^{-x^2/2a_1^2} e^{-y^2/2a_2^2} e^{-z^2/2a_3^2}, \tag{2.33}$$

where the widths a_i of the wave function in the three directions are given by

$$a_i^2 = \frac{\hbar}{m\omega_i}. \tag{2.34}$$

The density distribution is thus anisotropic if the three frequencies ω_1, ω_2 and ω_3 are not all equal, the greatest width being associated with the lowest frequency. The widths a_i may be written in a form analogous to (2.22)

$$a_i \approx 10.1 \left(\frac{100 \text{ Hz}}{f_i} \frac{1}{A}\right)^{1/2} \mu\text{m}, \tag{2.35}$$

in terms of the trap frequencies $f_i = \omega_i/2\pi$ and the mass number A, the number of nucleons in the nucleus of the atom.

In momentum space the wave function corresponding to (2.33) is obtained by taking its Fourier transform and is

$$\phi_0(\mathbf{p}) = \frac{1}{\pi^{3/4}(c_1 c_2 c_3)^{1/2}} e^{-p_x^2/2c_1^2} e^{-p_y^2/2c_2^2} e^{-p_z^2/2c_3^2}, \tag{2.36}$$

where

$$c_i = \frac{\hbar}{a_i} = \sqrt{m\hbar\omega_i}. \tag{2.37}$$

The density in momentum space corresponding to (2.32) is given by

$$n(\mathbf{p}) = N|\phi_0(\mathbf{p})|^2 = \frac{N}{\pi^{3/2} c_1 c_2 c_3} e^{-p_x^2/c_1^2} e^{-p_y^2/c_2^2} e^{-p_z^2/c_3^2}. \tag{2.38}$$

Since $c_i^2/m = \hbar\omega_i$, the distribution (2.38) has the form of a Maxwell distribution with different 'temperatures' $T_i = \hbar\omega_i/2k$ for the three directions.

Since the spatial distribution is anisotropic, the momentum distribution also depends on direction. By the uncertainty principle, a narrow spatial distribution implies a broad momentum distribution, as seen in the Fourier transform (2.36) where the widths c_i are proportional to the square root of the oscillator frequencies.

These density and momentum distributions may be contrasted with the corresponding expressions when the gas obeys classical statistics, at temperatures well above the Bose–Einstein condensation temperature. The density distribution is then proportional to $\exp[-V(\mathbf{r})/kT]$ and consequently it is given by

$$n(\mathbf{r}) = \frac{N}{\pi^{3/2} R_1 R_2 R_3} e^{-x^2/R_1^2} e^{-y^2/R_2^2} e^{-z^2/R_3^2}. \tag{2.39}$$

Here the widths R_i are given by

$$R_i^2 = \frac{2kT}{m\omega_i^2}, \tag{2.40}$$

and they therefore depend on temperature. Note that the ratio R_i/a_i equals $(2kT/\hbar\omega_i)^{1/2}$, which under typical experimental conditions is much greater than unity. Consequently the condition for semi-classical behaviour is well satisfied, and one concludes that the thermal cloud is much broader than the condensate, which below T_c emerges as a narrow peak in the spatial distribution with a weight that increases with decreasing temperature.

Above T_c the density $n(\mathbf{p})$ in momentum space is isotropic in equilibrium, since it is determined only by the temperature and the particle mass, and in the classical limit it is given by

$$n(\mathbf{p}) = C e^{-p^2/2mkT}, \tag{2.41}$$

where the constant C is independent of momentum. The width of the momentum distribution is thus $\sim (mkT)^{1/2}$, which is $\sim (kT/\hbar\omega_i)^{1/2}$ times the zero-temperature width $(m\hbar\omega_i)^{1/2}$. At temperatures comparable with the transition temperature one has $kT \sim N^{1/3}\hbar\omega_i$ and therefore the factor $(kT/\hbar\omega_i)^{1/2}$ is of the order of $N^{1/6}$. The density and velocity distributions of the thermal cloud are thus much broader than those of the condensate.

If a thermal cloud is allowed to expand to a size much greater than its original one, the resulting cloud will be spherically symmetric due to the isotropy of the velocity distribution. This is quite different from the anisotropic shape of an expanding cloud of condensate. In early experiments the anisotropy of clouds after expansion provided strong supporting evidence for the existence of a Bose–Einstein condensate.

Interactions between the atoms alter the sizes of clouds somewhat, as we shall see in Sec. 6.2. A repulsive interaction expands the zero-temperature condensate cloud by a numerical factor which depends on the number of particles and the interatomic potential, typical values being in the range between 2 and 10, while an attractive interaction can cause the cloud to collapse. Above T_c, where the cloud is less dense, interactions hardly affect the size of the cloud.

2.3.1 The semi-classical distribution

Quantum-mechanically, the density of non-interacting bosons is given by

$$n(\mathbf{r}) = \sum_\nu f_\nu |\phi_\nu(\mathbf{r})|^2, \qquad (2.42)$$

where f_ν is the occupation number for state ν, for which the wave function is $\phi_\nu(\mathbf{r})$. Such a description is unwieldy in general, since it demands a knowledge of the wave functions for the trapping potential in question. However, provided the de Broglie wavelengths of particles are small compared with the length scale over which the trapping potential varies significantly, it is possible to use a simpler description in terms of a semi-classical distribution function $f_\mathbf{p}(\mathbf{r})$. This is defined such that $f_\mathbf{p}(\mathbf{r}) d\mathbf{p} d\mathbf{r}/(2\pi\hbar)^3$ denotes the mean number of particles in the phase-space volume element $d\mathbf{p} d\mathbf{r}$. The physical content of this approximation is that locally the gas may be regarded as having the same properties as a bulk gas. We have used this approximation to discuss the high-temperature limit of Boltzmann statistics, but it may also be used under conditions when the gas is degenerate. The distribution function in equilibrium is therefore given by

$$f_\mathbf{p}(\mathbf{r}) = f_\mathbf{p}^0(\mathbf{r}) = \frac{1}{e^{[\epsilon_\mathbf{p}(\mathbf{r}) - \mu]/kT} - 1}. \qquad (2.43)$$

Here the particle energies are those of a classical free particle at point $\mathbf{r}$,

$$\epsilon_\mathbf{p}(\mathbf{r}) = \frac{p^2}{2m} + V(\mathbf{r}), \qquad (2.44)$$

where $V(\mathbf{r})$ is the external potential.

This description may be used for particles in excited states, but it is inappropriate for the ground state, which has spatial variations on length scales comparable with those over which the trap potential varies significantly. Also, calculating properties of the system by integrating over momentum states does not properly take into account the condensed state, but properties of particles in excited states are well estimated by the semi-classical result. Thus, for example, to determine the number of particles in excited states, one integrates the semi-classical distribution function (2.43) divided by $(2\pi\hbar)^3$ over $\mathbf{p}$ and $\mathbf{r}$. The results for T_c agree with those obtained by the methods described in Sec. 2.2 above, where the effect of the potential was included through the density of states. To demonstrate this for a harmonic trap is left as an exercise (Problem 2.1).

We now consider the density of particles which are not in the condensate. This is given by

$$n_{\text{ex}}(\mathbf{r}) = \int \frac{d\mathbf{p}}{(2\pi\hbar)^3} \frac{1}{e^{[\epsilon_{\mathbf{p}}(\mathbf{r}) - \mu]/kT} - 1}. \tag{2.45}$$

We evaluate the integral (2.45) by introducing the variable $x = p^2/2mkT$ and the quantity $z(\mathbf{r})$ defined by the equation

$$z(\mathbf{r}) = e^{[\mu - V(\mathbf{r})]/kT}. \tag{2.46}$$

For $V(\mathbf{r}) = 0$, z reduces to the fugacity. One finds

$$n_{\text{ex}}(\mathbf{r}) = \frac{2}{\sqrt{\pi}\lambda_T^3} \int_0^\infty dx \frac{x^{1/2}}{z^{-1}e^x - 1}, \tag{2.47}$$

where $\lambda_T = (2\pi\hbar^2/mkT)^{1/2}$ is the thermal de Broglie wavelength, Eq. (1.2). Integrals of this type occur frequently in expressions for properties of ideal Bose gases, so we shall consider a more general one. They are evaluated by expanding the integrand in powers of z, and one finds

$$\int_0^\infty dx \frac{x^{\gamma-1}}{z^{-1}e^x - 1} = \sum_{n=1}^\infty \int_0^\infty dx\, x^{\gamma-1} e^{-nx} z^n$$
$$= \Gamma(\gamma) g_\gamma(z), \tag{2.48}$$

where

$$g_\gamma(z) = \sum_{n=1}^\infty \frac{z^n}{n^\gamma}. \tag{2.49}$$

For $z = 1$, the sum in (2.49) reduces to $\zeta(\gamma)$, in agreement with (2.18).

The integral in (2.47) corresponds to $\gamma = 3/2$, and therefore

$$n_{\text{ex}}(\mathbf{r}) = \frac{g_{3/2}(z(\mathbf{r}))}{\lambda_T^3}. \tag{2.50}$$

In Fig. 2.2 we show for a harmonic trap the density of excited particles in units of $1/\lambda_T^3$ for a chemical potential equal to the minimum of the potential.

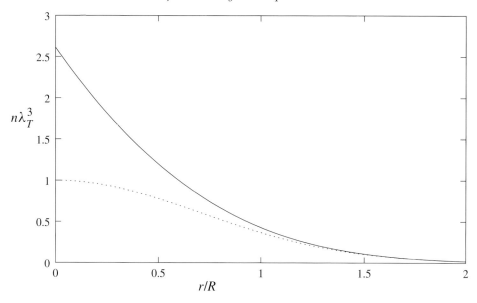

Fig. 2.2. The spatial distribution of non-condensed particles, Eq. (2.50), for an isotropic trap, $V(r) = m\omega_0^2 r^2/2$, with $R = (2kT/m\omega_0^2)^{1/2}$. The dotted line is a Gaussian distribution, corresponding to the first term in the sum (2.49).

This gives the distribution of excited particles at the transition temperature or below. For comparison the result for the classical Boltzmann distribution, which corresponds to the first term in the series (2.49), is also exhibited for the same value of μ. Note that in the semi-classical approximation the density has a cusp at the origin, whereas in a more precise treatment this would be smoothed over a length scale of order λ_T. For a harmonic trap above the transition temperature, the total number of particles is related to the chemical potential by

$$N = g_3(z(0)) \left(\frac{kT}{\hbar\bar{\omega}}\right)^3, \tag{2.51}$$

as one can verify by integrating (2.45) over space.

2.4 Thermodynamic quantities

In this section we determine thermodynamic properties of ideal Bose gases and calculate the energy, entropy, and other properties of the condensed phase. We explore how the temperature dependence of the specific heat for temperatures close to T_c depends on the parameter α characterizing the density of states.

2.4.1 Condensed phase

The energy of the macroscopically occupied state is taken to be zero, and therefore only excited states contribute to the total energy of the system. Consequently in converting sums to integrals it is not necessary to include an explicit term for the condensate, as it is when calculating the total number of particles. Below T_c, the chemical potential vanishes, and the internal energy is given by

$$E = C_\alpha \int_0^\infty d\epsilon \, \epsilon^{\alpha-1} \frac{\epsilon}{e^{\epsilon/kT} - 1} = C_\alpha \Gamma(\alpha+1)\zeta(\alpha+1)(kT)^{\alpha+1}, \quad (2.52)$$

where we have used the integral (2.18). The specific heat $C = \partial E/\partial T$ is therefore given by[1]

$$C = (\alpha+1)\frac{E}{T}. \quad (2.53)$$

Since the specific heat is also given in terms of the entropy S by $C = T\partial S/\partial T$, we find

$$S = \frac{C}{\alpha} = \frac{\alpha+1}{\alpha}\frac{E}{T}. \quad (2.54)$$

Note that below T_c the energy, entropy, and specific heat do not depend on the total number of particles. This is because only particles in excited states contribute, and consequently the number of particles in the macroscopically occupied state is irrelevant for these quantities.

Expressed in terms of the total number of particles N and the transition temperature T_c, which are related by Eq. (2.19), the energy is given by

$$E = Nk\alpha \frac{\zeta(\alpha+1)}{\zeta(\alpha)} \frac{T^{\alpha+1}}{T_c^\alpha}, \quad (2.55)$$

where we have used the property of the gamma function that $\Gamma(z+1) = z\Gamma(z)$. As a consequence, the specific heat is given by

$$C = \alpha(\alpha+1)\frac{\zeta(\alpha+1)}{\zeta(\alpha)}Nk\left(\frac{T}{T_c}\right)^\alpha, \quad (2.56)$$

while the entropy is

$$S = (\alpha+1)\frac{\zeta(\alpha+1)}{\zeta(\alpha)}Nk\left(\frac{T}{T_c}\right)^\alpha. \quad (2.57)$$

Let us compare the results above with those in the classical limit. At high

---

[1] The specific heat C is the temperature derivative of the internal energy, subject to the condition that the trap parameters are unchanged. For particles in a box, C is thus the specific heat at constant volume.

temperatures, the Bose–Einstein distribution becomes a Boltzmann distribution, and therefore

$$N = C_\alpha \int_0^\infty d\epsilon\, \epsilon^{\alpha-1} e^{(\mu-\epsilon)/kT} \tag{2.58}$$

and

$$E = C_\alpha \int_0^\infty d\epsilon\, \epsilon^\alpha e^{(\mu-\epsilon)/kT}. \tag{2.59}$$

On integrating Eq. (2.59) by parts, we obtain

$$E = \alpha N kT, \tag{2.60}$$

which implies that the high-temperature specific heat is

$$C = \alpha N k. \tag{2.61}$$

For a homogeneous gas in three dimensions, for which $\alpha = 3/2$, the result (2.61) is $C = 3Nk/2$, and for a harmonic-oscillator potential in three dimensions $C = 3Nk$. Both these results are in agreement with the equipartition theorem. The ratio of the specific heat in the condensed state to its value in the classical limit is thus given by

$$\frac{C(T)}{\alpha N k} = (\alpha + 1) \frac{\zeta(\alpha+1)}{\zeta(\alpha)} \left(\frac{T}{T_c}\right)^\alpha. \tag{2.62}$$

At T_c the ratio is approximately 1.28 for a uniform gas in three dimensions ($\alpha = 3/2$), and 3.60 for a three-dimensional harmonic-oscillator potential ($\alpha = 3$).

For later applications we shall require explicit expressions for the pressure and entropy of a homogeneous Bose gas. For an ideal gas in three dimensions, the pressure is given by $p = 2E/3V$, irrespective of statistics. For the condensed Bose gas this result may be derived by using the fact that $p = -(\partial E/\partial V)_S$, with the energy given by (2.55) for $\alpha = 3/2$. According to Eq. (2.23) T_c scales as $n^{2/3}$, and one finds

$$p = nk \frac{\zeta(5/2)}{\zeta(3/2)} \frac{T^{5/2}}{T_c^{3/2}} = \zeta(5/2) \left(\frac{m}{2\pi\hbar^2}\right)^{3/2} (kT)^{5/2}. \tag{2.63}$$

From Eq. (2.57), the entropy per particle is seen to be

$$\frac{S}{N} = k \frac{5}{2} \frac{\zeta(5/2)}{\zeta(3/2)} \left(\frac{T}{T_c}\right)^{3/2}. \tag{2.64}$$

These results will be used in the discussion of sound modes in Sec. 10.4.

2.4.2 Normal phase

Let us now consider the leading corrections to the classical result (2.61) for the specific heat. The general expression for the total number of particles is

$$N = C_\alpha \int_0^\infty d\epsilon\, \epsilon^{\alpha-1} \frac{1}{e^{(\epsilon-\mu)/kT} - 1}, \tag{2.65}$$

while that for the total energy is

$$E = C_\alpha \int_0^\infty d\epsilon\, \epsilon^\alpha \frac{1}{e^{(\epsilon-\mu)/kT} - 1}. \tag{2.66}$$

At high temperatures the mean occupation numbers are small. We may therefore use in these two equations the expansion $(e^x - 1)^{-1} \simeq e^{-x} + e^{-2x}$ valid for large x, and obtain

$$N \simeq C_\alpha \int_0^\infty d\epsilon\, \epsilon^{\alpha-1} [e^{(\mu-\epsilon)/kT} + e^{2(\mu-\epsilon)/kT}] \tag{2.67}$$

and

$$E \simeq C_\alpha \int_0^\infty d\epsilon\, \epsilon^\alpha [e^{(\mu-\epsilon)/kT} + e^{2(\mu-\epsilon)/kT}], \tag{2.68}$$

from which we may eliminate the chemical potential μ by solving (2.67) for $\exp(\mu/kT)$ and inserting the result in (2.68). This yields

$$\frac{E}{\alpha N kT} \simeq 1 - \frac{\zeta(\alpha)}{2^{\alpha+1}} \left(\frac{T_c}{T}\right)^\alpha, \tag{2.69}$$

where we have used (2.19) to express N/C_α in terms of T_c. The specific heat is then given by

$$C \simeq \alpha N k \left[1 + (\alpha - 1)\frac{\zeta(\alpha)}{2^{\alpha+1}}\left(\frac{T_c}{T}\right)^\alpha\right]. \tag{2.70}$$

This approximate form is useful even at temperatures only slightly above T_c.

2.4.3 Specific heat close to T_c

Having calculated the specific heat at high temperatures and at temperatures less than T_c we now determine its behaviour near T_c. We shall see that it exhibits a discontinuity at T_c if α exceeds 2. By contrast, for a uniform Bose gas (for which α equals 3/2) the specific heat at constant volume is continuous at T_c, but its derivative with respect to temperature is discontinuous.

We shall consider the energy E as a function of the two variables T and μ. These are constrained by the condition that the total number of particles be equal to N. The change in energy, δE, may then be written as $\delta E = (\partial E/\partial T)_\mu \delta T + (\partial E/\partial \mu)_T \delta \mu$. The term proportional to δT is the same just above and just below the transition, and therefore it gives contributions to the specific heat that are continuous at T_c. The source of the singular behaviour is the term proportional to $\delta \mu$, since the chemical potential μ is identically zero for temperatures less than T_c and becomes non-zero (in fact, negative) above the transition. To determine the nature of the singularity it is sufficient to consider temperatures just above T_c, and evaluate the change (from zero) of the chemical potential, $\delta \mu$, to lowest order in $T - T_c$. The non-zero value $\delta \mu$ of the chemical potential results in a contribution to the internal energy given by $(\partial E/\partial \mu)\delta \mu$. The derivative may be calculated by taking the derivative of Eq. (2.66) and integrating by parts, and one finds $\partial E/\partial \mu = \alpha N$. The discontinuity in the specific heat is therefore

$$\Delta C = C(T_{c+}) - C(T_{c-}) = \alpha N \frac{\partial \mu}{\partial T}\bigg|_{T=T_{c+}}, \qquad (2.71)$$

where the derivative is to be evaluated for fixed particle number.

We determine the dependence of μ on T just above T_c by calculating the derivative $(\partial \mu/\partial T)_N$, using the identity

$$\left(\frac{\partial \mu}{\partial T}\right)_N = -\left(\frac{\partial \mu}{\partial N}\right)_T \left(\frac{\partial N}{\partial T}\right)_\mu = -\left(\frac{\partial N}{\partial T}\right)_\mu \left(\frac{\partial N}{\partial \mu}\right)_T^{-1}, \qquad (2.72)$$

which follows from the fact that

$$dN = \left(\frac{\partial N}{\partial T}\right)_\mu dT + \left(\frac{\partial N}{\partial \mu}\right)_T d\mu = 0 \qquad (2.73)$$

when the particle number is fixed. The derivatives are evaluated by differentiating the expression (2.65) which applies at temperatures at or above T_c and integrating by parts. The results at T_c are

$$\left(\frac{\partial N}{\partial \mu}\right)_T = \frac{\zeta(\alpha-1)}{\zeta(\alpha)} \frac{N}{kT_c} \qquad (2.74)$$

and

$$\left(\frac{\partial N}{\partial T}\right)_\mu = \alpha \frac{N}{T_c}, \qquad (2.75)$$

which yield

$$\left(\frac{\partial \mu}{\partial T}\right)_N = -\alpha \frac{\zeta(\alpha)}{\zeta(\alpha-1)} k. \qquad (2.76)$$

We have assumed that α is greater than 2, since otherwise the expansion is not valid.

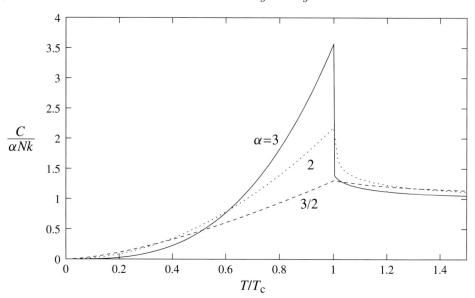

Fig. 2.3. The specific heat C, in units of $\alpha N k$, as a function of the reduced temperature T/T_c for different values of α.

For $T - T_c \ll T_c$ this yields

$$\mu \simeq -\alpha \frac{\zeta(\alpha)}{\zeta(\alpha-1)} k(T - T_c), \tag{2.77}$$

which, when inserted into Eq. (2.71), gives the specific heat discontinuity

$$\Delta C = -\alpha^2 \frac{\zeta(\alpha)}{\zeta(\alpha-1)} Nk. \tag{2.78}$$

For a harmonic-oscillator potential, corresponding to $\alpha = 3$, the jump in the specific heat is

$$\Delta C = -9 \frac{\zeta(3)}{\zeta(2)} Nk \approx -6.58 Nk. \tag{2.79}$$

For $\alpha \leq 2$, the expansion given above is not permissible, and the investigation of the specific heat for that case is the subject of Problem 2.4. We exhibit in Fig. 2.3 the temperature-dependent specific heat for different values of α. Note that the specific heat of an ideal, uniform Bose gas at constant pressure diverges as the temperature approaches T_c from above (Problem 2.5).

2.5 Effect of finite particle number

For a cloud of N bosons in a three-dimensional harmonic trap the transition temperature as given by (2.20) is proportional to $N^{1/3}$. We now consider the leading correction to this result due to the finiteness of the number of particles. As we shall see, this is independent of N, and is therefore of relative order $N^{-1/3}$. For a cloud containing 10^6 atoms it lowers the transition temperature by about 1%, while for 10^4 atoms the effect is somewhat larger, of order 5%.

The correction to T_c originates in the zero-point motion, which increases the energy of the lowest single-particle state by an amount (see Eq. (2.8))

$$\Delta \epsilon_{\min} = \frac{\hbar}{2}(\omega_1 + \omega_2 + \omega_3) = \frac{3\hbar\omega_m}{2}, \tag{2.80}$$

where

$$\omega_m = (\omega_1 + \omega_2 + \omega_3)/3 \tag{2.81}$$

is the algebraic mean of the frequencies. Thus the shift in the chemical potential at the transition temperature is

$$\Delta \mu = \Delta \epsilon_{\min}. \tag{2.82}$$

To determine the shift ΔT_c in the transition temperature for particles trapped by a three-dimensional harmonic-oscillator potential we use the inverse of the result (2.76) to relate the shift in transition temperature to the change in the chemical potential. When condensation sets in, the chemical potential is equal to the expression (2.82) when the zero-point energy is taken into account. Since $\alpha = 3$ for the three-dimensional harmonic trap, the shift in transition temperature is

$$\Delta T_c = -\frac{\zeta(2)}{3\zeta(3)}\frac{\Delta\mu}{k} = -\frac{\zeta(2)}{2\zeta(3)}\frac{\hbar\omega_m}{k}. \tag{2.83}$$

Inserting T_c from (2.20) we obtain the result

$$\frac{\Delta T_c}{T_c} = -\frac{\zeta(2)}{2[\zeta(3)]^{2/3}}\frac{\omega_m}{\bar{\omega}}N^{-1/3} \approx -0.73\frac{\omega_m}{\bar{\omega}}N^{-1/3}, \tag{2.84}$$

which shows that the fractional change in the transition temperature is proportional to $N^{-1/3}$.

For an anisotropic trap with axial symmetry, where $\omega_3 = \lambda\omega_1 = \lambda\omega_2$, the ratio $\omega_m/\bar{\omega}$ becomes $(2+\lambda)/3\lambda^{1/3}$. Since λ may be as small as 10^{-3}, anisotropy can enhance significantly the effects of finite particle number.

2.6 Lower-dimensional systems

There are physical situations where some degrees of freedom are frozen out, and quantum gases behave as one- or two-dimensional systems. Here we shall briefly describe some of them. One example of a two-dimensional system is spin-polarized hydrogen adsorbed on a liquid ^{4}He surface at temperatures low enough that atoms are in the lowest-energy state for motion perpendicular to the surface. To investigate whether or not Bose–Einstein condensation occurs, we calculate the number of particles in excited states when the chemical potential is equal to the energy $\epsilon_{\min}$ of the single-particle ground state. This is given by

$$N_{\text{ex}} = \sum_{i(\epsilon_i > \epsilon_{\min})} \frac{1}{e^{(\epsilon_i - \epsilon_{\min})/kT} - 1}. \tag{2.85}$$

In two dimensions, the smoothed density of states is $g(\epsilon) = L^2 m/2\pi\hbar^2$, where L^2 is the surface area. If we replace the sum in Eq. (2.85) by an integral and take the lower limit to be zero, the integral does not exist because of the divergence at low energies. This shows that we cannot use the simple prescription employed in three dimensions. In a two-dimensional box, the ground state has energy $\epsilon_{\min} \sim \hbar^2/mL^2$, since its wavelength is comparable to the linear extent of the area. Also, the separation of the lowest excited state from the ground state is of the same order. To estimate the number of excited particles, we replace the sum in Eq. (2.85) by an integral, and cut the integral off at an energy $\epsilon_{\min}$. This gives

$$N_{\text{ex}} \sim \frac{kT}{\epsilon_{\min}} \ln\left(\frac{kT}{\epsilon_{\min}}\right). \tag{2.86}$$

The number of excited particles becomes equal to the total number of particles at a temperature given by

$$kT_c \sim \frac{\hbar^2}{mL^2} \frac{N}{\ln N} = \hbar^2 \frac{\sigma}{m \ln N}, \tag{2.87}$$

where $\sigma = N/L^2$ is the number of particles per unit area. The transition temperature is therefore lower by a factor $\sim \ln N$ than the temperature at which the particle spacing is comparable to the thermal de Broglie wavelength. If we take the limit of a large system, but keep the areal density of particles constant, the transition temperature tends to zero. Despite the fact that fluctuations tend to destroy Bose–Einstein condensation in two-dimensional systems, many fascinating phenomena have been found for helium adsorbed on surfaces, among them the Kosterlitz–Thouless transition

[4]. Research on spin-polarized hydrogen on surfaces is being pursued at a number of centres, and a review is given in Ref. [5].

Another possibility for realizing systems that are effectively one- or two-dimensional is to use gases in very anisotropic traps [6]. Imagine an anisotropic harmonic trap, with $\omega_3 \ll \omega_1, \omega_2$, and let us assume that the thermal energy kT is small compared with $\hbar\omega_1$ and $\hbar\omega_2$. Particles are then in the lowest-energy states for motion in the x and y directions, and when the chemical potential is equal to the energy of the ground state, the number of particles in excited states for motion in the z direction is given by

$$N_{\text{ex}} = \sum_{n=1}^{\infty} \frac{1}{e^{n\hbar\omega_3/kT} - 1}. \tag{2.88}$$

Like the sum for the two-dimensional Bose gas above, this expression would diverge if one replaced it by an integral and took the lower limit on n to be zero. However, since the lowest excitation energy is $\hbar\omega_3$, a better estimate is obtained by cutting the integral off at $n = 1$, which corresponds to an energy $\hbar\omega_3$. In this approximation one finds

$$N_{\text{ex}} \simeq \frac{kT}{\hbar\omega_3} \ln\left(\frac{kT}{\hbar\omega_3}\right), \tag{2.89}$$

which is the same as Eq. (2.86), except that ϵ_{min} there is replaced by $\hbar\omega_3$. The temperature at which the number of particles in excited states must become macroscopic is therefore given by equating N_{ex} to the total number of particles N, and therefore

$$kT_{\text{c}} \simeq \hbar\omega_3 \frac{N}{\ln N}. \tag{2.90}$$

As N increases, T_{c} increases if the properties of the trap are fixed. This is similar to what would happen if one increased the number of particles in a two-dimensional 'box' of fixed area.

A wide range of physical phenomena are predicted to occur in very anisotropic traps. Recent experiments where low-dimensional behaviour has been observed in such traps are described in Ref. [7], where references to earlier theoretical and experimental work may be found.

Problems

PROBLEM 2.1 Use the semi-classical distribution function (2.43) to calculate the number of particles in the condensate for an isotropic harmonic-oscillator potential. Indicate how the calculation may be generalized to an anisotropic harmonic oscillator.

PROBLEM 2.2 Consider a gas of N identical bosons, each of mass m, in the quadrupole trap potential

$$V(x,y,z) = A(x^2 + y^2 + 4z^2)^{1/2},$$

where A is a positive constant (the physics of this trap will be explained in Sec. 4.1). Determine the density of single-particle states as a function of energy and calculate the transition temperature, the depletion of the condensate as a function of temperature, and the jump in the specific heat at the transition temperature.

PROBLEM 2.3 Determine the Bose–Einstein condensation temperature and the temperature dependence of the depletion of the condensate for N identical bosons of mass m in an isotropic potential given by the power law $V(\mathbf{r}) = Cr^\nu$, where C and ν are positive constants.

PROBLEM 2.4 Prove that the discontinuity in the specific heat at T_c, obtained in (2.78) for $\alpha > 2$, disappears when $\alpha < 2$ by using the identity

$$N - C_\alpha \Gamma(\alpha)\zeta(\alpha)(kT)^\alpha = C_\alpha \int_0^\infty d\epsilon\, \epsilon^{\alpha-1} \left[\frac{1}{e^{(\epsilon-\mu)/kT} - 1} - \frac{1}{e^{\epsilon/kT} - 1} \right]$$

at temperatures just above T_c. [Hint: Simplify the integrand by using the approximation $(e^x - 1)^{-1} \simeq 1/x$.]

PROBLEM 2.5 Consider a uniform non-interacting gas of N identical bosons of mass m in a volume V. Use the method employed in Problem 2.4 to calculate the chemical potential as a function of temperature and volume at temperatures just above T_c. Show that the specific heat at constant pressure, C_p, diverges as $(T - T_c)^{-1}$ when the temperature approaches T_c from above. [Hint: The thermodynamic identity $C_p = C_V - T(\partial p/\partial T)_V^2/(\partial p/\partial V)_T$ may be useful.]

References

[1] L. D. Landau and E. M. Lifshitz, *Statistical Physics*, Part 1, Third edition, (Pergamon, Oxford, 1980), §62.

[2] M. R. Andrews, M.-O. Mewes, N. J. van Druten, D. S. Durfee, D. M. Kurn, and W. Ketterle, *Science* **273**, 84 (1996).

[3] W. Ketterle, D. S. Durfee, and D. M. Stamper-Kurn, in *Bose–Einstein Condensation in Atomic Gases*, Proceedings of the Enrico Fermi International School of Physics, Vol. CXL, ed. M. Inguscio, S. Stringari, and C. E. Wieman, (IOS Press, Amsterdam, 1999), p. 67.

[4] D. J. Bishop and J. D. Reppy, *Phys. Rev. Lett.* **48**, 1727 (1978); *Phys. Rev. B* **22**, 5171 (1980).

[5] J. T. M. Walraven, in *Fundamental Systems in Quantum Optics*, ed. J. Dalibard, J.-M. Raimond, and J. Zinn-Justin (North-Holland, Amsterdam, 1992), p. 485.

[6] V. Bagnato and D. Kleppner, *Phys. Rev. A* **44**, 7439 (1991).

[7] A. Görlitz, J. M. Vogels, A. E. Leanhardt, C. Raman, T. L. Gustavson, J. R. Abo-Shaeer, A. P. Chikkatur, S. Gupta, S. Inouye, T. P. Rosenband, D. E. Pritchard, and W. Ketterle, *Phys. Rev. Lett.* **87**, 130 402 (2001).

3
Atomic properties

Atomic properties of the alkali atoms play a key role in experiments on cold atomic gases, and we shall discuss them briefly in the present chapter. Basic atomic structure is the subject of Sec. 3.1. Two effects exploited to trap and cool atoms are the influence of a magnetic field on atomic energy levels, and the response of an atom to radiation. In Sec. 3.2 we describe the combined influence of the hyperfine interaction and the Zeeman effect on the energy levels of an atom, and in Sec. 3.3 we review the calculation of the atomic polarizability. In Sec. 3.4 we summarize and compare some energy scales.

3.1 Atomic structure

The total spin of a Bose particle must be an integer, and therefore a boson made up of fermions must contain an even number of them. Neutral atoms contain equal numbers of electrons and protons, and therefore the statistics that an atom obeys is determined solely by the number of neutrons N: if N is even, the atom is a boson, and if it is odd, a fermion. Since the alkalis have odd atomic number Z, boson alkali atoms have odd mass numbers A. In Table 3.1 we list N, Z, and the nuclear spin quantum number I for some alkali atoms and hydrogen. We also give the nuclear magnetic moment μ, which is defined as the expectation value of the z component of the magnetic moment operator in the state where the z component of the nuclear spin, denoted by $m_I \hbar$, has its maximum value, $\mu = \langle I, m_I = I | \mu_z | I, m_I = I \rangle$. To date, essentially all experiments on Bose–Einstein condensation have been made with states having total electronic spin 1/2, and the majority of them have been made with states having nuclear spin $I = 3/2$ (^{87}Rb, ^{23}Na, and ^{7}Li). Successful experiments have also been carried out with hydrogen ($I = 1/2$) and ^{85}Rb atoms ($I = 5/2$). In addition, Bose–Einstein

Table 3.1. *The proton number Z, the neutron number N, the nuclear spin I, the nuclear magnetic moment μ (in units of the nuclear magneton $\mu_N = e\hbar/2m_p$), and the hyperfine splitting $\nu_{hf} = \Delta E_{hf}/h$ for hydrogen and some alkali isotopes. For completeness, the two fermion isotopes ^{6}Li and ^{40}K are included.*

Isotope	Z	N	I	μ/μ_N	ν_{hf} (MHz)
^{1}H	1	0	1/2	2.793	1420
^{6}Li	3	3	1	0.822	228
^{7}Li	3	4	3/2	3.256	804
^{23}Na	11	12	3/2	2.218	1772
^{39}K	19	20	3/2	0.391	462
^{40}K	19	21	4	−1.298	−1286
^{41}K	19	22	3/2	0.215	254
^{85}Rb	37	48	5/2	1.353	3036
^{87}Rb	37	50	3/2	2.751	6835
^{133}Cs	55	78	7/2	2.579	9193

condensation has been achieved for ^{4}He atoms, with nuclear spin $I = 0$, in a metastable electronic state with $S = 1$.

The ground-state electronic structure of alkali atoms is simple: all electrons but one occupy closed shells, and the remaining one is in an s orbital in a higher shell. In Table 3.2 we list the ground-state electronic configurations for alkali atoms. The nuclear spin is coupled to the electronic spin by the hyperfine interaction. Since the electrons have no orbital angular momentum ($L = 0$), there is no magnetic field at the nucleus due to the orbital motion, and the coupling arises solely due to the magnetic field produced by the electronic spin. The coupling of the electronic spin, $S = 1/2$, to the nuclear spin I yields the two possibilities $F = I \pm 1/2$ for the quantum number F for the total spin, according to the usual rules for addition of angular momentum.

In the absence of an external magnetic field the atomic levels are split by the hyperfine interaction. The coupling is represented by a term H_{hf} in the Hamiltonian of the form

$$H_{hf} = A\mathbf{I}\cdot\mathbf{J}, \qquad (3.1)$$

where A is a constant, while $\mathbf{I}$ and $\mathbf{J}$ are the operators for the nuclear spin and the electronic angular momentum, respectively, in units of $\hbar$. The operator for the total angular momentum is equal to

$$\mathbf{F} = \mathbf{I} + \mathbf{J}. \qquad (3.2)$$

Table 3.2. *The electron configuration and electronic spin for selected isotopes of alkali atoms and hydrogen. For clarity, the inner-shell configuration for Rb and Cs is given in terms of the noble gases Ar, $1s^2 2s^2 2p^6 3s^2 3p^6$, and Kr, $(Ar)3d^{10} 4s^2 4p^6$.*

Element	Z	Electronic spin	Electron configuration
H	1	1/2	$1s$
Li	3	1/2	$1s^2 2s^1$
Na	11	1/2	$1s^2 2s^2 2p^6 3s^1$
K	19	1/2	$1s^2 2s^2 2p^6 3s^2 3p^6 4s^1$
Rb	37	1/2	$(Ar)3d^{10} 4s^2 4p^6 5s^1$
Cs	55	1/2	$(Kr)4d^{10} 5s^2 5p^6 6s^1$

By squaring this expression we may express **I·J** in terms of the quantum numbers I, J, and F determining the squares of the angular momentum operators and the result is

$$\mathbf{I \cdot J} = \frac{1}{2}[F(F+1) - I(I+1) - J(J+1)]. \tag{3.3}$$

Alkali and hydrogen atoms in their ground states have $J = S = 1/2$. The splitting between the levels $F = I + 1/2$ and $F = I - 1/2$ is therefore given by

$$\Delta E_{\text{hf}} = h\nu_{\text{hf}} = (I + \frac{1}{2})A. \tag{3.4}$$

Measured values of the hyperfine splitting are given in Table 3.1.

As a specific example, consider an alkali atom with $I = 3/2$ in its ground state ($J = S = 1/2$). The quantum number F may be either 1 or 2, and $\mathbf{I \cdot J} = -5/4$ and $3/4$, respectively. The corresponding shifts of the ground-state multiplet are given by $E_1 = -5A/4$ (three-fold degenerate) and $E_2 = 3A/4$ (five-fold degenerate). The energy difference $\Delta E_{\text{hf}} = E_2 - E_1$ due to the hyperfine interaction (3.1) is thus

$$\Delta E_{\text{hf}} = 2A, \tag{3.5}$$

in agreement with the general expression (3.4).

A first-order perturbation treatment of the magnetic dipole interaction between the outer s electron and the nucleus yields the expression [1, §121]

$$\Delta E_{\text{hf}} = \frac{\mu_0}{4\pi} \frac{16\pi}{3} \mu_B \mu \frac{(I+1/2)}{I} |\psi(0)|^2. \tag{3.6}$$

The quantity $\mu_B = e\hbar/2m_e$ (the Bohr magneton) is the magnitude of the magnetic moment of the electron, and μ is the magnetic moment of the nucleus, which is of order the nuclear magneton $\mu_N = e\hbar/2m_p = (m_e/m_p)\mu_B$. Here m_e is the electron mass and m_p the proton mass. Throughout we denote the elementary charge, the absolute value of the charge of the electron, by e. The quantity $\psi(0)$ is the valence s-electron wave function at the nucleus. It follows from Eq. (3.6) that for atoms with a positive nuclear magnetic moment, which is the case for all the alkali isotopes listed in Table 3.1 except ^{40}K, the state with lowest energy has $F = I - 1/2$. For negative nuclear magnetic moment the state with $F = I + 1/2$ has the lower energy.

The hyperfine line for the hydrogen atom has a measured frequency of 1420 MHz, and it is the famous 21-cm line exploited by radio astronomers. Let us compare this frequency with the expression (3.6). For hydrogen, μ is the proton magnetic moment $\mu_p \approx 2.793\mu_N$ and $|\psi(0)|^2 = 1/\pi a_0^3$, where $a_0 = \hbar^2/m_e e_0^2$ is the Bohr radius, with $e_0^2 = e^2/4\pi\epsilon_0$. Thus (3.6) becomes

$$\Delta E_{\rm hf} = \frac{32}{3}\frac{\mu_0}{4\pi}\frac{\mu_B \mu_p}{a_0^3}. \qquad (3.7)$$

The magnitude of $\Delta E_{\rm hf}$ is of order $(m_e/m_p)\alpha_{\rm fs}$ in atomic units (a. u.), where $\alpha_{\rm fs} = e_0^2/\hbar c$, c being the velocity of light, is the fine structure constant and the atomic unit of energy is e_0^2/a_0. The calculated result (3.7) differs from the experimental one by less than 1%, the bulk of the difference being due to the reduced-mass effect and the leading radiative correction to the electron g factor, $g = 2(1 + \alpha_{\rm fs}/2\pi)$.

For multi-electron atoms, Eq. (3.6) shows that the hyperfine splitting depends on the electron density at the nucleus due to the valence electron state. If interactions between electrons could be neglected, the electron orbitals would be hydrogenic, and the hyperfine interaction would scale as $(Z/n)^3$, where n is the principal quantum number for the radial wave function of the outermost s electron. However, the electrons in closed shells screen the charge of the nucleus, and the outermost electron sees a reduced charge. As a consequence the hyperfine splitting increases less rapidly with Z, as is illustrated for alkali atoms in Table 3.1.

The variation in hyperfine splitting exhibited in Table 3.1 may be roughly understood in terms of the measured nuclear magnetic moment μ and the valence electron probability density at the nucleus. If Z is large compared with unity, but still small compared with $\alpha_{\rm fs}^{-1}$, the inverse fine structure constant, so that relativistic effects may be neglected, one may show within the quasi-classical WKB approximation that the characteristic scale of densities for the outermost electron at the origin is proportional to Z [1, §71]. An empirical measure of the valence electron density at the nucleus may be obtained by dividing $\nu_{\rm hf}$ by $\mu(I+1/2)/I$. This ratio is independent of nuclear properties, since the hyperfine splitting is proportional to $\mu(I+1/2)/I$. If one divides further by a factor of Z, one finds from Table 3.1 that the resulting ratios differ by no more than 20 % for the elements from sodium to cesium, in good agreement with the result of the quasi-classical method.

3.2 The Zeeman effect

To take into account the effect of an external magnetic field on the energy levels of an atom we must add to the hyperfine Hamiltonian (3.1) the Zeeman energies arising from the interaction of the magnetic moments of the electron and the nucleus with the magnetic field. If we take the magnetic field **B** to be in the z direction, the total Hamiltonian is thus

$$H_{\rm spin} = A\mathbf{I}\cdot\mathbf{J} + CJ_z + DI_z. \tag{3.8}$$

The constants C and D are given by

$$C = g\mu_{\rm B}B, \tag{3.9}$$

and

$$D = -\frac{\mu}{I}B, \tag{3.10}$$

where in writing Eq. (3.9) we have assumed that the electronic orbital angular momentum L is zero and its spin S is 1/2. For ^{87}Rb, μ equals $2.751\mu_{\rm N}$, and $D = -1.834\mu_{\rm N}B$, according to Table 3.1. Since $|C/D| \sim m_{\rm p}/m_{\rm e} \approx 2000$, for most applications D may be neglected. At the same level of approximation the g factor of the electron may be put equal to 2.

Because of its importance we first consider a nuclear spin of 3/2. In order to obtain the level scheme for an atom in an external magnetic field, we diagonalize $H_{\rm spin}$ in a basis consisting of the eight states $|m_I, m_J\rangle$, where $m_I = 3/2, 1/2, -1/2, -3/2$, and $m_J = 1/2, -1/2$. The hyperfine interaction may be expressed in terms of the usual raising and lowering operators $I_\pm = I_x \pm iI_y$ and $J_\pm = J_x \pm iJ_y$ by use of the identity

$$\mathbf{I}\cdot\mathbf{J} = I_zJ_z + \frac{1}{2}(I_+J_- + I_-J_+). \tag{3.11}$$

The Hamiltonian (3.8) conserves the z component of the total angular momentum, and therefore it couples only states with the same value of the sum $m_I + m_J$, since the raising (lowering) of m_J by 1 must be accompanied by the lowering (raising) of m_I by 1. This reflects the invariance of the interaction under rotations about the z axis.

The energies of the states $|3/2, 1/2\rangle$ and $|-3/2, -1/2\rangle$ are easily calculated, since these states do not mix with any others. They are

$$E(\frac{3}{2}, \frac{1}{2}) = \frac{3}{4}A + \frac{1}{2}C + \frac{3}{2}D \tag{3.12}$$

and

$$E(-\frac{3}{2}, -\frac{1}{2}) = \frac{3}{4}A - \frac{1}{2}C - \frac{3}{2}D, \tag{3.13}$$

which are linear in the magnetic field.

Since the Hamiltonian conserves the z component of the total angular momentum, the only states that mix are pairs like $|m_I, -1/2\rangle$ and $|m_I - 1, 1/2\rangle$. Therefore to calculate the energies of the other states we need to diagonalize only 2×2 matrices. Let us first consider the matrix for $m_I + m_J = 1$, corresponding to the states $|3/2, -1/2\rangle$ and $|1/2, 1/2\rangle$. The matrix elements of the Hamiltonian (3.8) are

$$\begin{pmatrix} -\frac{3}{4}A - \frac{1}{2}C + \frac{3}{2}D & \frac{\sqrt{3}}{2}A \\ \frac{\sqrt{3}}{2}A & \frac{1}{4}A + \frac{1}{2}C + \frac{1}{2}D \end{pmatrix},$$

and the eigenvalues are

$$E = -\frac{A}{4} + D \pm \sqrt{\frac{3}{4}A^2 + \frac{1}{4}(A + C - D)^2}. \tag{3.14}$$

In the absence of a magnetic field ($C = D = 0$) the eigenvalues are seen to be $-5A/4$ and $3A/4$, in agreement with the energies of the $F = 1$ and $F = 2$ states calculated earlier. For the states $|-3/2, 1/2\rangle$ and $|-1/2, -1/2\rangle$ the matrix is obtained from the one above by the substitution $C \to -C$ and $D \to -D$. The matrix for the states $|1/2, -1/2\rangle$, $|-1/2, 1/2\rangle$ is

$$\begin{pmatrix} -\frac{1}{4}A + \frac{1}{2}(C - D) & A \\ A & -\frac{1}{4}A - \frac{1}{2}(C - D) \end{pmatrix},$$

and the eigenvalues are

$$E = -\frac{A}{4} \pm \sqrt{A^2 + \frac{1}{4}(C - D)^2}. \tag{3.15}$$

The eigenvalues resulting from the matrix diagonalization are plotted in Fig. 3.1. As we remarked earlier, $|D|$ is much less than both C and $|A|$ at the fields attainable. Therefore to a good approximation we may set D equal to

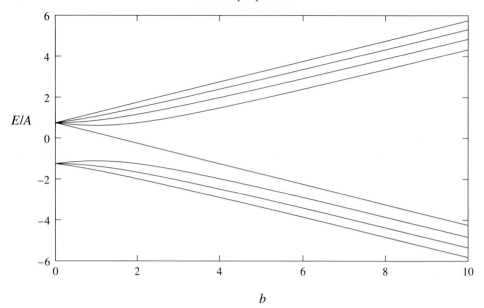

Fig. 3.1. Energies of hyperfine levels of an alkali atom with $I = 3/2$ and $A > 0$ in a magnetic field. The energy is measured in units of $A = \Delta E_{\text{hf}}/2$, and the dimensionless magnetic field $b = C/A = 4\mu_B B/\Delta E_{\text{hf}}$ (see Eq. (3.16)).

zero, and we also set $g = 2$. The dimensionless magnetic field b is defined for arbitrary nuclear spin by

$$b = \frac{C}{A} = \frac{(2I+1)\mu_B B}{\Delta E_{\text{hf}}}. \tag{3.16}$$

The two straight lines correspond to the energies of the states $|3/2, 1/2\rangle$ and $|-3/2, -1/2\rangle$, which do not mix with other states.

When D is neglected, the energy levels are given for $m_I + m_J = \pm 2$ by

$$E(\tfrac{3}{2},\tfrac{1}{2}) = A\left(\tfrac{3}{4} + \tfrac{b}{2}\right) \quad \text{and} \quad E(-\tfrac{3}{2},-\tfrac{1}{2}) = A\left(\tfrac{3}{4} - \tfrac{b}{2}\right), \tag{3.17}$$

for $m_I + m_J = \pm 1$ by

$$E = A\left(-\frac{1}{4} \pm \sqrt{\frac{3}{4} + \frac{1}{4}(1+b)^2}\right) \tag{3.18}$$

and

$$E = A\left(-\frac{1}{4} \pm \sqrt{\frac{3}{4} + \frac{1}{4}(1-b)^2}\right), \tag{3.19}$$

and for $m_I + m_J = 0$ by

$$E = A\left(-\frac{1}{4} \pm \sqrt{1 + \frac{b^2}{4}}\right). \tag{3.20}$$

At high magnetic fields, $b \gg 1$, the leading contributions to these expressions are $\pm Ab/2 = \pm C/2$ corresponding to the energy eigenvalues $\pm \mu_B B$ associated with the electronic spin. These calculations may easily be generalized to other values of the nuclear spin.

Many experiments on alkali atoms are carried out in relatively low magnetic fields, for which the Zeeman energies are small compared with the hyperfine splitting. To first order in the magnetic field, the energy may be written as

$$E(F, m_F) = E(F) + m_F g_F \mu_B B, \tag{3.21}$$

where g_F is the Landé g factor and $E(F)$ is the energy for $B = 0$. For $F = I + 1/2$, the electron spin is aligned essentially parallel to the total spin, and therefore the g factor is negative. Consequently the state with $m_F = F$ has the highest energy. For $F = I - 1/2$, the electron spin is predominantly antiparallel to the total spin, and the state with $m_F = -F$ has the highest energy. Calculation of the g factors is left as an exercise, Problem 3.1.

One state of particular importance experimentally is the *doubly polarized state* $|m_I = I, m_J = 1/2\rangle$, in which the nuclear and electronic spin components have the largest possible values along the direction of the magnetic field. Another is the *maximally stretched state*, which corresponds to quantum numbers $F = I - 1/2, m_F = -(I - 1/2)$ in zero magnetic field. These states have negative magnetic moments and therefore, according to the discussion in Sec. 4.1, they can be trapped by magnetic fields. In addition, they have long inelastic relaxation times, as we shall explain in Sec. 5.4.2. For a nuclear spin 3/2 the doubly polarized state is $|m_I = 3/2, m_J = 1/2\rangle$, which has $F = 2, m_F = 2$, and in zero magnetic field the maximally stretched state is $F = 1, m_F = -1$.

For hydrogen, the nuclear spin is $I = 1/2$. The eigenvalues of (3.8) are determined in precisely the same manner as for a nuclear spin $I = 3/2$, and the result is

$$E(\frac{1}{2}, \frac{1}{2}) = A\left(\frac{1}{4} + \frac{b}{2}\right) \quad \text{and} \quad E(-\frac{1}{2}, -\frac{1}{2}) = A\left(\frac{1}{4} - \frac{b}{2}\right), \tag{3.22}$$

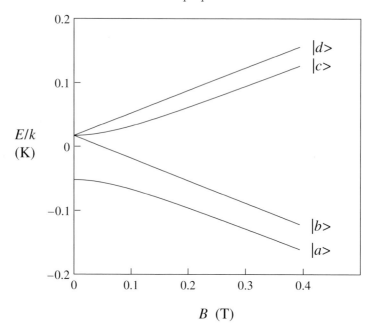

Fig. 3.2. Energies of the ground-state multiplet of a hydrogen atom as a function of the magnetic field.

and for the states with $m_I + m_J = 0$

$$E = -A\left(\frac{1}{4} \pm \frac{1}{2}\sqrt{1+b^2}\right). \qquad (3.23)$$

For $I = 1/2$, A is equal to the hyperfine splitting ΔE_{hf}, and the dimensionless magnetic field $b = C/A = 2\mu_B B/\Delta E_{\text{hf}}$ according to (3.16). The energies, converted to equivalent temperatures, are plotted as functions of the magnetic field B in Fig. 3.2.

The four states corresponding to these eigenvalues are conventionally labelled $a, b, c,$ and d in order of increasing energy. When expressed in terms of the basis $|m_I, m_J\rangle$ they are

$$|a\rangle = \cos\theta \, |\tfrac{1}{2}, -\tfrac{1}{2}\rangle - \sin\theta \, |-\tfrac{1}{2}, \tfrac{1}{2}\rangle, \qquad (3.24)$$

$$|b\rangle = |-\tfrac{1}{2}, -\tfrac{1}{2}\rangle, \qquad (3.25)$$

$$|c\rangle = \cos\theta \, |-\tfrac{1}{2}, \tfrac{1}{2}\rangle + \sin\theta \, |\tfrac{1}{2}, -\tfrac{1}{2}\rangle, \qquad (3.26)$$

and

$$|d\rangle = |\tfrac{1}{2}, \tfrac{1}{2}\rangle. \qquad (3.27)$$

The dependence of the mixing angle θ on the magnetic field is given by $\tan 2\theta = 1/b = \Delta E_{\mathrm{hf}}/2\mu_{\mathrm{B}} B$.

3.3 Response to an electric field

When an atom is subjected to an electric field $\mathcal{E}$ it acquires an electric dipole moment, and its energy levels are shifted. This effect is exploited extensively in experiments on cold dilute gases for trapping and cooling atoms using the strong electric fields generated by lasers. Such electric fields are time dependent, but to set the scale of effects we begin by considering static fields. For the hydrogen atom we may estimate the order of magnitude of the polarizability by arguing that the average position of the electron will be displaced by an amount comparable to the atomic size $\sim a_0$ if an external electric field comparable in strength to the electric field in the atom $\mathcal{E} \sim e/(4\pi\epsilon_0 a_0^2)$ is applied. Here $a_0 = \hbar^2/m_e e_0^2$ is the Bohr radius and $e_0^2 = e^2/4\pi\epsilon_0$. The polarizability α relates the expectation value $<\mathbf{d}>$ of the electric dipole moment to the electric field according to the definition

$$<\mathbf{d}> = \alpha \mathcal{E}, \tag{3.28}$$

and therefore it is given in order of magnitude by

$$\alpha \sim \frac{ea_0}{e/4\pi\epsilon_0 a_0^2} = 4\pi\epsilon_0 a_0^3. \tag{3.29}$$

More generally, the polarizability is a tensor, but for the ground states of alkali atoms and hydrogen, which are S states, it is a scalar, since it does not depend on the direction of the field. In order to avoid exhibiting the factor of $4\pi\epsilon_0$ we define the quantity

$$\tilde{\alpha} = \frac{\alpha}{4\pi\epsilon_0}. \tag{3.30}$$

The estimate (3.29) then leads to the result

$$\tilde{\alpha} \sim a_0^3. \tag{3.31}$$

The energy of an atom in an electric field may be evaluated quantitatively using perturbation theory. In an electric field which is spatially uniform on the scale of the atom, the interaction between the atom and the electric field may be treated in the dipole approximation, and the interaction Hamiltonian is

$$H' = -\mathbf{d}\cdot\mathcal{E}, \tag{3.32}$$

where

$$\mathbf{d} = -e \sum_j \mathbf{r}_j \qquad (3.33)$$

is the electric dipole moment operator for the atomic electrons. Here the $\mathbf{r}_j$ are the position operators for the electrons relative to the atomic nucleus, and the sum is over all electrons in the atom. In the absence of external fields, most atomic states are eigenstates of the parity operator to a very good approximation, since the only deviations are due to the weak interaction, which violates parity. From symmetry it then follows that the dipole moment of the atom in the absence of a field vanishes, and consequently the first-order contribution to the energy also vanishes. The first non-vanishing term in the expansion of the energy in powers of the electric field is of second order, and for the ground state this is given by

$$\Delta E = -\sum_n \frac{|\langle n|H'|0\rangle|^2}{E_n - E_0}. \qquad (3.34)$$

In this expression the energies in the denominator are those for the unperturbed atom, and the sum is over all excited states, including those in the continuum. Because all energy denominators are positive for the ground state, the second-order contribution to the ground-state energy is negative.

The energy of the state may also be calculated in terms of the polarizability. To do this we make use of the fact that the change in the energy due to a change in the electric field is given by

$$dE = -<\mathbf{d}> \cdot d\boldsymbol{\mathcal{E}}, \qquad (3.35)$$

and therefore, when the expectation value of the dipole moment is given by (3.28), the field-dependent contribution to the energy is

$$\Delta E = -\frac{1}{2}\alpha \mathcal{E}^2. \qquad (3.36)$$

Comparing Eqs. (3.34) and (3.36) we see that

$$\alpha = -\frac{\partial^2 \Delta E}{\partial \mathcal{E}^2} = \sum_n \frac{2|\langle n|d_i|0\rangle|^2}{E_n - E_0}, \qquad (3.37)$$

where for definiteness we have taken the electric field to be in the i direction.

For hydrogen the polarizability may be calculated exactly, and it is given by $\tilde{\alpha} = 9a_0^3/2$, in agreement with our qualitative estimate above and with experiment. Upper and lower bounds on the polarizability may be obtained by simple methods, as described in Problem 3.3.

The polarizabilities of alkali atoms are larger than that of hydrogen by factors which range from 30 to 90. To understand the magnitude of the numerical factor and its variation throughout the alkali series in simple terms we introduce the concept of the *oscillator strength* for a transition from a state k to a state l. This dimensionless quantity is defined by

$$f_{kl}^i = \frac{2m_c(E_k - E_l)}{e^2\hbar^2}|\langle k|d_i|l\rangle|^2 \tag{3.38}$$

for the i component of the dipole moment. This is the squared modulus of the dipole matrix element measured in terms of the electronic charge and a length equal to $1/2\pi$ times the wavelength of a free electron with energy equal to that of the transition. The polarizability of an atom in its ground state may then be written as

$$\alpha = 4\pi\epsilon_0\tilde{\alpha} = \frac{e^2}{m_e}\sum_n \frac{f_{n0}^i}{\omega_{n0}^2}, \tag{3.39}$$

where $\omega_{n0} = (E_n - E_0)/\hbar$. In atomic units ($a_0^3$ for $\tilde{\alpha}$ and e_0^2/a_0 for energies) this result may be written

$$\tilde{\alpha} = \sum_n \frac{f_{n0}^i}{(E_n - E_0)^2}. \tag{3.40}$$

For an atom with Z electrons the oscillator strengths obey the Thomas–Reiche–Kuhn or f sum rule [2, p. 181]

$$\sum_n f_{n0} = Z. \tag{3.41}$$

In alkali atoms, by far the main contribution to the polarizability comes from the valence electron. Electrons in other states contribute relatively little because the corresponding excitations have high energies. In addition, for the valence electron the bulk of the oscillator strength is in the *resonance lines* in the optical part of the spectrum. These are due to nP–nS transitions, which are doublets due to the spin–orbit interaction. The best-known example is the 3P to 3S transition in sodium, which gives the yellow Fraunhofer D lines in the spectrum of the Sun. To an excellent approximation, the valence electron states are those of a single electron moving in a static potential due to the nucleus and the electrons in the core. The contribution to the sum rule from the valence electron transitions is then unity. If we further neglect all transitions except that for the resonance line, the total oscillator strength for the resonance line is unity, and the polarizability is

Table 3.3. *Wavelengths and energies of resonance lines of alkali atoms and hydrogen. The wavelengths of both members of the doublet are given, and for H and Li they are the same to within the number of figures quoted in the table. The energies given correspond to the average of the energies for the transitions to the spin–orbit doublet, weighted according to their statistical weights (4 for $P_{3/2}$ and 2 for $P_{1/2}$).*

Atom	Wavelength (nm)	ΔE_{res} (eV)	ΔE_{res} (a. u.)
H	121.6	10.20	0.375
Li	670.8	1.848	0.0679
Na	589.0, 589.6	2.104	0.0773
K	766.5, 769.9	1.615	0.0594
Rb	780.0, 794.8	1.580	0.0581
Cs	852.1, 894.3	1.432	0.0526

given by

$$\tilde{\alpha} \approx \frac{1}{(\Delta E_{\text{res}})^2}, \qquad (3.42)$$

where ΔE_{res} is the energy difference associated with the resonance line measured in atomic units ($e_0^2/a_0 \approx 27.2$ eV). In Table 3.3 we list values of the wavelengths and energy differences for the resonance lines of alkali atoms and hydrogen.

The measured value of $\tilde{\alpha}$ for Li is 164 in atomic units, while (3.42) yields 217. For Na the measured value is 163 [3], while Eq. (3.42) gives 167. For K, Rb, and Cs the measured values are 294, 320, and 404, and Eq. (3.42) evaluated using the energy of the resonance line averaged over the two members of the doublet yields 284, 297, and 361, respectively. Thus we see that the magnitude of the polarizability and its variation through the alkali series may be understood simply in terms of the dominant transition being the resonance line.

The resonance lines in alkali atoms have energies much less than the Lyman-α line in hydrogen because they are due to transitions between valence electron states with the same principal quantum number, e.g., 3P–3S for Na. If the potential in which the electron moved were purely Coulombic, these states would be degenerate. However the s-electron wave function penetrates the core of the atom to a greater extent than does that of the p electron, which is held away from the nucleus by virtue of its angular momentum. Consequently screening of the nuclear charge by core electrons is more effective for a p electron than for an s electron, and as a result the s state has a lower energy than the p state.

For the heavier alkali atoms, the experimental value of the polarizability exceeds that given by the simple estimate (3.42), and the difference increases with increasing Z. This is because core electrons, which have been neglected in making the simple estimate, contribute significantly to the polarizability. For hydrogen, the line that plays a role analogous to that of the resonance line in the alkalis is the Lyman-α line, whose energy is $(3/8)e_0^2/a_0$, and the estimate (3.42) gives a polarizability of $64/9 \approx 7.1$. This is nearly 60% more than the actual value, the reason being that this transition has an oscillator strength much less than unity. The estimate for Li is 30% high for a similar reason.

Oscillating electric fields

Next we turn to time-dependent electric fields. We assume that the electric field is in the i direction and varies in time as $\mathcal{E}(t) = \mathcal{E}_0 \cos \omega t$. Therefore the perturbation is given by

$$H' = -d_i \mathcal{E}_0 \cos \omega t = -\frac{d_i \mathcal{E}_0}{2}(e^{i\omega t} + e^{-i\omega t}). \tag{3.43}$$

By expanding the wave function ψ in terms of the complete set of unperturbed states u_n with energies E_n,

$$\psi = \sum_n a_n u_n e^{-iE_n t/\hbar}, \tag{3.44}$$

we obtain from the time-dependent Schrödinger equation the usual set of coupled equations for the expansion coefficients a_n,

$$i\hbar \dot{a}_n = \sum_k \langle n|H'|k\rangle a_k(t) e^{i\omega_{nk} t}, \tag{3.45}$$

where $\omega_{nk} = (E_n - E_k)/\hbar$, and the dot on a denotes the derivative with respect to time.

Let us consider an atom initially in an eigenstate m of the unperturbed Hamiltonian, and imagine that the perturbation is turned on at time $t = 0$. The expansion coefficients a_n for $n \neq m$ are then obtained to first order in the perturbation H' by inserting (3.43) into (3.45) and replacing a_k on the right hand side by its value δ_{km} when the perturbation is absent,

$$a_n^{(1)} = -\frac{1}{2i\hbar} \int_0^t dt' \langle n|d_i \mathcal{E}_0|m\rangle [e^{i(\omega_{nm}+\omega)t'} + e^{i(\omega_{nm}-\omega)t'}]. \tag{3.46}$$

For simplicity we assume here that the frequency is not equal to any of the transition frequencies. In Sec. 4.2 we shall relax this assumption. By carrying out the integration over time one finds

$$a_n^{(1)} = \frac{\langle n|d_i\mathcal{E}_0|m\rangle}{2\hbar}\left[\frac{e^{i(\omega_{nm}+\omega)t}-1}{\omega_{nm}+\omega} + \frac{e^{i(\omega_{nm}-\omega)t}-1}{\omega_{nm}-\omega}\right] \quad (3.47)$$

for $n \neq m$. To determine the coefficient a_m for the initial state, which yields the energy shift, we write it as $a_m = e^{i\phi_m}$, where ϕ_m is a complex phase, and we insert (3.47) into (3.45). To second order in the perturbation the result is

$$\hbar\dot\phi_m = \langle m|d_i|m\rangle\mathcal{E}_0\cos\omega t$$
$$+ \frac{\mathcal{E}_0^2}{2\hbar}\sum_{n\neq m}|\langle n|d_i|m\rangle|^2 e^{-i\omega_{nm}t}\cos\omega t\left[\frac{e^{i(\omega_{nm}+\omega)t}-1}{\omega_{nm}+\omega} + \frac{e^{i(\omega_{nm}-\omega)t}-1}{\omega_{nm}-\omega}\right]. \quad (3.48)$$

On the right hand side of (3.48) we have replaced $e^{-i\phi_m}$ by unity, since we work only to second order in the strength of the electric field. The matrix element $\langle m|d_i|m\rangle$ vanishes by symmetry when the state m is an eigenstate of the parity operator.

The shift in the energy is given by $\hbar$ times the average rate at which the phase of the state decreases in time. We therefore average (3.48) over time and obtain

$$\hbar<\dot\phi_m>_t = \frac{\mathcal{E}_0^2}{4\hbar}\sum_n\left(\frac{1}{\omega_{nm}+\omega} + \frac{1}{\omega_{nm}-\omega}\right)|\langle n|d_i|m\rangle|^2. \quad (3.49)$$

Here $<\cdots>_t$ denotes an average over one oscillation period of the electric field. We write the time-averaged energy shift $\Delta E = -\hbar<\dot\phi_m>_t$ for the ground state of the atom ($m = 0$) in a form analogous to Eq. (3.36) for static fields,

$$\Delta E = -\frac{1}{2}\alpha(\omega)<\mathcal{E}(t)^2>_t, \quad (3.50)$$

with $<\mathcal{E}(t)^2>_t = \mathcal{E}_0^2/2$. The frequency-dependent polarizability is thus

$$\alpha(\omega) = \sum_n|\langle n|d_i|0\rangle|^2\left(\frac{1}{E_n-E_0+\hbar\omega} + \frac{1}{E_n-E_0-\hbar\omega}\right)$$
$$= \sum_n\frac{2(E_n-E_0)|\langle n|d_i|0\rangle|^2}{(E_n-E_0)^2-(\hbar\omega)^2} \quad (3.51)$$

or

$$\alpha(\omega) = \frac{e^2}{m_e}\sum_n\frac{f_{n0}^i}{\omega_{n0}^2-\omega^2}. \quad (3.52)$$

In the limit $\omega \to 0$ this agrees with the result (3.39) for static fields.

Table 3.4. *Characteristic atomic energies E_i for sodium together with the corresponding frequencies E_i/h, and temperatures E_i/k. The quantity $\Delta E_{\rm res}$ is the energy of the resonance line due to the transition between the 3P and 3S levels, $\Delta E_{\rm so}$ is the spin–orbit splitting in the 3P doublet, $\Delta E_{\rm hf}$ the hyperfine splitting in the ground state, $\hbar \Gamma_e$ the linewidth of the resonance line, while $\mu_{\rm B} B$ and $\mu_{\rm N} B$ are the Zeeman energies. The magnetic field has been chosen to be $B = 0.1$ T.*

Quantity	Energy (eV)	Frequency (Hz)	Temperature (K)
$\Delta E_{\rm res}$	2.1	5.1×10^{14}	2.4×10^4
$\Delta E_{\rm so}$	2.1×10^{-3}	5.2×10^{11}	2.5×10^1
$\Delta E_{\rm hf}$	7.3×10^{-6}	1.8×10^9	8.5×10^{-2}
$\mu_{\rm B} B$	5.8×10^{-6}	1.4×10^9	6.7×10^{-2}
$\hbar \Gamma_e$	4.1×10^{-8}	1.0×10^7	4.8×10^{-4}
$\mu_{\rm N} B$	3.2×10^{-9}	7.6×10^5	3.7×10^{-5}

3.4 Energy scales

As a prelude to our discussion in Chapter 4 of trapping and cooling processes we give in the following some characteristic atomic energy scales.

Since the Zeeman energies $\mu_{\rm B} B$ and $\mu_{\rm N} B$ differ by three orders of magnitude, the interaction of the nuclear spin with the external field may generally be neglected. The magnitude of the hyperfine splitting, $\Delta E_{\rm hf}$, is comparable with the Zeeman energy $\mu_{\rm B} B$ for $B = 0.1$ T.

As we shall see in the next chapter, laser cooling exploits transitions between atomic levels which are typically separated by an energy of the order of one electron volt (eV). The two resonance lines in sodium are due to transitions from the 3P level to the 3S level with wavelengths of 589.0 nm and 589.6 nm corresponding to energy differences $\Delta E_{\rm res}$ of 2.1 eV (cf. Table 3.3). The splitting of the lines is due to the spin–orbit interaction. The spin–orbit splitting $\Delta E_{\rm so}$ involves – apart from quantum numbers associated with the particular states in question – an average of the derivative of the potential. For hydrogen one has $\Delta E_{\rm so} \sim \hbar^2 e_0^2 / m_e^2 c^2 a_0^3 = \alpha_{\rm fs}^2 e_0^2 / a_0 \sim 10^{-3}$ eV, where $\alpha_{\rm fs}$ is the fine structure constant. For sodium the splitting of the resonance line doublet is 2.1×10^{-3} eV.

Yet another energy scale that plays a role in the cooling processes is the intrinsic width of atomic levels. An atom in an excited state decays to lower states by emitting radiation. The rate of this process may be determined by using second-order perturbation theory, the perturbation being the interaction between the atom and the quantized radiation field. Alternatively, within a semi-classical approach, one may calculate the rate of absorption and stimulated emission of photons by treating the electric field classically. From this, the rate of spontaneous emission processes is obtained by using the relationship between the Einstein A and B coefficients. For a state n whose decay is allowed in the dipole approximation the rate of decay Γ_{nm} to state m by spontaneous emission is found to be [2, p. 168]

$$\Gamma_{nm} = \frac{4}{3}\omega_{nm}^3 \frac{\sum_i |\langle n|d_i|m\rangle|^2}{4\pi\epsilon_0 \hbar c^3}, \tag{3.53}$$

where $\hbar\omega_{nm}$ is the energy difference between the levels in question, while $\langle n|d_i|m\rangle$ is the dipole matrix element. This result is identical with the damping rate of a classical electric dipole moment of magnitude equal to the square root of $\sum_i |\langle n|d_i|m\rangle|^2$ and oscillating with frequency ω_{nm}. The total decay rate of an excited state n is therefore given by

$$\Gamma_e = \sum_m \Gamma_{nm}, \tag{3.54}$$

where the sum is over all states m with energy less than that of the initial state. Estimating (3.53) for the 2P–1S Lyman-α transition in hydrogen, with $\omega_{nm} \approx e_0^2/\hbar a_0$ and $|\langle n|d_i|m\rangle| \approx e a_0$, yields

$$\Gamma_e \approx \left(\frac{e_0^2}{\hbar c}\right)^3 \frac{e_0^2}{\hbar a_0}. \tag{3.55}$$

The rate of spontaneous emission is thus a factor of order $(\alpha_{\text{fs}})^3$ or 4×10^{-7} times atomic transition frequencies.

For the resonance lines in alkali atoms, the possible final states are all members of the ground-state multiplet, and therefore we may write the decay rate (3.54) as

$$\Gamma_e \approx \frac{2}{3} f_{\text{res}} \frac{e_0^2}{\hbar c} \frac{\hbar \omega_{\text{res}}^2}{m_e c^2}, \tag{3.56}$$

where $\omega_{\text{res}} = \Delta E_{\text{res}}/\hbar$ is the resonance-line frequency and f_{res} is the total oscillator strength from the excited state to all members of the ground-state multiplet. This strength, like the total strength from a member of the ground-state multiplet to all members of the excited-state multiplet, is close to unity, and for similar reasons, and therefore the decay rate of the excited state is given approximately by

$$\Gamma_e \approx \frac{2}{3} \frac{e_0^2}{\hbar c} \frac{\hbar \omega_{\text{res}}^2}{m_e c^2}. \tag{3.57}$$

For the resonance line in sodium, $\omega_{\text{res}} = 3.2 \times 10^{15}$ s^{-1} and the estimate (3.57) gives $\Gamma_e = 6.4 \times 10^7$ s^{-1}, which agrees closely with the measured value $\Gamma_e = 6.3 \times 10^7$ s^{-1}.

In Table 3.4 (see p. 55) we list for sodium the characteristic energies, frequencies, and temperatures discussed above.

Problems

PROBLEM 3.1 Calculate the Zeeman splitting of the hyperfine levels for the ^{87}Rb atom in low magnetic fields and determine the Landé g factors.

PROBLEM 3.2 Use a high-field perturbation treatment to obtain explicit expressions for the Zeeman-split hyperfine levels of ^{87}Rb and ^{133}Cs in a strong magnetic field and compare the results with Fig. 3.1.

PROBLEM 3.3 The static polarizability of a hydrogen atom in its ground state may be calculated in an approximate way from Eq. (3.34). First include only the unperturbed states $|nlm\rangle$ associated with the next-lowest unperturbed energy level, which has $n = 2$, and show that if the electric field is in the z direction, the only non-vanishing matrix element is $\langle 210|z|100\rangle$. Calculate the matrix element and use it to obtain an upper bound on the second-order correction to the energy, ΔE. Determine a lower bound on ΔE by replacing all the energy differences $E_n - E_0$ in Eq. (3.34) by the difference between the energy of the lowest excited state and that of the ground state, and use closure to evaluate the sum. The exact expression, valid to second order in $\mathcal{E}$, is $\Delta E/4\pi\epsilon_0 = -(9/4)a_0^3\mathcal{E}^2$.

References

[1] L. D. Landau and E. M. Lifshitz, *Quantum Mechanics*, Third edition, (Pergamon, Oxford, 1977).

[2] See, e.g., B. H. Bransden and C. J. Joachain, *Physics of Atoms and Molecules*, (Longman, New York, 1983).

[3] Experimental values of polarizabilities are taken from the compilation by A. Derevianko, W. R. Johnson, M. S. Safronova, and J. F. Babb, *Phys. Rev. Lett.* **82**, 3589 (1999).

4
Trapping and cooling of atoms

The advent of the laser opened the way to the development of powerful new methods for manipulating and cooling atoms which were exploited in the realization of Bose–Einstein condensation in alkali atom vapours. To set the stage we describe a typical experiment, which is shown schematically in Fig. 4.1 [1]. A beam of sodium atoms emerges from an oven at a temperature of about 600 K, corresponding to a speed of about 800 m s^{-1}, and is then passed through a so-called Zeeman slower, in which the velocity of the atoms is reduced to about 30 m s^{-1}, corresponding to a temperature of about 1 K. In the Zeeman slower, a laser beam propagates in the direction opposite that of the atomic beam, and the radiation force produced by absorption of photons retards the atoms. Due to the Doppler effect, the frequency of the atomic transition in the laboratory frame is not generally constant, since the atomic velocity varies. However, by applying an inhomogeneous magnetic field designed so that the Doppler and Zeeman effects cancel, the frequency of the transition in the rest frame of the atom may be held fixed. On emerging from the Zeeman slower the atoms are slow enough to be captured by a magneto-optical trap (MOT), where they are further cooled by interactions with laser light to temperatures of order 100 µK. Another way of compensating for the changing Doppler shift is to increase the laser frequency in time, which is referred to as 'chirping'. In other experiments the MOT is filled by transferring atoms from a second MOT where atoms are captured directly from the vapour. After a sufficiently large number of atoms (typically 10^{10}) have accumulated in the MOT, a magnetic trap is turned on and the laser beams are turned off: the atoms are then confined by a purely magnetic trap. At this stage, the density of atoms is relatively low, and the gas is still very non-degenerate, with a phase-space density of order 10^{-6}. The final step in achieving Bose–Einstein condensation is

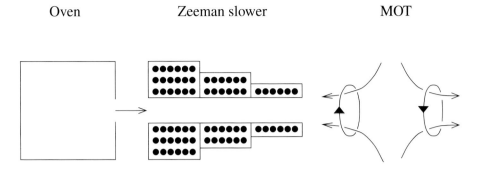

Fig. 4.1. A typical experiment to cool and trap alkali atoms.

evaporative cooling, in which relatively energetic atoms leave the system, thereby lowering the average energy of the remaining atoms.

In this chapter we describe the physics of cooling and trapping atoms. We begin with magnetic traps (Sec. 4.1). Subsequently, as a prelude to a discussion of laser cooling and trapping, we consider the effects of laser radiation on atoms and describe optical traps (Sec. 4.2). We then discuss, in Sec. 4.3, the simple theory of laser cooling and, in Sec. 4.4, the magneto-optical trap. In Sec. 4.5 an account is given of the Sisyphus cooling process. Section 4.6 is devoted to evaporative cooling. Atomic hydrogen is different from alkali atoms, in that it cannot be cooled by lasers, and the final section, Sec. 4.7, is devoted to experiments on hydrogen. For a more extensive description of many of the topics treated in this chapter, see Ref. [2]. Another useful source is the summer school lectures [3].

4.1 Magnetic traps

Magnetic trapping of neutral atoms is due to the Zeeman effect, which we described in Chapter 3: the energy of an atomic state depends on the magnetic field, and therefore an atom in an inhomogeneous field experiences a spatially-varying potential. For simplicity, let us begin by assuming that the energy of a state is linear in the magnetic field. As one can see from our earlier discussion in Chapter 3 (see, e.g., Fig. 3.1), this is true generally for the doubly polarized states, and for other states it is a good approximation provided the level shifts produced by the magnetic field are either very small or very large compared with the hyperfine splitting. The energy of an atom in a particular state i may then be written as

$$E_i = C_i - \mu_i B, \tag{4.1}$$

where μ_i is the magnetic moment of the state and C_i is a constant. The magnetic contribution to the energy thus provides a potential energy $-\mu_i B$ for the atom. If the magnetic moment is positive, the atom experiences a force tending to drive it to regions of higher field, while if it is negative, the force is towards regions of lower field. For this reason, states with a positive magnetic moment are referred to as *high-field seekers*, and those with a negative one as *low-field seekers*.

The energy depth of magnetic traps is determined by the Zeeman energy, $\mu_i B$. Atomic magnetic moments are of order the Bohr magneton, $\mu_B = e\hbar/2m_e$, which in temperature units is approximately 0.67 K/T. Since laboratory magnetic fields are generally considerably less than 1 tesla, the depth of magnetic traps is much less than a kelvin, and therefore atoms must be cooled in order to be trapped magnetically.

The task of constructing a magnetic trap is thus to design magnetic field configurations with either a local minimum in the magnitude of the magnetic field, or a local maximum. The latter possibility is ruled out by a general theorem, published surprisingly recently, that a local maximum in $|\mathbf{B}|$ is impossible in regions where there are no electrical currents [4]. Thus the case of interest is that of a local minimum, and consequently the only atomic states that can be trapped by magnetic fields alone are low-field seekers. Magnetic field configurations with a local minimum in $|\mathbf{B}|$ have been important over the past few decades for trapping *charged* particles as a step in the continuing quest to realize nuclear fusion in hot plasmas. Then the trapping results not from the intrinsic magnetic moment of the particle, but rather from the magnetic moment associated with its cyclotron motion. However, despite the very different physical conditions in the two cases, the requirements in terms of the magnetic field configuration are quite similar, and the design of traps for cold atoms has been significantly influenced by work on plasmas. Field configurations with a minimum in $|\mathbf{B}|$ may be divided into two classes: ones where the minimum of the field is zero, and those where it is non-zero. We shall now describe these in turn.

4.1.1 The quadrupole trap

A simple magnetic field configuration in which the magnetic field vanishes at some point is the standard quadrupole one, in which the magnetic field varies linearly with distance in all directions. Such a magnetic field may be produced by, e.g., a pair of opposed Helmholtz coils, as in standard focusing magnets. For definiteness, let us consider a situation with axial symmetry about the z direction. If we denote the magnetic field gradients along the

x and y axes by B', the gradient along the z axis must be $-2B'$, since the divergence of the magnetic field vanishes, $\boldsymbol{\nabla}\cdot\mathbf{B} = 0$. The magnetic field in the vicinity of the minimum, whose position we choose to be at the origin of the coordinate system, is thus given by

$$\mathbf{B} = B'(x, y, -2z). \tag{4.2}$$

The magnitude of the field is given by $B = B'(x^2 + y^2 + 4z^2)^{1/2}$, and thus it varies linearly with distance from the minimum, but with a slope that depends on direction.

The quadrupole trap suffers from one important disadvantage. In the above discussion of the effective potential produced by a magnetic field, we assumed implicitly that atoms remain in the same quantum state. This is a good approximation provided the magnetic field experienced by an atom changes slowly with time, since the atom then remains in the same quantum state relative to the instantaneous direction of the magnetic field: it is said to follow the magnetic field variations adiabatically. However, a moving atom experiences a time-dependent magnetic field, which will induce transitions between different states. In particular, atoms in low-field seeking states may make transitions to high-field seeking ones, and thereby be ejected from the trap. The effects of the time-dependent magnetic field become serious if its frequency is comparable with or greater than the frequencies of transitions between magnetic sublevels. The latter are of order $\mu_B B$, and therefore vanish if $B = 0$. Thus trap losses can be appreciable in the vicinity of the zero-field point: the quadrupole trap effectively has a 'hole' near the node in the field, and this limits the time for which atoms can be stored in it.

This disadvantage of the simple quadrupole trap may be circumvented in a number of ways. One of these is to 'plug the hole' in the trap. In the first successful experiment to realize Bose–Einstein condensation this was done by applying an oscillating bias magnetic field, as we shall explain in the next subsection. An alternative approach, adopted by the MIT group of Ketterle and collaborators in early experiments [5], is to apply a laser field in the region of the node in the magnetic field. The resulting radiation forces repel atoms from the vicinity of the node, thereby reducing losses. The physics of this mechanism will be described in Sec. 4.2. Instead of using traps having a node in the magnetic field, one can remove the 'hole' by working with magnetic field configurations that have a non-zero field at the minimum. These will be described in Sec. 4.1.3.

4.1.2 The TOP trap

As mentioned in Chapter 1, Bose–Einstein condensation in dilute gases was first achieved in experiments using a modified quadrupole trap known as the time-averaged orbiting potential (TOP) trap. In this trap one superimposes on the quadrupole field a rotating, spatially-uniform, magnetic field [6]. For definiteness we consider the geometry used in the original experiment, where the oscillating magnetic field has components $B_0 \cos \omega t$ in the x direction, and $B_0 \sin \omega t$ in the y direction [7]. The instantaneous field is therefore given by

$$\mathbf{B} = (B'x + B_0 \cos \omega t, B'y + B_0 \sin \omega t, -2B'z). \tag{4.3}$$

Thus the effect of the oscillating bias field is to move the instantaneous position of the node in the magnetic field. The frequency of the bias field is chosen to be low compared with the frequencies of transitions between magnetic substates. This condition ensures that an atom will remain in the same quantum state relative to the instantaneous magnetic field, and therefore will not undergo transitions to other hyperfine states and be lost from the trap. Under these conditions the effect of the bias field may be described in terms of an oscillatory component of the energy of an atom. If the frequency of the bias field is chosen to be much greater than that of the atomic motions, an atom moves in an effective potential given by the time average of the instantaneous potential over one rotation period of the field. In experiments, frequencies of atomic motions are typically $\sim 10^2$ Hz, frequencies of transitions between magnetic substates are of order $\sim 10^6$ Hz or more, and the frequency of the bias field is typically in the kilohertz range.

To determine the effective potential we first evaluate the instantaneous strength of the magnetic field, which is given by

$$\begin{aligned} B(t) &= [(B_0 \cos \omega t + B'x)^2 + (B_0 \sin \omega t + B'y)^2 + 4B'^2 z^2]^{1/2} \\ &\simeq B_0 + B'(x \cos \omega t + y \sin \omega t) \\ &\quad + \frac{B'^2}{2B_0}[x^2 + y^2 + 4z^2 - (x \cos \omega t + y \sin \omega t)^2], \end{aligned} \tag{4.4}$$

where the latter form applies for small distances from the node of the quadrupole field, $r \ll |B_0/B'|$. The time average, $<B>_t$, of the magnitude of the magnetic field over a rotation period of the field is defined by

$$<B>_t = \frac{\omega}{2\pi} \int_0^{2\pi/\omega} dt\, B(t). \tag{4.5}$$

By performing the time average, we find from (4.4) that

$$<B>_t \simeq B_0 + \frac{B'^2}{4B_0}(x^2 + y^2 + 8z^2). \tag{4.6}$$

The important feature of this result is that the time-averaged field never vanishes, and consequently there is no longer a 'hole' in the trap. The magnetic contribution to the energy of an atom in a magnetic substate i is thus given for small r by

$$\begin{aligned} E_i(<B>_t) &\simeq E_i(B_0) - \mu_i(B_0)(<B>_t - B_0) \\ &\simeq E_i(B_0) - \mu_i(B_0)\frac{B'^2}{4B_0}(x^2 + y^2 + 8z^2), \end{aligned} \tag{4.7}$$

where

$$\mu_i = -\left.\frac{\partial E_i}{\partial B}\right|_{B_0} \tag{4.8}$$

is the projection of the magnetic moment in the direction of the magnetic field. For doubly polarized states, in which the nuclear and electron spins have the largest possible projections on the magnetic field direction and are in the same direction, the magnetic moment is independent of the magnetic field (see Sec. 3.2). The oscillating bias field thus converts the linear dependence of magnetic field strength on distance from the node in the original quadrupole field to a quadratic one, corresponding to an anisotropic harmonic-oscillator potential. The angular frequencies for motion in the three coordinate directions are

$$\omega_x^2 = \omega_y^2 = -\mu_i \frac{B'^2}{2mB_0}, \tag{4.9}$$

and

$$\omega_z^2 = 8\omega_x^2 = -8\mu_i \frac{B'^2}{2mB_0}. \tag{4.10}$$

Different choices for the rotation axis of the bias field give traps with different degrees of anisotropy (see Problem 4.3).

Another force which can be important in practice is gravity. This gives rise to a potential which is linear in distance. If the potential produced by the TOP trap were purely harmonic, the only effect of gravity would be to displace the minimum in the harmonic-oscillator potential. However, the harmonic behaviour of the TOP trap potential extends only to a distance of order $l = B_0/|B'|$ from the origin, beyond which the bias magnetic field has little effect. If the gravitational force is strong enough to displace the minimum a distance greater than or of order $B_0/|B'|$, the total potential no longer has the form (4.7) when expanded about the new minimum. Gravity modifies the trapping potential appreciably if the gravitational force mg, where g is the acceleration due to gravity, exceeds that due to the magnetic trap at a distance $B_0/|B'|$ from the origin. From Eq. (4.7) one sees that the force constant of the trap is of order $|\mu_i|B'^2/B_0$, and therefore the force at a distance $B_0/|B'|$ from the origin is of order $|\mu_i B'|$. Thus gravity is important if

$$|\mu_i B'| \lesssim mg. \tag{4.11}$$

By appropriate choice of magnetic field strengths and of the direction of the axis of a magnetic trap relative to that of the gravitational force it is possible to make the minimum of the potential due to both magnetic and gravitational forces lie at a point such that the force constants of the trap are not in the usual ratio for a TOP trap in the absence of gravity.

4.1.3 Magnetic bottles and the Ioffe–Pritchard trap

An inhomogeneous magnetic field with a minimum in the magnetic field at a non-zero value may be generated by a configuration based on two Helmholtz coils with identical currents circulating in the *same* direction, in contrast to the simple quadrupole field, which is generated by Helmholtz coils with the currents in the two coils in *opposite* directions. Let us investigate the form of the magnetic field in the vicinity of the point midway between the coils on their symmetry axis, which we take to be the origin of our coordinate system. We denote the coordinate in the direction of the symmetry axis by z, and the distance from the axis by ρ. Since there are no currents in the vicinity of the origin, the magnetic field may be derived from a scalar potential Φ, $\mathbf{B} = -\nabla\Phi$. We assume the current coils to be rotationally symmetric about the z axis, and thus the magnetic field is independent of the azimuthal angle. Since the field is an even function of the coordinates, the potential must be an odd one, and therefore an expansion of the potential in terms of spherical harmonics Y_{lm} can contain only terms with odd order l. Because of the rotational invariance about the symmetry axis, the potential is a function only of the distance from the origin, $r = (z^2 + \rho^2)^{1/2}$, and of $\cos\theta = z/r$. The potential satisfies Laplace's equation, and it may be written simply as

$$\Phi = \sum_l A_l r^l P_l(z/r), \tag{4.12}$$

where the P_l are Legendre polynomials and the A_l are coefficients. There can be no terms with inverse powers of r since the field is finite at the origin.

In the immediate vicinity of the origin we may restrict ourselves to the first two terms in this expansion, and therefore one has

$$\Phi = A_1 r P_1(\cos\theta) + A_3 r^3 P_3(\cos\theta) \qquad (4.13)$$

$$= A_1 z + A_3 (\frac{5}{2} z^3 - \frac{3}{2} z r^2) \qquad (4.14)$$

$$= A_1 z + A_3 (z^3 - \frac{3}{2} z \rho^2), \qquad (4.15)$$

and the magnetic field has components

$$B_z = -A_1 - 3A_3(z^2 - \frac{1}{2}\rho^2), \quad B_\rho = 3A_3 z\rho, \quad \text{and} \quad B_\varphi = 0, \qquad (4.16)$$

where φ is the azimuthal angle. If A_1 and A_3 have the same sign, the magnetic field increases in magnitude with increasing $|z|$. Such a field configuration is referred to as a 'magnetic bottle' in plasma physics. Provided its energy is not too high, a charged particle gyrating about the field will be reflected as it moves towards regions of higher field, and thereby contained. Neutral particles, unlike charged ones, can move freely perpendicular to the direction of the magnetic field, and therefore to trap them magnetically the magnetic field must increase in directions perpendicular to the magnetic field as well as parallel to it. To second order in z and ρ the magnitude of the magnetic field is given by

$$B = A_1 + 3A_3(z^2 - \frac{1}{2}\rho^2), \qquad (4.17)$$

and therefore the magnetic field does not have a local minimum at the origin.

The problem of creating a magnetic field with a local minimum in B arose in the context of plasma physics, and one way of doing this was proposed by Ioffe [8]. It is clear from the expansion of the potential given above that this cannot be done if the magnetic field is axially symmetric: there are only two adjustable coefficients, and there is no way to create a local minimum. The suggestion Ioffe made was to add to the currents creating the magnetic bottle other currents that break the rotational invariance about the symmetry axis of the bottle. In plasma physics the configuration used is that shown in Fig. 4.2, where the additional currents are carried by conductors parallel to the symmetry axis (so-called 'Ioffe bars'). If the magnitude of the current is the same for each bar, it follows from symmetry under rotation about the axis of the system that the potential produced by the currents in the bars must correspond to a potential having components with degree m equal to 2 or more. Thus the lowest-order spherical harmonics that can contribute are $Y_{2,\pm 2} \propto (\rho/r)^2 e^{\pm i 2\varphi}$, and the corresponding solutions of Laplace's equation

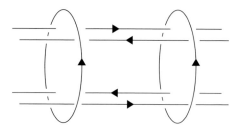

Fig. 4.2. Currents that generate the Ioffe–Pritchard configuration for the magnetic field.

are $r^2 Y_{2,\pm 2} \propto \rho^2 e^{\pm i2\varphi}$. The leading term in the expansion of the potential near the origin must therefore be of the form

$$\Phi = \frac{\rho^2}{2}\left[Ce^{i2\varphi} + C^*e^{-i2\varphi}\right], \qquad (4.18)$$

since terms with higher values of l have higher powers of r. Here C is a constant determined by the current in the bars and their geometry. If the zero of the azimuthal angle is taken to lie midway between two adjacent conductors, C must be real, and by symmetry the potential function must be proportional to $x^2 - y^2$, since on the x and y axes the magnetic field lies along the axes. The components of the field due to the Ioffe bars are therefore

$$B_x = -Cx, \quad B_y = Cy, \quad \text{and} \quad B_z = 0. \qquad (4.19)$$

When this field is added to that of the magnetic bottle the magnitude of the total magnetic field is given to second order in the coordinates by

$$B = A_1 + 3A_3(z^2 - \frac{1}{2}\rho^2) + \frac{C^2}{2A_1}\rho^2, \qquad (4.20)$$

and consequently the magnitude of the field has a local minimum at the origin if $C^2 > 3A_1A_3$, that is for sufficiently strong currents in the bars. A convenient feature of this trap is that by adjusting the current in the coils relative to that in the bars it is possible to make field configurations with different degrees of curvature in the axial and radial directions.

The use of such magnetic field configurations to trap neutral atoms was first proposed by Pritchard [9], and in the neutral atom literature this trap is commonly referred to as the Ioffe–Pritchard trap. A variant of it which is convenient in practice is the clover-leaf trap, where the non-axially-symmetric field is produced by coils configured as shown in Fig. 4.3 [1]. The windings look very different from the Ioffe bars, but since the azimuthal symmetry of the currents corresponds to $m = 2$ or more, as in the Ioffe

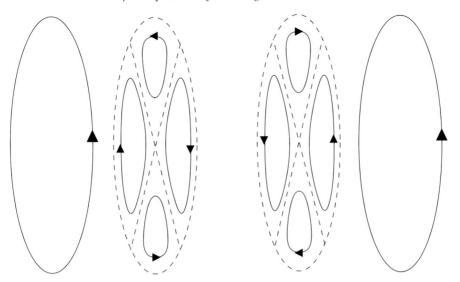

Fig. 4.3. Schematic view of the currents in the clover-leaf trap.

case, the field near the origin must have the same symmetry. To first order in the coordinates the component of the magnetic field in the axial direction vanishes, and the first non-vanishing contributions are of third order. A virtue of the clover-leaf configuration relative to the original Ioffe one is that experimental access to the region near the field minimum is easier due to the absence of the current-carrying bars.

4.2 Influence of laser light on an atom

Many techniques for trapping and cooling atoms exploit the interaction of atoms with radiation fields, especially those of lasers. As a prelude to the applications later in this chapter, we give here a general description of the interaction of an atom with a radiation field. In Sec. 3.3 we calculated the polarizability of atoms with non-degenerate ground states, and here we shall generalize the treatment to allow for the lifetime of the excited state, and arbitrary directions for the electric field.

The interaction between an atom and the electric field is given in the dipole approximation by

$$H' = -\mathbf{d}\cdot\boldsymbol{\mathcal{E}}, \qquad (4.21)$$

where $\mathbf{d}$ is the electric dipole moment operator and $\boldsymbol{\mathcal{E}}$ is the electric field vector. In a static electric field the change ΔE_g in the ground-state energy

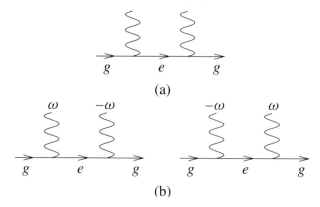

Fig. 4.4. Diagrammatic representation of second-order contributions to the energy of an atom in its ground state for (a) a static electric field and (b) a time-dependent field. The straight lines correspond to the atom and the wavy ones to the interaction with the electric field.

of an atom is given to second order in the electric field by

$$\Delta E_g = -\sum_e \frac{|\langle e|H'|g\rangle|^2}{E_e - E_g} = -\frac{1}{2}\alpha\mathcal{E}^2, \qquad (4.22)$$

where

$$\alpha = 2\sum_e \frac{|\langle e|\mathbf{d}\cdot\hat{\epsilon}|g\rangle|^2}{E_e - E_g} \qquad (4.23)$$

is the atomic polarizability. Here $\hat{\epsilon}$ is a unit vector in the direction of the electric field, and we label the ground state by g and the excited states by e. This contribution to the energy may be represented by the diagram shown in Fig. 4.4 (a). The interaction vertices give factors $\langle e|(-\mathbf{d}\cdot\mathcal{E})|g\rangle$ and $\langle g|(-\mathbf{d}\cdot\mathcal{E})|e\rangle$, and the line for the intermediate atomic state gives a factor $1/(E_e - E_g)$ to the summand, as one may confirm by comparing with the explicit calculation in Sec. 3.3. To describe a time-dependent electric field with frequency ω we write the electric field as $\mathcal{E}(\mathbf{r}, t) = \mathcal{E}_\omega e^{-i\omega t} + \mathcal{E}_{-\omega} e^{i\omega t}$. Since the electric field is real, the condition

$$\mathcal{E}_{-\omega} = \mathcal{E}_\omega^* \qquad (4.24)$$

must be satisfied. In Sec. 3.3 we calculated the energy shift due to this time-dependent electric field by conventional perturbation theory. It is instructive, however, to derive the result (3.50) in an alternative fashion, by use of diagrammatic perturbation theory. The second-order contribution to the energy may be expressed as the sum of two terms which are represented by the diagrams shown in Fig. 4.4 (b). These are identical apart from the reversal of the order in which the dipolar perturbations with positive and negative frequencies occur. The term varying as $e^{-i\omega t}$ corresponds in a quantum-mechanical description of the radiation field to absorption of a photon, and that varying as $e^{i\omega t}$ to emission. The terms where either the positive-frequency component of the field or the negative-frequency one act twice do not contribute to the energy shift, since the final state differs from the initial one by the addition or removal of two photons (these are the terms which averaged to zero in the perturbation calculation of Sec. 3.3). By generalizing the approach used above for a static field, one finds for the energy shift

$$\begin{aligned}\Delta E_g &= \sum_e \langle g|\mathbf{d}\cdot\boldsymbol{\mathcal{E}}_\omega|e\rangle \frac{1}{E_g - E_e + \hbar\omega} \langle e|\mathbf{d}\cdot\boldsymbol{\mathcal{E}}_{-\omega}|g\rangle \\ &+ \sum_e \langle g|\mathbf{d}\cdot\boldsymbol{\mathcal{E}}_{-\omega}|e\rangle \frac{1}{E_g - E_e - \hbar\omega} \langle e|\mathbf{d}\cdot\boldsymbol{\mathcal{E}}_\omega|g\rangle \\ &= \sum_e |\langle e|\mathbf{d}\cdot\hat{\boldsymbol{\epsilon}}|g\rangle|^2 \left(\frac{1}{E_g - E_e - \hbar\omega} + \frac{1}{E_g - E_e + \hbar\omega}\right) |\mathcal{E}_\omega|^2 \\ &= -\alpha(\omega)|\mathcal{E}_\omega|^2 \\ &= -\frac{1}{2}\alpha(\omega) <\mathcal{E}(\mathbf{r},t)^2>_t, \end{aligned} \quad (4.25)$$

where $<\cdots>_t$ denotes a time average, and the dynamical polarizability is given by

$$\begin{aligned}\alpha(\omega) &= \sum_e |\langle e|\mathbf{d}\cdot\hat{\boldsymbol{\epsilon}}|g\rangle|^2 \left(\frac{1}{E_e - E_g + \hbar\omega} + \frac{1}{E_e - E_g - \hbar\omega}\right) \\ &= \sum_e \frac{2(E_e - E_g)|\langle e|\mathbf{d}\cdot\hat{\boldsymbol{\epsilon}}|g\rangle|^2}{(E_e - E_g)^2 - (\hbar\omega)^2}, \end{aligned} \quad (4.26)$$

in agreement with Eqs. (3.50) and (3.51). Note that the only difference from the static case is that the intermediate energy denominators are shifted by $\pm\hbar\omega$ to take into account the non-zero frequency of the electric field.

In many situations of interest the frequency of the radiation is close to that of an atomic resonance, and it is then a good approximation to neglect all transitions except the resonant one. In addition, in the expression for the polarizability one may take only the term with the smallest energy denominator. The polarizability then reduces to a single term

$$\alpha(\omega) \approx \frac{|\langle e|\mathbf{d}\cdot\hat{\boldsymbol{\epsilon}}|g\rangle|^2}{E_e - E_g - \hbar\omega}. \quad (4.27)$$

In the above discussion we implicitly assumed that the excited state has an infinitely long lifetime. However, in reality it will decay by spontaneous emission of photons. This effect can be taken into account phenomenologically by attributing to the excited state an energy with both real and imaginary parts. If the excited state has a lifetime $1/\Gamma_e$, corresponding to the e-folding time for the occupation probability of the state, the corresponding e-folding time for the *amplitude* will be twice this, since the probability is equal to the squared modulus of the amplitude. If, in the usual way, the amplitude of the excited state is taken to vary as $\exp(-iE_e t/\hbar)$, exponential decay of the amplitude with a lifetime $2/\Gamma_e$ corresponds to an imaginary contribution to the energy of the state equal to $-i\hbar\Gamma_e/2$. The polarizability is then

$$\alpha(\omega) \approx \frac{|\langle e|\mathbf{d}\cdot\hat{\boldsymbol{\epsilon}}|g\rangle|^2}{E_e - i\hbar\Gamma_e/2 - E_g - \hbar\omega}. \tag{4.28}$$

Quite generally, the energy of the ground state is a complex quantity, and we shall write the energy shift as

$$\Delta E_g = V_g - i\hbar\Gamma_g/2. \tag{4.29}$$

This has the form of an effective potential acting on the atom, the real part corresponding to a shift of the energy of the state, and the imaginary part to a finite lifetime of the ground state due to transitions to the excited state induced by the radiation field, as described above. The shift of the energy level is given by

$$V_g = -\frac{1}{2}\alpha'(\omega) <\mathcal{E}(\mathbf{r},t)^2>_t, \tag{4.30}$$

where

$$\alpha'(\omega) \approx \frac{(E_e - E_g - \hbar\omega)|\langle e|\mathbf{d}\cdot\hat{\boldsymbol{\epsilon}}|g\rangle|^2}{(E_e - E_g - \hbar\omega)^2 + (\hbar\Gamma_e/2)^2} \tag{4.31}$$

is the real part of α. These shifts are sometimes referred to as ac Stark shifts, since the physics is the same as for the usual Stark effect except that the electric field is time-dependent.

It is convenient to introduce the *detuning*, which is the difference between the laser frequency and the frequency $\omega_{eg} = (E_e - E_g)/\hbar$ of the atomic transition:

$$\delta = \omega - \omega_{eg}. \tag{4.32}$$

Positive δ is referred to as *blue* detuning, and negative δ as *red* detuning. The energy shift is given by

$$V_g = \frac{\hbar\Omega_R^2 \delta}{\delta^2 + \Gamma_e^2/4}. \tag{4.33}$$

Here we have introduced the *Rabi frequency*, which is the magnitude of the perturbing matrix element $|\langle e|\mathbf{d}\cdot\boldsymbol{\mathcal{E}}_\omega|g\rangle|$ expressed as a frequency:

$$\Omega_R = |\langle e|\mathbf{d}\cdot\boldsymbol{\mathcal{E}}_\omega|g\rangle|/\hbar. \tag{4.34}$$

Ground-state energy shifts are thus positive for blue detuning and negative for red detuning.

The rate of loss of atoms from the ground state is given by

$$\Gamma_g = -\frac{2}{\hbar}\,\mathrm{Im}\,\Delta E_g = \frac{1}{\hbar}\alpha''(\omega)<\mathcal{E}(\mathbf{r},t)^2>_t, \tag{4.35}$$

where α'' is the imaginary part of α,

$$\alpha''(\omega) \approx \frac{\hbar\Gamma_e/2}{(E_e - E_g - \hbar\omega)^2 + (\hbar\Gamma_e/2)^2}|\langle e|\mathbf{d}\cdot\hat{\boldsymbol{\epsilon}}|g\rangle|^2. \tag{4.36}$$

Thus the rate of transitions from the ground state has a Lorentzian dependence on frequency in this approximation.

The perturbative treatment given above is valid provided the admixture of the excited state into the ground state is small. To lowest order in the perturbation, this admixture is of order the matrix element of the perturbation divided by the excitation energy of the intermediate state. If the decay of the intermediate state may be neglected, the magnitude of the energy denominator is $|\hbar(\omega_{eg} - \omega)| = \hbar|\delta|$ and, with allowance for decay, the effective energy denominator has a magnitude $\hbar(\delta^2 + \Gamma_e^2/4)^{1/2}$. The condition for validity of perturbation theory is therefore $|\langle e|\mathbf{d}\cdot\boldsymbol{\mathcal{E}}_\omega|g\rangle| \ll \hbar(\delta^2 + \Gamma_e^2/4)^{1/2}$ or $\Omega_R \ll (\delta^2 + \Gamma_e^2/4)^{1/2}$. For larger electric fields it is necessary to go beyond simple perturbation theory but, fortunately, under most conditions relevant for experiments on Bose–Einstein condensation, electric fields are weak enough that the perturbative approach is adequate.

4.2.1 Forces on an atom in a laser field

Experiments on clouds of dilute gases exploit the forces on atoms in a laser field in a variety of ways. Before discussing specific applications, we describe the origin of these forces. The energy shift of an atom may be regarded as an effective potential V in which the atom moves. This way of viewing the problem is sometimes referred to as the *dressed atom picture*, since the energy of interest is that of an atom with its accompanying perturbations in the radiation field, not just an isolated atom. It is the analogue of the concept of an elementary excitation, or quasiparticle, that has been so powerful in understanding the properties of solids and quantum liquids. If the time-averaged electric field varies with position, the shift of the energy due to the field gives rise to a force

$$\mathbf{F}_{\text{dipole}} = -\boldsymbol{\nabla} V(\mathbf{r}) = \frac{1}{2}\alpha'(\omega)\boldsymbol{\nabla} <\mathcal{E}(\mathbf{r},t)^2>_t \qquad (4.37)$$

on an atom. Here $<\cdots>_t$ denotes a time average. This result may be understood as being due to the interaction of the induced dipole moment of the atom with a spatially-varying electric field, and it is often referred to as the *dipole force*. More generally, terms due to higher moments of the electric charge distribution such as the quadrupole moment will also give rise to forces, but these will usually be much less than the dipole force.

At low frequencies the polarizability is positive, and the dipole moment is in the same direction as the electric field. However, at frequencies above those of transitions contributing importantly to the polarizability, the induced dipole moment is in the opposite direction, as one can see by inspection of Eq. (4.31). It is illuminating to consider a frequency close to a resonance, in which case we may use the approximate form (4.31) for the real part of the polarizability. From this one can see that for frequencies below the resonance the force is towards regions of higher electric field, while for ones above it the force is towards regions of lower field. As can be seen from Eqs. (4.37) and (4.31), the magnitude of the force can be of order ω_{eg}/Γ_e times larger than for static fields of the same strength.

As we remarked above, the radiation force repelling atoms from regions of high electric field at frequencies above resonance has been used to reduce loss of atoms at the centre of a quadrupole trap [5]. A blue-detuned laser beam passing through the trap centre gave a repulsive contribution to the energy of atoms, which were thereby prevented from penetrating into the dangerous low-field region, where spin flips could occur.

The change in sign of the force at resonance may be understood in terms of a classical picture of an electron moving in a harmonic potential under the influence of an electric field. The equation of motion for the atomic dipole moment $\mathbf{d} = -e\mathbf{r}$, where $\mathbf{r}$ is the coordinate of the electron, is

$$\frac{d^2\mathbf{d}}{dt^2} + \omega_0^2\mathbf{d} = \frac{e^2}{m_e}\boldsymbol{\mathcal{E}}, \qquad (4.38)$$

with ω_0 being the frequency of the harmonic motion. For an electric field which oscillates in time as $\exp(-i\omega t)$, we then obtain

$$(-\omega^2 + \omega_0^2)\mathbf{d} = \frac{e^2}{m_e}\boldsymbol{\mathcal{E}}. \qquad (4.39)$$

This shows that the polarizability, the ratio of the dipole moment to the electric field, is

$$\alpha(\omega) = \frac{e^2}{m_e(\omega_0^2 - \omega^2)}, \qquad (4.40)$$

which becomes negative for frequencies ω that exceed ω_0. If we compare this result with the quantum-mechanical one, Eq. (3.52), we see that it corresponds to having oscillator strength unity for a single transition at the oscillator frequency.

In addition to the contribution to the force on an atom due to energy-level shifts, which are associated with virtual transitions between atomic states, there is another one due to real transitions. Classically this is due to the radiation pressure on the atom. In quantum-mechanical language it is a consequence of the momentum of a photon being imparted to or removed from an atom during an absorption or an emission process. The rate of absorption of photons by an atom in the ground state is equal to the rate of excitation of the ground state, given by (4.35). Therefore, if the radiation field is a travelling wave with wave vector $\mathbf{q}$, the total force on the atom due to absorption processes is

$$\mathbf{F}_{\text{rad}} = \hbar \mathbf{q} \Gamma_g. \qquad (4.41)$$

As we shall describe, both this force and the dipole force (4.37) play an important role in laser cooling.

4.2.2 Optical traps

By focusing a laser beam it is possible to create a radiation field whose intensity has a maximum in space. If the frequency of the light is detuned to the red, the energy of a ground-state atom has a spatial minimum, and therefore it is possible to trap atoms. The depth of the trap is determined by the magnitude of the energy shift, given by Eq. (4.33).

One advantage of optical traps is that the potential experienced by an alkali atom in its ground state is essentially independent of the magnetic substate. This is due to the outermost electron in the ground state of alkali atoms being in an s state. The situation is quite different for trapping by magnetic fields, since the potential is then strongly dependent on the magnetic substate. With magnetic traps it is difficult to investigate the influence of the interaction energy on the spin degrees of freedom of an atomic cloud since the energy is dominated by the Zeeman term. By contrast, optical traps are well suited for this purpose.

Optical traps are also important in the context of Feshbach resonances. As we shall describe in Sec. 5.4.2, in the vicinity of such a resonance the effective interaction is a strong function of the magnetic field, and therefore it is desirable that the magnetic field be homogeneous. This may be achieved by applying a uniform magnetic field to atoms in an optical trap, but it is not possible with magnetic traps, since without inhomogeneity of the magnetic field there is no trapping.

To reduce heating of atoms by absorption of photons, the laser frequency in optical traps must be chosen to be away from atomic resonances. For example, a Bose–Einstein condensate of Na atoms has been held in a purely optical trap by Stamper-Kurn et al. [10]. The laser had a wavelength of 985 nm while that of the atomic transition is 589 nm. The resulting optical traps are shallow, with depths of order µK in temperature units (see Problem 4.4), and therefore atoms must be precooled in other sorts of traps before they can be held by purely optical forces.

4.3 Laser cooling: the Doppler process

The basic idea that led to the development of laser cooling may be understood by considering an atom subjected to two oppositely directed laser beams of the same angular frequency, ω, and the same intensity. Imagine that the frequency of the laser beams is tuned to lie just below the frequency, ω_{eg}, of an atomic transition between an excited state $|e\rangle$ and the ground state $|g\rangle$. For definiteness we assume that the laser beams are directed along the z axis.

To estimate the frictional force, we assume that the radiation field is sufficiently weak that the absorption may be calculated by perturbation theory. From Eqs. (4.35) and (4.36), the rate dN_{ph}/dt at which a single atom absorbs photons from one of the beams is given by

$$\frac{dN_{\text{ph}}}{dt} = CL(\omega), \qquad (4.42)$$

where

$$C = \frac{\pi}{\hbar^2}|\langle e|\mathbf{d}\cdot\hat{\boldsymbol{\epsilon}}|g\rangle|^2 <\mathcal{E}(\mathbf{r},t)^2> \qquad (4.43)$$

and L is the Lorentzian function

$$L(\omega) = \frac{\Gamma_e/2\pi}{(\omega-\omega_{eg})^2 + (\Gamma_e/2)^2}, \qquad (4.44)$$

which is normalized so that its integral over ω is unity. The lifetime of an atom in the ground state in the presence of one of the laser beams is $1/CL(\omega)$. An atom initially at rest will absorb as many left-moving photons as right-moving ones, and thus the total momentum change will be zero

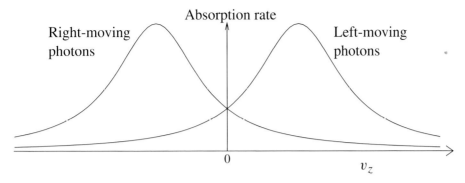

Fig. 4.5. Rate of absorption of photons from the two red-detuned laser beams as functions of the atomic velocity.

on average. However, for an atom moving to the right with velocity v_z, the frequency of the right-moving photons in the rest frame of the atom is decreased due to the Doppler effect, and to lowest order in the velocity it is given by $\omega - v_z q$, where $q = \omega/c$ is the wave number of the photons. The frequency thus lies further from the atomic resonance than it would for an atom at rest, and consequently the rate of absorption of right-moving photons is reduced, and is given approximately by

$$\frac{dN_{\text{right}}}{dt} = CL(\omega - v_z q). \tag{4.45}$$

For photons moving to the left the story is the opposite: the frequency of the photons in the rest frame of the atom is increased to $\omega + v_z q$, and consequently absorption of these photons is increased:

$$\frac{dN_{\text{left}}}{dt} = CL(\omega + v_z q). \tag{4.46}$$

The situation is represented schematically in Fig. 4.5, where we show the absorption of photons from the two beams as functions of the velocity of the atom.

Since the absorption of a photon is accompanied by transfer to the atom of momentum $\hbar q$ in the direction of propagation of the photon, the absorption of photons from the two laser beams produces a frictional force on the atom. The net rate of transfer of momentum to the atom is given by

$$\frac{dp_z}{dt} = -\gamma v_z, \tag{4.47}$$

where the friction coefficient γ is defined by

$$\gamma = \frac{\hbar q C}{v_z}[L(\omega + v_z q) - L(\omega - v_z q)] \simeq 2\hbar q^2 C \frac{dL(\omega)}{d\omega}, \tag{4.48}$$

and in the second form we have assumed that the atom moves sufficiently slowly that the Doppler shift is small compared with the larger of the linewidth and the detuning. The characteristic braking time, which determines the rate of loss of momentum by an atom, is given by

$$\frac{1}{\tau_{\text{fric}}} = -\frac{1}{p_z}\frac{dp_z}{dt} = \frac{\gamma}{m}. \tag{4.49}$$

For narrow lines, $dL/d\omega$ can be very large in magnitude if the detuning is of order Γ_e, and the frictional force is correspondingly large. Because of this, the configuration of oppositely directed laser beams is referred to as *optical molasses*. The frictional force is strong only for a limited range of velocities because, if the velocity of the atom exceeds the larger of Γ_e/q and $|\delta|/q$, the linear expansion fails, and the force is reduced.

We now estimate the lowest atomic kinetic energies that one would expect to be attainable with the configuration described above. Absorption of photons by atoms, as well as giving rise to the frictional force, also heats them. An atom at rest is equally likely to absorb photons from either of the beams and, since absorption events are uncorrelated with each other, the momentum of the atom undergoes a random walk. In such a walk, the mean-square change in a quantity is the total number of steps in the walk times the square of the step size. The total number of photons absorbed per unit time is given by

$$\frac{dN_{\text{ph}}}{dt} = 2CL(\omega), \tag{4.50}$$

since for small velocities the Doppler shifts may be neglected, and therefore the momentum diffusion coefficient, which is the rate of change of the mean-square momentum $\overline{p_z^2}$ of the atom due to absorption of photons, is given by

$$\left.\frac{d\overline{p_z^2}}{dt}\right|_{\text{abs}} = 2CL(\omega)(\hbar q)^2. \tag{4.51}$$

The emission of photons as the atom de-excites also contributes to the random walk of the momentum. Just how large this effect is depends on detailed assumptions about the emission pattern. If one makes the somewhat artificial assumption that the problem is purely one-dimensional, and that the photons are always emitted in the direction of the laser beams, the step size of the random walk for the z momentum of the atom is again $\hbar q$, and the total number of photons emitted is equal to the number absorbed. Thus the rate of change of the mean-square momentum is precisely the same as for absorption,

$$\left.\frac{d\overline{p_z^2}}{dt}\right|_{\text{em}} = \left.\frac{d\overline{p_z^2}}{dt}\right|_{\text{abs}}. \tag{4.52}$$

The total momentum diffusion coefficient, D_p, due to both absorption and emission of photons is thus given by

$$D_p = \left.\frac{d\overline{p_z^2}}{dt}\right|_{\text{heat}} = 4CL(\omega)(\hbar q)^2. \tag{4.53}$$

The kinetic energy of the atom in a steady state is determined by balancing the heating rate (4.53) with the cooling due to the frictional force which, from (4.47) and (4.49), is given by

$$\left.\frac{d\overline{p_z^2}}{dt}\right|_{\text{fric}} = -2\frac{\overline{p_z^2}}{\tau_{\text{fric}}}. \tag{4.54}$$

One thus arrives at the equation

$$\overline{p_z^2} = \frac{1}{2}D_p\tau_{\text{fric}}, \tag{4.55}$$

which shows that the root-mean-square momentum in a steady state is, roughly speaking, the momentum an atom would acquire during a random walk of duration equal to the braking time τ_{fric} of an atom in the ground state. Thus the mean kinetic energy and temperature T associated with the motion of the atom in the z direction are given by

$$m\overline{v_z^2} = kT = \hbar L(\omega)\left(\frac{dL(\omega)}{d\omega}\right)^{-1}. \tag{4.56}$$

The lowest temperature attainable by this mechanism is obtained by minimizing this expression with respect to ω, and is found from (4.44) to be

$$kT = \frac{\hbar\Gamma_e}{2}. \tag{4.57}$$

For other assumptions about the emission of photons from the excited state, the limiting temperature differs from this result by a numerical factor. In terms of the detuning parameter $\delta = \omega - \omega_{eg}$, the minimum temperature is attained for $\delta = -\Gamma_e/2$, corresponding to red detuning. For blue detuning, the atom is accelerated rather than braked, as one can see from the general formula for the force.

As an example, let us estimate the lowest temperature that can be achieved by laser cooling of sodium atoms. Since the width Γ_e corresponds to a temperature of 480 μK according to Table 3.4, we conclude that the minimum temperature attainable by the Doppler mechanism is ~ 240 µK. In our discussion above we have shown how cooling is achieved for one velocity component. With three pairs of mutually opposed laser beams all three components of the velocity may be cooled.

As we mentioned in the introduction to this chapter, in some experiments atoms are passed through a so-called *Zeeman slower* to reduce their velocities to values small enough for trapping in a magneto-optical trap to be possible. In the Zeeman slower a beam of atoms is subjected to a single laser beam propagating in the direction opposite that of the atoms. As we have seen earlier in this section, absorption and subsequent re-emission of radiation by atoms transfers momentum from the laser beam to the atoms, thereby slowing the atoms. However, if the frequency of the laser beam is resonant with the atomic frequency when atoms emerge from the oven, the slowing of the atoms and the consequent change of the Doppler shift will cause the transition to become non-resonant. Thus the velocity range over which laser light will be maximally effective in decelerating atoms is limited. In the Zeeman slower the effect of the decreasing atomic velocity on the frequency of the atomic transition is compensated by a Zeeman shift produced by an inhomogeneous magnetic field.

4.4 The magneto-optical trap

Radiation pressure may also be used to confine atoms in space. In the magneto-optical trap (MOT) this is done with a combination of laser beams and a spatially-varying magnetic field. The basic physical effect is that, because atomic energy levels depend on the magnetic field, the radiation pressure depends on position. By way of illustration, consider an atom with a ground state having zero total angular momentum, and an excited state with angular momentum quantum number $J = 1$. For simplicity we neglect the nuclear spin. Consider, for example, the quadrupole magnetic field (4.2). On the z axis, the magnetic field is in the z direction and it is linear in z. The magnetic substates of the excited state are specified by the quantum number m, in terms of which the projection of the angular momentum of the state along the z axis is $m\hbar$. Circularly polarized laser beams with equal intensity and frequency are imposed in the positive and negative z directions. The polarization of both beams is taken to be clockwise with respect to the direction of propagation, which means that the beam directed to the right (σ_+) couples the ground state to the $m = +1$ excited substate. On the other hand, the polarization of the beam directed to the left has the opposite sense

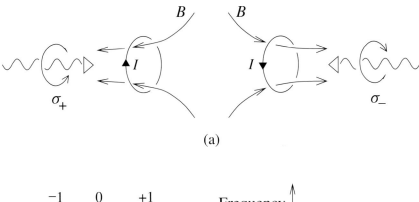

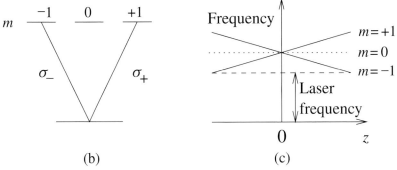

Fig. 4.6. (a) The magneto-optical trap. (b) The relevant transitions. (c) Influence of a spatially-varying magnetic field on the atomic transitions. (After Ref. [11].)

(σ_-) with respect to the z axis, and thus induces transitions to the $m = -1$ substate. The situation is shown schematically in Fig. 4.6.

Let us assume that the laser frequency is detuned to the red. At $z = 0$ the two laser beams are absorbed equally by the atom, and thus there is no net radiation force on the atom. However, if the atom is displaced to positive values of z, the frequency of the transition to the $m = -1$ substate is reduced, and is thus closer to the laser frequency, while the reverse holds for the $m = +1$ substate. This results in an increased absorption rate for σ_- photons, which are moving to the left, and a decreased rate for σ_+ photons, which are moving to the right. Consequently there is a force towards the origin, where the two transitions have the same frequency. Similar arguments apply for negative z. By applying six laser beams, two counterpropagating beams along each axis, one can make a three-dimensional trap.

The use of MOT's is a universal feature of experiments on cold alkali atoms. Not only do they trap atoms, but they also cool them, since the Doppler mechanism described above and the Sisyphus process described below operate in them. The fact that the atomic frequency depends on position, due to the inhomogeneous magnetic field, implies that efficient cooling is possible for atoms with a range of velocities.

In practice, MOT's are more complicated than the simple schematic model described above. One reason for this is that the ground state of an alkali atom has more than one hyperfine state. As an example let us consider Na, which has $F = 2$ and $F = 1$ hyperfine levels of the $3S_{1/2}$ ground state. The excited $3P_{3/2}$ state has hyperfine levels with total angular momentum quantum numbers $F' = 0, 1, 2,$ and 3. If laser light resonant with the $F = 2 \to F' = 3$ transition is applied, some atoms will be excited non-resonantly to the $F' = 2$ state, from which they will decay either to the $F = 2$ or $F = 1$ levels of the ground state. Since there is no radiation resonant with the $F = 1 \to F' = 2$ transition, the net effect is to build up the population of atoms in the $F = 1$ level compared with that in the $F = 2$ level. This process is referred to as *optical pumping*. If this depletion of atoms in the $F = 2$ level is not hindered, the MOT will cease to work effectively because of the small number of atoms in the $F = 2$ level, referred to as the *bright* state, which is the one involved in the transition the MOT is working on. To remove atoms from the $F = 1$ level radiation resonant with the $F = 1 \to F' = 2$ transition is applied. This is referred to as *repumping*.

As we shall explain below, for evaporative cooling to be effective it is necessary to achieve a sufficiently high density of atoms. In a standard MOT there are a number of effects which limit the density to values too low for evaporative cooling to be initiated. One of these is that the escaping fluorescent radiation produces a force on atoms which counteracts the trapping force of the MOT. A second is that if the density is sufficiently high, the cloud becomes opaque to the trapping light. Both of these can be mitigated by reducing the amount of repumping light so that only a small fraction of atoms are in the substate relevant for the transition the MOT operates on. This reduces the effective force constant of the MOT, but the density that can be attained with a given number of atoms increases. In the experiment of Ketterle *et al.* [12] repumping light was applied preferentially in the outer parts of the cloud, thereby giving rise to strong frictional forces on atoms arriving from outside, while in the interior of the cloud radiation forces were reduced because of the depletion of atoms in the bright state. Such a trap is referred to as a *dark-spot MOT*, and densities achievable with it are of order 100 times higher than with a conventional MOT. The dark-spot MOT made it possible to create clouds with densities high enough for evaporative cooling to be efficient, and it was a crucial element in the early experiments on Bose–Einstein condensation in alkali atom vapours.

4.5 Sisyphus cooling

It was encouraging that the temperatures achieved in early experiments on laser cooling appeared to agree with the estimates made for the Doppler mechanism. However, subsequent studies showed that temperatures below the Doppler value could be realized, and that this happened for large detunings, not for a detuning equal to $\Gamma_e/2$ as the Doppler theory predicts. This was both gratifying, since it is commonly the case that what can be achieved in practice falls short of theoretical prediction, and disquieting, since the measurements demonstrated that the cooling mechanisms were not understood [13]. In addition, experimental results depended on the polarization of the laser beams. This led to the discovery of new mechanisms which can cool atoms to temperatures corresponding to a thermal energy of order the so-called *recoil energy*,

$$E_r = \frac{(\hbar q)^2}{2m}. \tag{4.58}$$

This is the energy imparted to an atom at rest when it absorbs a photon of momentum $\hbar q$, and it corresponds to a temperature

$$T_r = \frac{E_r}{k} = \frac{(\hbar q)^2}{2mk}. \tag{4.59}$$

These temperatures lie several orders of magnitude below the lowest temperature achievable by the Doppler mechanism, since the recoil energy is $\hbar^2\omega^2/2mc^2$. The atomic transitions have energies on the scale of electron volts, while the rest-mass energy of an atom is $\sim A$ GeV, where A is the mass number of the atom. The recoil energy is therefore roughly $5 \times 10^{-10}(\hbar\omega/1\text{eV})^2/A$ eV, and the corresponding temperature is $\sim 6 \times 10^{-6}(\hbar\omega/1\text{eV})^2/A$ K, which is of order 0.1–1 μK.

The new cooling mechanisms rely on two features not taken into account in the Doppler theory. First, alkali atoms are not simple two-level systems, and their ground states have substates which are degenerate in the absence of a magnetic field, as we saw in Chapter 3. Second, the radiation field produced by two opposed laser beams is inhomogeneous. To understand how one of these mechanisms, the so-called Sisyphus process, works, consider two counterpropagating linearly-polarized laser beams of equal intensity. For definiteness, we assume that the beam propagating in the positive z direction is polarized along the x axis, while the one propagating in the negative z direction is polarized in the y direction. The electric field is thus of the form

$$\boldsymbol{\mathcal{E}}(z,t) = \boldsymbol{\mathcal{E}}(z)e^{-i\omega t} + \boldsymbol{\mathcal{E}}^*(z)e^{i\omega t}, \tag{4.60}$$

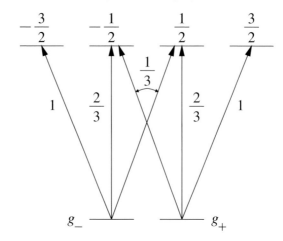

Fig. 4.7. Dipole transitions between a doublet ground state and a quadruplet excited state. The numbers indicate the square of the Clebsch–Gordan coefficients for the transitions.

where

$$\mathcal{E}(z) = \mathcal{E}_0(\hat{e}_x e^{iqz} + \hat{e}_y e^{-iqz}). \tag{4.61}$$

In writing this equation we have chosen the origin of the coordinate system to eliminate the arbitrary phase difference between the two counterpropagating beams. Thus the polarization of the radiation field varies with z, and the polarization vector is

$$\hat{\epsilon} = \frac{1}{\sqrt{2}}(\hat{e}_x + \hat{e}_y e^{-2iqz}). \tag{4.62}$$

This varies regularly in the z direction with a period π/q, which is one-half of the optical wavelength, $\lambda = 2\pi/q$. At $z = 0$ the electric field is linearly polarized at $45°$ to the x axis, and at $z = \lambda/4$ it is again linearly polarized, but at an angle $-45°$ to the x axis. At $z = \lambda/8$ the electric field is circularly polarized with negative sense (σ_-) about the z axis, while at $z = 3\lambda/8$ it is circularly polarized with positive sense (σ_+). At an arbitrary point the intensities of the positively and negatively circularly-polarized components of the radiation field vary as $(1 \mp \sin 2qz)/2$, as may be seen by expressing (4.62) in terms of the polarization vectors $(\hat{e}_x \pm i\hat{e}_y)/\sqrt{2}$.

As a simple example that illustrates the physical principles, consider now the energy of an atom with a doublet ($J_g = 1/2$) ground state coupled to a quadruplet ($J_e = 3/2$) excited state, as shown schematically in Fig. 4.7. This would correspond to the transition from a $^2S_{1/2}$ to a $^2P_{3/2}$ state for an alkali atom if the nuclear spin were neglected. Due to interaction with

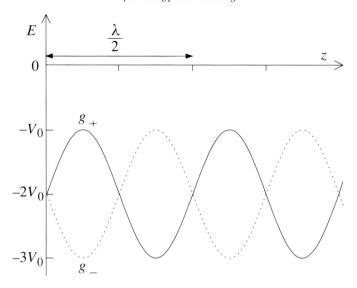

Fig. 4.8. Energy of substates of an atom as a function of position for a red-detuned radiation field ($\delta < 0$). The zero of the energy scale is taken to be the energy of the atom in the absence of radiation.

the laser field, the energies of the substates are shifted. For each sublevel of the ground state there are two contributions to the energy shift, one for each of the two circularly-polarized components of the radiation field. The contribution of a particular transition to the energy shift is proportional to the product of the intensity of the appropriate component of the radiation field times the square of the corresponding Clebsch–Gordan coefficient. The latter factors are indicated on the diagram. Since the intensities of the two circularly-polarized components of the radiation field vary in space, the energy shifts induced by the radiation field do so too. Thus at $z = 0$ the shifts of the two lower substates are the same, while at $z = \lambda/8$, where the radiation is completely circularly-polarized in the negative sense, the g_+ state couples only to the upper substate with magnetic quantum number $-1/2$ (with the square of the Clebsch–Gordan coefficient equal to $1/3$), while the g_- state couples only to the upper substate with magnetic quantum number $-3/2$ (with the square of the Clebsch–Gordan coefficient equal to 1). The shift of the g_- substate is thus three times as large as that of the g_+ substate. At a general point in space the energy shift of an atom is

$$V^\pm = V_0(-2 \pm \sin 2qz), \qquad (4.63)$$

as sketched in Fig. 4.8. The magnitude of V_0 is found by adding the contributions from the two transitions illustrated in Fig. 4.7: the energy shifts

$V^\pm$ of the two states $g_\pm$ are proportional to

$$\Omega_R^2(1 \mp \sin 2qz) + \frac{1}{3}\Omega_R^2(1 \pm \sin 2qz) = \frac{2}{3}\Omega_R^2(2 \mp \sin 2qz), \qquad (4.64)$$

where Ω_R is the Rabi frequency of the transition from the g_+ sublevel to the $m = 3/2$ excited level at $z = 0$. Combining this with Eq. (4.33) one sees that the prefactor in (4.63) is given by

$$V_0 = -\frac{2}{3}\frac{\hbar\Omega_R^2 \delta}{\delta^2 + \Gamma_e^2/4}, \qquad (4.65)$$

which is positive for red detuning. The periodic potential acting on an atom subjected to counterpropagating laser beams is referred to as an *optical lattice*.

A second key ingredient in understanding cooling in this configuration is that the rate at which atoms are optically pumped between the two lower substates depends on position. Consider a point where the radiation is circularly polarized in the positive sense. Under the influence of the radiation field, an atom in the g_+ substate will make transitions to the upper substate with $m = 3/2$, from which it will decay to the g_+ state again, since there are no other possibilities for dipole transitions. By contrast, an atom in the g_- substate will be excited to the $m = 1/2$ upper substate, from which it can decay by dipole radiation either to the g_- substate, with probability $1/3$, or to the g_+ substate, with probability $2/3$. The net effect is thus to pump atoms from the g_- substate, which at this point has the higher energy of the two substates, into the g_+ one. Where the radiation field is linearly polarized there is no net pumping, while where it is circularly polarized with negative sense, atoms are pumped into the g_- substate. At any point, the rate of pumping atoms from the g_- substate to the g_+ substate is proportional to the intensity of the circularly-polarized component of the radiation with positive sense, that is to $(1 - \sin 2qz)/2$, and the rate of pumping from the g_+ substate to the g_- substate is proportional to $(1 + \sin 2qz)/2$. For red detuning, pumping thus tends to move atoms from the substate with higher energy to that with lower energy, while, for blue detuning, pumping rates and energy shifts are anticorrelated and atoms tend to accumulate in the higher-energy substate. We assume that the radiation field is sufficiently weak that spontaneous emission processes from the excited state to the ground state are more rapid than the corresponding induced process, and consequently the characteristic time τ_p for pumping is of order the time for the radiation field to excite an atom, which we estimated earlier (see (4.35) and (4.36)):

$$\frac{1}{\tau_{\rm p}} \sim \frac{\Omega_{\rm R}^2 \Gamma_e}{\delta^2 + \Gamma_e^2/4}. \tag{4.66}$$

The cooling mechanism may be understood by considering atoms with a thermal spread of velocities. Where the radiation is linearly polarized, there is no net tendency to pump atoms between the two substates. However, if an atom is moving in a direction such that its energy increases, the rate at which it is pumped to the other substate also increases. Thus there is a tendency for atoms in the substate with the higher energy to be pumped into the substate with lower energy. Consider an atom moving away from a point where the energies of the two substates are equal. If it is moving into a region where its radiation-induced energy shift is greater, it will, by virtue of conservation of the total energy of the atom, tend to lose kinetic energy. In addition, there will be an increasing tendency for the atom to be pumped into the other substate. Conversely, if an atom is moving into a region where its energy shift is smaller, it tends to gain kinetic energy, but the rate of optical pumping to the higher-energy substate is reduced. Because of the correlation between pumping rates and energy shifts, there is a net tendency for an atom to lose kinetic energy irrespective of its direction of motion. Since optical pumping tends to repopulate the lower-energy substate at any point in space, the process of losing kinetic energy followed by optical pumping will be repeated, thereby leading to continual cooling of the atoms. This mechanism is referred to as *Sisyphus cooling*, since it reminded Dalibard and Cohen-Tannoudji of the Greek myth in which Sisyphus was condemned to eternal punishment in Tartarus, where he vainly laboured to push a heavy rock up a steep hill.

The friction coefficient

To estimate temperatures that can be achieved by this process we adopt an approach analogous to that used in the discussion of Doppler cooling, and calculate a friction coefficient, which describes energy loss, and a momentum diffusion coefficient, which takes into account heating [14]. If in the characteristic pumping time an atom moved from a point where the radiation field is linearly polarized to one where it is circularly polarized, the rate of energy loss would be $dE/dt \sim -V_0/\tau_{\rm p}$. However, for an atom moving with velocity v, the distance moved in a time $\tau_{\rm p}$ is $v\tau_{\rm p}$, and we shall consider the situation when this is small compared with the scale of modulations of the radiation field, the optical wavelength $\lambda = 2\pi/q$. For an atom starting at $z = 0$, the net pumping rate is reduced by a factor $\sim v\tau_{\rm p}/\lambda$, while the extra energy lost due to the motion of the atom from its original position is $\sim V_0 v \tau_p / \lambda$. The total energy loss rate is thus reduced by a factor $\sim (v\tau_p/\lambda)^2$ compared with the naive estimate. The energy loss rate averaged over possible starting points is reduced by a similar factor, as the kinetic theory calculation below will demonstrate. Such factors are familiar from the kinetic theory of gases, where rates of dissipation processes in the hydrodynamic regime are given by the typical collision rate reduced by a factor proportional to the square of the mean free path of an atom divided by the length scale over which the density, temperature, or velocity of the gas vary. Thus the effective energy loss rate is

$$\frac{dE}{dt} \sim -\frac{V_0 v^2 \tau_p}{\lambda^2} = -\frac{E}{\tau_{\text{cool}}}, \qquad (4.67)$$

where

$$\frac{1}{\tau_{\text{cool}}} \sim \frac{V_0 \tau_p}{\lambda^2 m} \qquad (4.68)$$

is a characteristic cooling time. As we shall show below, this result may also be obtained from kinetic theory. Substituting the expressions (4.33) and (4.66) into this equation one finds

$$\frac{1}{\tau_{\text{cool}}} \sim \frac{-\delta}{\Gamma} \frac{E_r}{\hbar}, \qquad (4.69)$$

where the recoil energy is defined in (4.58). A noteworthy feature of the cooling rate is that it does not depend on the strength of the radiation field: the energy shift is proportional to the intensity, while the pumping time is inversely proportional to it. This should be contrasted with the Doppler process, for which the cooling rate is proportional to the intensity.

Kinetic theory approach

It is instructive to derive this result in a more formal manner starting from the kinetic equation for the atoms. The evolution of the distribution function is governed by the Boltzmann equation:

$$\frac{\partial f_{\mathbf{p}}^{\pm}}{\partial t} + \frac{\partial \epsilon_{\mathbf{p}}}{\partial \mathbf{p}} \cdot \frac{\partial f_{\mathbf{p}}^{\pm}}{\partial \mathbf{r}} - \frac{\partial \epsilon_{\mathbf{p}}}{\partial \mathbf{r}} \cdot \frac{\partial f_{\mathbf{p}}^{\pm}}{\partial \mathbf{p}} = \left.\frac{\partial f_{\mathbf{p}}^{\pm}}{\partial t}\right|_{\text{pump}}, \qquad (4.70)$$

where the Hamiltonian for the particle, its energy, is given by $\epsilon_{\mathbf{p}} = p^2/2m + V$, the potential being that due to the radiation field. The right hand side of this equation, usually referred to as the collision term, takes into account pumping of atoms between substates by the radiation field. These may be written

$$\left.\frac{\partial f_{\mathbf{p}}^+}{\partial t}\right|_{\text{pump}} = -\left.\frac{\partial f_{\mathbf{p}}^-}{\partial t}\right|_{\text{pump}} = -\Gamma_{-+}f_{\mathbf{p}}^+ + \Gamma_{+-}f_{\mathbf{p}}^-. \qquad (4.71)$$

Here $f_{\mathbf{p}}^{\pm}$ is the distribution function for atoms in the substates $+$ and $-$ as a function of the atomic momentum $\mathbf{p} = m\mathbf{v}$, and Γ_{-+} and Γ_{+-} are the pumping rates, which depend on the velocity of the atom. The first equality in (4.71) follows from the fact that an atom lost from one substate is pumped to the other one.

Let us first assume that the intensity of the radiation is independent of space. In a steady state the Boltzmann equation reduces to $\partial f_{\mathbf{p}}^{\pm}/\partial t\big|_{\text{pump}} = 0$, and therefore

$$\frac{f_{\mathbf{p}}^+}{f_{\mathbf{p}}^-} = \frac{\Gamma_{+-}}{\Gamma_{-+}} \qquad (4.72)$$

or

$$\frac{f_{\mathbf{p}}^+ - f_{\mathbf{p}}^-}{f_{\mathbf{p}}^+ + f_{\mathbf{p}}^-} = \frac{\Gamma_{+-} - \Gamma_{-+}}{\Gamma_{+-} + \Gamma_{-+}}. \qquad (4.73)$$

In this situation the average energy of the atoms remains constant, because the net pumping rate vanishes. However, when the radiation field is inhomogeneous, the distribution function will change due to atoms moving between points where the radiation field is different. Provided the radiation field varies little over the distance that an atom moves in a pumping time, the distribution function will remain close to that for a steady state locally, which we denote by $\bar{f}_{\mathbf{p}}^{\pm}$. This is given by Eq. (4.72) evaluated for the spatial dependence of the pumping rates given above, that is

$$\frac{\bar{f}_{\mathbf{p}}^+ - \bar{f}_{\mathbf{p}}^-}{\bar{f}_{\mathbf{p}}^+ + \bar{f}_{\mathbf{p}}^-} = -\sin 2qz, \quad \text{(local steady state)}. \qquad (4.74)$$

For $z = \lambda/8$, atoms are pumped completely into the g_- state, while for $z = -\lambda/8$, they are pumped into the g_+ state.

To calculate the deviation of the distribution function from the local equilibrium result we make use of the fact that under experimental conditions the pumping time τ_{p} is short compared with the time it takes an atom to move a distance λ, over which the population of atoms varies significantly. We may then insert the local steady-state solution (4.74) on the left hand side of the Boltzmann equation (4.70), just as one does in calculations of transport coefficients in gases. The last term on this side of the equation, which is due to the influence of the 'washboard' potential (4.63) on the atoms, is of order $V^{\pm}/kT$ times the second term, which comes from the spatial gradient of the local equilibrium distribution function. As we shall show, the lowest temperatures attainable are such that $kT \gtrsim |V^\pm|$, and therefore we shall neglect this term. We now linearize the right hand side of the Boltzmann equation about the local steady state, and write $\delta f_\mathbf{p}^\pm = f_\mathbf{p}^\pm - \bar{f}_\mathbf{p}^\pm$. Thus the deviations from the *local* steady-state solution satisfy the equation

$$\mathbf{v} \cdot \boldsymbol{\nabla}(\bar{f}_\mathbf{p}^+ - \bar{f}_\mathbf{p}^-) = -(\Gamma_{+-} + \Gamma_{-+})(\delta f_\mathbf{p}^+ - \delta f_\mathbf{p}^-). \tag{4.75}$$

On the right hand side of the equation we have omitted a term proportional to the deviation $\delta f_\mathbf{p}^+ + \delta f_\mathbf{p}^-$ of the total density from its local steady-state value since this is small as a consequence of the requirement that the pressure be essentially constant. Thus the deviation of the distribution function from its steady-state value is given by

$$\delta f_\mathbf{p}^+ - \delta f_\mathbf{p}^- = -\mathbf{v} \cdot \boldsymbol{\nabla}(\bar{f}_\mathbf{p}^+ - \bar{f}_\mathbf{p}^-)\tau_\mathrm{p} = 2qv_z f_\mathbf{p}\tau_\mathrm{p} \cos 2qz, \tag{4.76}$$

since the characteristic pumping time is given by $1/\tau_\mathrm{p} = \Gamma_{+-} + \Gamma_{-+}$. Here $f_\mathbf{p} = f_\mathbf{p}^+ + f_\mathbf{p}^-$ is the distribution function for atoms irrespective of which sublevel they are in.

We now evaluate the average force on an atom. For an atom in a particular sublevel the force is $-\boldsymbol{\nabla} V^\pm$, and therefore the total force on atoms with momentum $\mathbf{p}$ is $-\boldsymbol{\nabla} V^+ f_\mathbf{p}^+ - \boldsymbol{\nabla} V^- f_\mathbf{p}^- = -2qV_0(f_\mathbf{p}^+ - f_\mathbf{p}^-)\hat{z}\cos 2qz$. Since the local steady-state result for the distribution function (4.74) varies as $\sin 2qz$, the spatial average of the force vanishes for this distribution, and the leading contributions to the force come from the deviations from the local steady state, Eq. (4.76). One then finds that the spatial average of the force on an atom is

$$F_z|_\mathrm{av} = -2q^2 V_0 v_z \tau_\mathrm{p} = -4\pi q V_0 \frac{v_z \tau_\mathrm{p}}{\lambda}. \tag{4.77}$$

The latter expression indicates that this is of order the force due to atomic energy shifts, $\sim qV_0$, reduced by a factor of the mean free path for pumping, $v\tau_\mathrm{p}$, divided by the wavelength of light, which gives the spatial scale of variations in the distribution function. The average rate of change of the energy of an atom is therefore given by $F_z|_\mathrm{av} v_z = -2q^2 V_0 v_z^2 \tau_\mathrm{p}$ and the characteristic cooling rate is thus

$$\frac{1}{\tau_\mathrm{cool}} = \frac{4q^2 V_0 \tau_\mathrm{p}}{m}, \tag{4.78}$$

in agreement with our earlier estimate (4.68). The remarkable effectiveness of Sisyphus cooling is a consequence of the almost complete reversal of sublevel populations on a short length scale, the optical wavelength.

Temperature of atoms

There are a number of effects contributing to the energy diffusion coefficient, among them fluctuations in the number of photons absorbed from the two beams, and the different directions of the emitted photons which we considered in the discussion of Doppler cooling. In Sisyphus cooling there is an additional effect due to the fact that the periodic potential accelerates atoms, and this is the dominant contribution. The force on an atom is of order $2qV_0 = 4\pi V_0/\lambda$, and therefore the momentum imparted to an atom in the characteristic pumping time, which plays the role of a mean free time in this process, is $\sim 4\pi V_0 \tau_\mathrm{p}/\lambda$. Diffusion coefficients are of order the square of the step size divided by the time for a step, and therefore the momentum diffusion coefficient is of order

$$D \sim \left(\frac{4\pi V_0 \tau_\mathrm{p}}{\lambda}\right)^2 \frac{1}{\tau_\mathrm{p}} = \frac{(4\pi)^2 V_0^2 \tau_\mathrm{p}}{\lambda^2}. \tag{4.79}$$

The mean-square momentum in a steady state is of order that produced by the diffusion process in a characteristic cooling time, where τ_cool is given by (4.78), that is $\overline{p_z^2} \sim D\tau_\mathrm{cool}$. Thus the characteristic energy of an atom in a steady state is of order

$$E \sim D\tau_\mathrm{cool}/m \sim V_0. \tag{4.80}$$

This remarkably simple result would appear to indicate that by reducing the laser power atoms could be cooled to arbitrarily low temperatures. However, there are limits to the validity of the classical description of the atomic motion. We have spoken as though atoms have a definite position and velocity, but according to Heisenberg's uncertainty principle the momentum of an atom confined to within a distance l, has an uncertainty of magnitude $\sim \hbar/l$, and thus a kinetic energy of at least $\sim \hbar^2/2ml^2$. The energy of an atom confined to within one minimum of the effective potential is of order $\hbar^2/2m\lambda^2 \sim E_\mathrm{r}$. If this exceeds the depth of modulation of the washboard potential produced by the radiation, a classical treatment fails. A quantum-mechanical calculation shows that for $V_0 \lesssim E_\mathrm{r}$ the energy of the atoms rises with decreasing radiation intensity, and thus the minimum particle energies that can be achieved are of order E_r. Detailed one-dimensional calculations confirm the order-of-magnitude estimates and scaling laws given by our simplified treatment. For example, in the limit of large detuning, $|\delta| \gg \Gamma_e$, the minimum kinetic energy achievable is of order $30E_\mathrm{r}$ [15]. The existence of substantial numerical factors should come as no surprise, since in our simple approach we were very cavalier and neglected all numerical factors. The single most important source of the large numerical factor is the proper quantum-mechanical treatment of the motion of atoms in the periodic potential, which shows that the lowest temperatures are attained for $V_0 \sim 50 E_\mathrm{r}$. The theory has been extended to higher dimensions, and for three-dimensional optical lattices Castin and Mølmer find a minimum kinetic energy of about $40 E_\mathrm{r}$ [16].

To conclude this section, we remark that methods have been developed to cool atoms to kinetic energies less than the recoil energy. This is done by collecting atoms in states weakly coupled to the laser radiation and having a small spread in velocity [15]. None of these methods have yet been exploited in the context of Bose–Einstein condensation.

4.6 Evaporative cooling

The temperatures reached by laser cooling are impressively low, but they are not low enough to produce Bose–Einstein condensation in gases at the densities that are realizable experimentally. In the experiments performed to date, Bose–Einstein condensation of alkali gases is achieved by using evaporative cooling after laser cooling. The basic physical effect in evaporative cooling is that, if particles escaping from a system have an energy higher than the average energy of particles in the system, the remaining particles are cooled. Widespread use of this effect has been made in low-temperature physics, where evaporation of liquified gases is one of the commonly used techniques for cooling. In the context of rarefied gases in traps it was first proposed by Hess [17]. For a more extensive account of evaporative cooling we refer to the review [18]. Imagine atoms with a thermal distribution of energies in a trap, as illustrated schematically in Fig. 4.9. If one makes a 'hole' in the trap high up on the sides of the trap, only atoms with an energy at least equal to the energy of the trap at the hole will be able to escape. In practice one can make such a hole by applying radio-frequency (rf) radiation that flips the spin state of an atom from a low-field seeking one to a high-field seeking one, thereby expelling the atom from the trap. Since the resonant frequency depends on position as a consequence of the Zeeman effect and the inhomogeneity of the field, as described in Chapter 3, the position of the 'hole' in the trap may be selected by tuning the frequency of the rf radiation. As atoms are lost from the trap and cooling proceeds, the frequency is steadily adjusted to allow loss of atoms with lower and lower energy.

A simple example that illustrates the effect is to imagine a gas with atoms which have an average energy $\bar{\epsilon}$. If the average energy of an evaporated

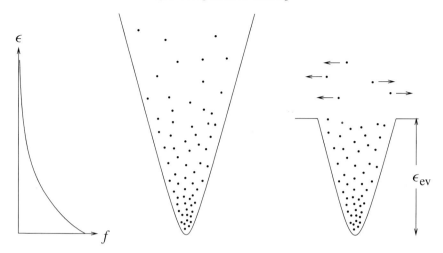

Fig. 4.9. Evaporative cooling. The curve on the left shows the equilibrium Maxwell–Boltzmann distribution proportional to $\exp(-\epsilon/kT)$. The energy $\epsilon_{\rm ev}$ is the threshold value for evaporation, see Eq. (4.93).

particle is $(1+\beta)\bar{\epsilon}$, the change in the average particle energy may be obtained from the condition that the total energy of all particles be constant. If the change in the number of particles is denoted by dN, the energy removed by the evaporated particles is $(1+\beta)\bar{\epsilon}dN$. For particle loss, dN is negative. By energy conservation, the total energy of the atoms remaining in the trap is $E+(1+\beta)\bar{\epsilon}dN$ and their number is $N+dN$. Thus the average energy per atom in the trap after the evaporation is

$$\bar{\epsilon} + d\bar{\epsilon} = \frac{E + (1+\beta)\bar{\epsilon}dN}{N + dN}, \tag{4.81}$$

or

$$\frac{d\ln\bar{\epsilon}}{d\ln N} = \beta. \tag{4.82}$$

Thus, if β is independent of N,

$$\frac{\bar{\epsilon}}{\bar{\epsilon}(0)} = \left(\frac{N}{N(0)}\right)^{\beta}, \tag{4.83}$$

where $\bar{\epsilon}(0)$ and $N(0)$ are initial values. In this simple picture, the average energy per particle of the gas in the trap decreases as a power of the particle number.

The relationship between the average energy and the temperature depends on the trapping potential. To obtain simple results, we consider potentials which are homogeneous functions of the coordinates with degree ν,

$$V(\mathbf{r}) = K(\hat{\mathbf{r}})r^\nu, \tag{4.84}$$

where $K(\hat{\mathbf{r}})$ is a coefficient which may depend on direction. A further variable is the effective dimensionality d of the trap, which specifies the number of dimensions for which the trapping potential is operative. For example, a potential $V = m\omega_\perp^2(x^2 + y^2)/2$, with no restoring force in the z direction, has $d = 2$. The discussion of magnetic traps given in Sec. 4.1 shows that the assumption of homogeneity is reasonable for many traps used in practice. For an ideal gas one can show that (Problem 4.7)

$$2\bar{\epsilon}_{\text{kin}} = \frac{3}{d}\nu\bar{V}, \tag{4.85}$$

where ϵ_{kin} is the kinetic energy of a particle, and V is the trapping potential. The bar denotes a thermal average. Throughout most of the evaporation process, the effects of quantum degeneracy are modest, so we may treat the gas as classical. Thus

$$\bar{\epsilon}_{\text{kin}} = \frac{3}{2}kT, \tag{4.86}$$

and

$$\bar{\epsilon} = \left(\frac{3}{2} + \frac{d}{\nu}\right)kT. \tag{4.87}$$

This shows that the temperature of the gas depends on the particle number in the same way as the average energy, and it therefore follows from Eq. (4.82) that

$$\frac{d\ln T}{d\ln N} = \beta. \tag{4.88}$$

This result shows that the higher the energy of the evaporating particles, the more rapidly the temperature falls for loss of a given number of particles. However, the rate of evaporation depends on the energy threshold and on the rate of elastic collisions between atoms in the gas, since collisions are responsible for scattering atoms into states at energies high enough for evaporation to occur. Thus the higher the threshold, the lower the evaporation rate. The threshold may not be chosen to be arbitrarily high because there are other processes by which particles are lost from the trap without cooling the remaining atoms. These limit the time available for evaporation, and therefore the threshold energy must be chosen as a compromise between the conflicting requirements of obtaining the greatest cooling per particle evaporated, and at the same time not losing the sample by processes other than evaporation.

One of these other processes is collisions with foreign gas atoms. A second is inelastic collisions, in which two atoms collide and are scattered to other hyperfine states. As we shall explain in greater detail in Sec. 5.4.1, some processes can proceed via the central part of the two-body interaction, and the loss rate due to them is generally so high that one must work with hyperfine states such as the doubly polarized state and, for atoms with a positive nuclear magnetic moment, the maximally stretched state $(F = I - 1/2, m_F = -F)$ which cannot decay by this route. However, even these states can scatter inelastically via the magnetic dipole–dipole interaction between electron spins. Experiments with Bose–Einstein condensation in atomic hydrogen in magnetic traps are performed on states of H↑, and dipolar losses are a dominant mechanism for loss of atoms and heating, as we shall describe in the following section. In experiments on alkali atoms dipolar losses are generally less important, and then formation of diatomic molecules can be a significant loss mechanism. Two atoms cannot combine directly because it is impossible to get rid of the binding energy of the molecule. The most effective way of satisfying the conservation laws for energy and momentum under typical experimental conditions is for a third atom to participate in the reaction.

A simple model

To estimate the cooling that may be achieved by evaporation, we consider a simple model [19]. We assume that the rate at which particles are removed from the trap by evaporation is given by

$$\frac{dN}{dt}\bigg|_{\mathrm{ev}} = -\frac{N}{\tau_{\mathrm{ev}}}, \qquad (4.89)$$

and that the rate of particle loss due to other processes is

$$\frac{dN}{dt}\bigg|_{\mathrm{loss}} = -\frac{N}{\tau_{\mathrm{loss}}}, \qquad (4.90)$$

where τ_{ev} and τ_{loss} are decay times for the two types of mechanisms. Depending upon which particular other process is dominant, the loss rate $1/\tau_{\mathrm{loss}}$ depends on the density in different ways. For scattering by foreign gas atoms it is independent of the density n of atoms that are to be cooled, and proportional to the density of background gas. For dipolar losses, it varies as n, and for three-body ones, as n^2. If we further assume that the average energy of particles lost by the other processes is the same as the average energy of particles in the gas and neglect heating of the gas by inelastic processes, only particles lost by evaporation change the average energy of particles remaining in the trap. The fraction of particles lost by evaporation is

$$\left.\frac{dN}{dt}\right|_{\text{ev}} \Big/ \left(\left.\frac{dN}{dt}\right|_{\text{ev}} + \left.\frac{dN}{dt}\right|_{\text{loss}}\right) = \frac{1/\tau_{\text{ev}}}{1/\tau_{\text{ev}} + 1/\tau_{\text{loss}}} = \frac{\tau_{\text{loss}}}{\tau_{\text{loss}} + \tau_{\text{ev}}}, \quad (4.91)$$

and therefore the temperature change is obtained by multiplying the expression (4.88) in the absence of losses by this factor. Thus

$$\frac{d\ln T}{d\ln N} = \beta \frac{\tau_{\text{loss}}}{\tau_{\text{loss}} + \tau_{\text{ev}}} \equiv \beta'. \quad (4.92)$$

As we argued earlier, the evaporation time increases rapidly as the average energy of the evaporated particles increases, and therefore there is a particular average energy of the evaporated particles for which the temperature drop for loss of a given number of particles is maximal. To model the evaporation process, we assume that any atom with energy greater than some threshold value ϵ_{ev} is lost from the system. The evaporation rate is therefore equal to the rate at which atoms are scattered to states with energies in excess of ϵ_{ev}. We shall assume that ϵ_{ev} is large compared with kT and, therefore, in an equilibrium distribution the fraction of particles with energies greater than ϵ_{ev} is exponentially small. The rate at which particles are promoted to these high-lying states may be estimated by using the principle of detailed balance, since in a gas in thermal equilibrium it is equal to the rate at which particles in these states are scattered to other states. The rate at which atoms with energy in excess of ϵ_{ev} collide in an equilibrium gas is

$$\left.\frac{dN}{dt}\right|_{\text{coll}} = \int d\mathbf{r} \int_{\epsilon_p > \epsilon_{\text{ev}}} \frac{d\mathbf{p}}{(2\pi\hbar)^3} \int \frac{d\mathbf{p}'}{(2\pi\hbar)^3} f_\mathbf{p} f_{\mathbf{p}'} |\mathbf{v}_\mathbf{p} - \mathbf{v}_{\mathbf{p}'}| \sigma, \quad (4.93)$$

where σ is the total elastic cross section, which we assume to be independent of the particle momenta. For $\epsilon_{\text{ev}} \gg kT$ we may replace the velocity difference by $\mathbf{v}_\mathbf{p}$, and the distribution functions are given by the classical Maxwell–Boltzmann result

$$f_\mathbf{p} = e^{-(p^2/2m + V - \mu)/kT}. \quad (4.94)$$

The integral over $\mathbf{p}'$ gives the total density of particles at the point under consideration, and the remaining integrals may be evaluated straightforwardly. The leading term for $\epsilon_{\text{ev}} \gg kT$ is given by

$$\left.\frac{dN}{dt}\right|_{\text{ev}} = -\left.\frac{dN}{dt}\right|_{\text{coll}} = -N n(0) \sigma \bar{v} \left(\frac{\epsilon_{\text{ev}}}{kT}\right) e^{-\epsilon_{\text{ev}}/kT}. \quad (4.95)$$

Here $n(0)$ is the particle density at the centre of the trap, where $V = 0$, while $\bar{v}$ is the mean thermal velocity given by

$$\bar{v} = \frac{\int_0^\infty dv v^3 \exp(-mv^2/2kT)}{\int_0^\infty dv v^2 \exp(-mv^2/2kT)} = \left(\frac{8kT}{\pi m}\right)^{1/2}. \tag{4.96}$$

It is convenient to introduce the collision time τ_{el} for elastic collisions, which we define by

$$\frac{1}{\tau_{\text{el}}} = n(0)\sigma \bar{v}_{\text{rel}}, \tag{4.97}$$

where $\bar{v}_{\text{rel}}$ is the mean relative velocity of particles in a gas, given by $\bar{v}_{\text{rel}} = \sqrt{2}\bar{v}$. In experiments, the elastic collision time is typically a few orders of magnitude less than the loss time. By combining Eqs. (4.89), (4.95), and (4.97) we conclude that the decay time for evaporation is given by

$$\frac{1}{\tau_{\text{ev}}} = \frac{1}{\tau_{\text{el}}} \left(\frac{\epsilon_{\text{ev}}}{\sqrt{2}kT}\right) e^{-\epsilon_{\text{ev}}/kT}. \tag{4.98}$$

Since the occupancy of single-particle states falls off rapidly at energies large compared with the thermal energy kT, the majority of particles leaving the cloud by evaporation have energies close to the threshold energy ϵ_{ev}, and it is a good approximation to replace the average energy $(1+\beta)\bar{\epsilon}$ of an evaporated particle by ϵ_{ev}. We may therefore write Eq. (4.92) as

$$\frac{d \ln T}{d \ln N} = \left(\frac{\epsilon_{\text{ev}}}{\bar{\epsilon}} - 1\right) \left(1 + \frac{\tau_{\text{el}}}{\tau_{\text{loss}}} \frac{\sqrt{2}kT}{\epsilon_{\text{ev}}} e^{\epsilon_{\text{ev}}/kT}\right)^{-1}. \tag{4.99}$$

This function first increases as the threshold energy increases above the average particle energy, and then falls off when the evaporation rate becomes comparable to the loss rate. The optimal choice of the threshold energy may easily be estimated by maximizing this expression, and it is given to logarithmic accuracy by $\epsilon_{\text{ev}} \sim kT \ln(\tau_{\text{loss}}/\tau_{\text{el}})$. This condition amounts to the requirement that the evaporation time and the loss time be comparable.

It is also of interest to investigate how the degree of degeneracy develops as particles are lost. In a trap with a power-law potential, the spatial extent of the cloud is $\sim T^{1/\nu}$ for each dimension for which the potential is effective, and therefore the volume varies as $\sim T^{d/\nu}$, where d is the effective dimensionality of the trap, introduced above Eq. (4.85). The mean thermal momentum varies as $\sim T^{1/2}$, and therefore the volume in momentum space varies as $\sim T^{3/2}$. Thus the phase-space volume scales as $T^{d/\nu+3/2}$, and the phase-space density ϖ scales as $N/T^{d/\nu+3/2}$. Thus

$$-\frac{d\ln\varpi}{d\ln N} = \beta'\left(\frac{d}{\nu} + \frac{3}{2}\right) - 1. \qquad (4.100)$$

This shows that the attainment of a large increase in phase-space density for loss of a given number of particles is aided by use of traps with low values of ν, and thus linear traps are better than harmonic ones.

In experiments, it is desirable that the elastic scattering time decrease as the evaporation proceeds. In this case one realizes what is referred to as *runaway evaporation*. In the opposite case, the rate of elastic collisions becomes less, and evaporation becomes relatively less important compared with losses. The scattering rate scales as the atomic density times a thermal velocity ($\propto T^{1/2}$), since the cross section is essentially constant, and thus one finds that

$$\frac{d\ln\tau_{\rm el}}{d\ln N} = \beta'\left(\frac{d}{\nu} - \frac{1}{2}\right) - 1. \qquad (4.101)$$

For runaway evaporation, this expression should be positive, and this sets a more stringent condition on β' than does the requirement of increasing phase-space density.

4.7 Spin-polarized hydrogen

As we described in Chapter 1, the first candidate considered for Bose–Einstein condensation in a dilute gas was spin-polarized hydrogen. However, the road to the experimental realization of Bose–Einstein condensation in hydrogen was a long one, and the successful experiment combined techniques from low-temperature physics with ones from atomic and optical physics.

The initial appeal of spin-polarized hydrogen was its having no two-body bound states. However, as has been strikingly demonstrated by the experiments on spin-polarized alkali gases, the absence of bound states is not a prerequisite for Bose–Einstein condensation. Polarized alkali atoms have many molecular bound states, but the rate at which molecules are produced is slow. This is because, as we described in the previous section, molecule formation is a three-body process and may thus be reduced by working at low enough densities. What is essential in realizing Bose–Einstein condensation in dilute systems is that the polarized atoms have a sufficiently long lifetime, irrespective of which particular loss processes operate.

Working with spin-polarized hydrogen presents a number of formidable experimental challenges. One is that laser cooling is impractical, since the lowest-frequency optical transition for the hydrogen atom in its ground state is the Lyman-α line (1S–2P), which has a wavelength of 122 nm in the ultraviolet. Even if it were practical, the temperatures attainable by laser cooling would not be particularly low because both the excited state linewidth and the recoil temperature (Eq. (4.59)), which determine the minimum temperatures for cooling by the Doppler and Sisyphus processes, are large due to the high frequency of the transition and the low mass of the hydrogen atom. In the experiments on hydrogen, the gas was first cooled cryogenically by heat exchange with a cold surface, and condensation was achieved by subsequent evaporative cooling, as described in Sec. 4.6.

The level structure of the ground state of hydrogen has been described in Sec. 3.2 (see Fig. 3.2). As we mentioned in Chapter 1, early experiments on spin-polarized hydrogen, in which hydrogen atoms were pressed against a surface by a spatially-varying magnetic field, employed states with the electron spin aligned opposite the direction of the magnetic field. These are the high-field seeking states (H$\downarrow$) labelled a and b in Fig. 3.2. The limited densities that could be obtained by these methods led to the development of purely magnetic traps. Since it is impossible to make a magnetic field with a local maximum in a current-free region, trapping is possible only if the electron spin is in the same direction as the field, H$\uparrow$, corresponding to states c and d in the figure.

The experimental techniques are described in detail in the review article [20] and the original papers [21, 22]. The heart of the experiment is a magnetic trap of the Ioffe–Pritchard type enclosed in a chamber whose walls can be cooled cryogenically. Atomic hydrogen is generated in an adjoining cell by applying an rf discharge to gaseous molecular hydrogen. The magnetic field in the source cell is higher than in the experimental cell, so atoms in the low-field seeking states c and d are driven into the experimental cell, where they can be trapped magnetically. The experimental chamber also contains helium gas. The helium atom has no electronic magnetic dipole moment and is unaffected by the magnetic field of the trap. Thus the helium acts as a medium for transporting heat away from the hydrogen gas in the trap to the cold walls of the experimental cell.

The state d is a doubly polarized one, and consequently, as we mentioned in Sec. 4.6 and will be elaborated in Sec. 5.4.1, it is relatively unaffected by atomic collisions, the only allowed processes being ones mediated by the magnetic dipole–dipole interaction. In the trapped gas, spin-exchange collisions $c+c \to d+b$ can occur, and the b atoms produced by this process, being high-field seekers, will return to the source. Thus the gas in the trap will consist only of d atoms.

After the hydrogen gas in the trap has been cooled, it is thermally isolated by removing the helium gas. This is done by cooling the walls of the experimental cell to a temperature less than 80 mK, so low that any atom arriving at the surface will be absorbed but will not subsequently be desorbed.

The gas in the trap is cooled further by evaporation, as described in Sec. 4.6. The scattering length for hydrogen atoms is typically 1–2 orders of magnitude smaller than for the alkalis, and therefore to achieve rapid enough evaporation it is necessary to use higher atomic densities for hydrogen than for alkalis. However, this is not a problem, since traps may be loaded to higher densities using cryogenic methods than is possible with MOT's. Bose–Einstein condensation sets in at a temperature of about 50 mK, which is higher than in experiments with alkalis because of the lower atomic mass and the higher atomic densities.

The main process for the destruction of d atoms is the dipolar one $d+d \rightarrow a+a$. Due to the heat which this generates, for hydrogen it is more difficult to realize condensates containing a large fraction of the total number of atoms than it is for alkali atoms, and in the first experiments the fraction of atoms in the condensate was estimated to be around 5%. However, the number of condensed particles (10^9), the peak condensate density (5×10^{15} cm^{-3}), and the size of the condensed cloud (5 mm long, and 15 μm in diameter) are impressive.

Detection of the condensate represented another challenge. Time-of-flight methods could not be used, because of the low condensate fraction. The technique employed was to measure the frequency shift of the 1S–2S transition (at the same frequency as the Lyman-α line), as illustrated schematically in Fig. 4.10. A great advantage of this line is that it is very narrow because the transition is forbidden in the dipole approximation. For the same reason it cannot be excited by a single photon, so the method employed was to use two-photon absorption of radiation from a laser operating at half the frequency of the transition. Absorption of one photon mixes into the wave function of the 1S state a component of a P state, which by absorption of a second photon is subsequently converted to the 2S state. The shift of the line is proportional to the density of hydrogen atoms, and thus the density of condensed atoms could be determined. By this method it was possible to identify components of the line due to the condensate and others due to the thermal cloud. We shall consider the theory of the line shift in Sec. 8.4.

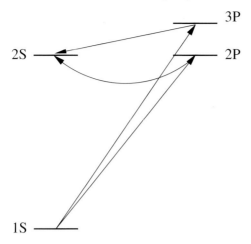

Fig. 4.10. Level scheme of the hydrogen atom to illustrate detection of atomic hydrogen by two-photon absorption.

Problems

PROBLEM 4.1 Consider two circular coils, each of radius a and with N turns. The coils have a common axis, which we take to be the z axis, and their centres are at $z = \pm d/2$. The current I in each coil has the same magnitude, but the opposite direction. Calculate the magnetic field in the vicinity of the origin.

PROBLEM 4.2 Find the classical oscillation frequencies for a ^{87}Rb atom in the state $|3/2, 1/2\rangle$ moving in the time-averaged magnetic field given by (4.6). The magnitude of the rotating magnetic field is $B_0 = 1$ mT, while the value of the radial field-gradient B' is 1.2 T m^{-1}. Compare your result with the value $\omega/2\pi = 7.5$ kHz for the frequency of the rotating magnetic field used experimentally [6].

PROBLEM 4.3 Find the time average of the magnitude of the magnetic field close to the origin for a TOP trap in which the quadrupole field is of the form $B'(x, y, -2z)$ and the bias field has the form $(B_0 \cos \omega t, 0, B_0 \sin \omega t)$. Contrast the result with Eq. (4.6) for the original TOP trap.

PROBLEM 4.4 Estimate the depth of an optical trap for Na atoms produced by a 5-mW laser beam of wavelength 1 µm when focused to a circular cross section of minimum diameter 10 µm.

PROBLEM 4.5 Determine the force constant of a magneto-optical trap using the simple model described in Sec. 4.4.

PROBLEM 4.6 Show that the lowest temperatures attainable by the Sisyphus process are of order $(m_e/m)/(e_0^2/\hbar c)$ times those for the Doppler process. How large is the numerical prefactor?

PROBLEM 4.7 Consider an ideal gas in three dimensions in a potential $V(\mathbf{r})$ which is a homogeneous function of the coordinates of degree ν, that is, $V(\lambda \mathbf{r}) = \lambda^\nu V(\mathbf{r})$. Assume that ν is positive. The effective dimensionality of the trap is denoted by d, and, for example, for the anisotropic harmonic oscillator potential (2.7), d is the number of the frequencies ω_i which are non-zero. By using the semi-classical distribution function, show for a classical gas that

$$\frac{2d}{3}\overline{\epsilon}_{\rm kin} = \nu \overline{V},$$

where the bar denotes a thermal average. The result also holds for degenerate Bose and Fermi gases, and is an example of the virial theorem.

References

[1] C. Townsend, W. Ketterle, and S. Stringari, *Physics World*, March 1997, p. 29.

[2] H. J. Metcalf and P. van der Straten, *Laser Cooling and Trapping*, (Springer, Berlin, 1999).

[3] *Bose–Einstein Condensation in Atomic Gases*, Proceedings of the Enrico Fermi International School of Physics, Vol. CXL, ed. M. Inguscio, S. Stringari, and C. E. Wieman, (IOS Press, Amsterdam, 1999).

[4] W. H. Wing, *Prog. Quantum Electronics* **8**, 181 (1984).

[5] K. B. Davis, M.-O. Mewes, M. R. Andrews, N. J. van Druten, D. S. Durfee, D. M. Kurn, and W. Ketterle, *Phys. Rev. Lett.* **75**, 3969 (1995).

[6] W. Petrich, M. H. Anderson, J. R. Ensher, and E. A. Cornell, *Phys. Rev. Lett.* **74**, 3352 (1995).

[7] M. H. Anderson, J. R. Ensher, M. R. Matthews, C. E. Wieman, and E. A. Cornell, *Science* **269**, 198 (1995).

[8] See, e.g., N. A. Krall and A. W. Trivelpiece, *Principles of Plasma Physics*, (McGraw-Hill, New York, 1973), p. 269 for a discussion of the plasma physics context. The original work is described by Yu. V. Gott, M. S. Ioffe, and V. G. Tel'kovskii, *Nucl. Fusion*, 1962 Suppl., Part 3, p. 1045 (1962).

[9] D. E. Pritchard, *Phys. Rev. Lett.* **51**, 1336 (1983).

[10] D. M. Stamper-Kurn, M. R. Andrews, A. P. Chikkatur, S. Inouye, H.-J. Miesner, J. Stenger, and W. Ketterle, *Phys. Rev. Lett.* **80**, 2027 (1998).

[11] W. D. Phillips, in *Fundamental Systems in Quantum Optics*, ed. J. Dalibard, J.-M. Raimond, and J. Zinn-Justin (North-Holland, Amsterdam, 1992), p. 165.

[12] W. Ketterle, K. B. Davis, M. A. Joffe, A. Martin, and D. E. Pritchard, *Phys. Rev. Lett.* **70**, 2253 (1993).

[13] C. Cohen-Tannoudji and W. D. Phillips, *Physics Today* **43**, October 1990, p. 33. Personal accounts of the history of laser cooling are given in the Nobel lectures of S. Chu, C. Cohen-Tannoudji, and W. D. Phillips in *Rev. Mod. Phys.* **70**, 685–741 (1998).

[14] Our discussion is a simplified version of the arguments given in the paper by J. Dalibard and C. Cohen-Tannoudji, *J. Opt. Soc. Am. B* **6**, 2023 (1989).

[15] Y. Castin, J. Dalibard, and C. Cohen-Tannoudji, in *Bose–Einstein Condensation*, ed. A. Griffin, D. W. Snoke, and S. Stringari, (Cambridge University Press, Cambridge, 1995), p. 173.

[16] Y. Castin and K. Mølmer, *Phys. Rev. Lett.* **74**, 3772 (1995).

[17] H. F. Hess, *Phys. Rev. B* **34**, 3476 (1986).

[18] W. Ketterle and N. J. van Druten, *Adv. At. Mol. Opt. Phys.* **37**, 181 (1996).

[19] O. J. Luiten, M. W. Reynolds, and J. T. M. Walraven, *Phys. Rev. A* **53**, 381 (1996).

[20] T. J. Greytak, in *Bose–Einstein Condensation*, ed. A. Griffin, D. W. Snoke, and S. Stringari, (Cambridge University Press, Cambridge, 1995), p. 131.

[21] D. G. Fried, T. C. Killian, L. Willmann, D. Landhuis, S. C. Moss, D. Kleppner, and T. J. Greytak, *Phys. Rev. Lett.* **81**, 3811 (1998).

[22] T. C. Killian, D. G. Fried, L. Willmann, D. Landhuis, S. C. Moss, T. J. Greytak, and D. Kleppner, *Phys. Rev. Lett.* **81**, 3807 (1998).

5
Interactions between atoms

From a theoretical point of view, one of the appealing features of clouds of alkali atom vapours is that particle separations, which are typically of order 10^2 nm, are large compared with the scattering length a which characterizes the strength of interactions. Scattering lengths for alkali atoms are of the order of $100 a_0$, where a_0 is the Bohr radius, and therefore alkali atom vapours are dilute, in the sense that the dominant effects of interaction are due to two-body encounters. It is therefore possible to calculate properties of the gas reliably from a knowledge of two-body scattering at low energies, which implies that information about atomic scattering is a key ingredient in work on Bose–Einstein condensates.

An alkali atom in its electronic ground state has several different hyperfine states, as we have seen in Secs. 3.1 and 3.2. Interatomic interactions give rise to transitions between these states and, as we described in Sec. 4.6, such processes are a major mechanism for loss of trapped atoms. In a scattering process, the internal states of the particles in the initial or final states are described by a set of quantum numbers, such as those for the spin, the atomic species, and their state of excitation. We shall refer to a possible choice of these quantum numbers as a *channel*.[1] At the temperatures of interest for Bose–Einstein condensation, atoms are in their electronic ground states, and the only relevant internal states are therefore the hyperfine states. Because of the existence of several hyperfine states for a single atom, the scattering of cold alkali atoms is a multi-channel problem.

Besides inelastic processes that lead to trap loss, coupling between channels also gives rise to Feshbach resonances, in which a low-energy bound state in one channel strongly modifies scattering in another channel. Feshbach resonances make it possible to tune both the magnitude and the sign

---

[1] Other authors use the word *channel* in a different sense, to describe the physical processes leading from one particular internal state (the initial state) to another (the final one).

of the effective atom–atom interaction, and they have become a powerful tool for investigating cold atoms.

For all but the very lightest atoms, it is impossible from theory alone to evaluate scattering properties of cold atoms because the atom–atom interaction potentials cannot be calculated with sufficient accuracy. In addition, many properties relevant for cold-atom studies are not directly accessible to experiment. Consequently, in deriving information about two-body scattering it is usually necessary to extract information about the interaction from one class of measurements, and then to use theory to predict properties of interest. Following the development of laser cooling, understanding of low-energy atomic collisions has increased enormously. In particular the use of photoassociation spectroscopy and the study of Feshbach resonances have made it possible to deduce detailed information on scattering lengths.

In the present chapter we give an introduction to atom–atom scattering. For further details we refer the reader to the review [1], the lectures [2], and the original literature. We first give dimensional arguments to show that, because of the van der Waals interaction, scattering lengths for alkali atoms are much larger than the atomic size (Sec. 5.1). We then review elements of scattering theory for a single channel, and discuss the concept of effective interactions (Sec. 5.2). In Sec. 5.3 we determine the scattering length for a simple model potential consisting of a hard-core repulsion and a long-range van der Waals attraction, which varies as r^{-6}, where r is the atomic separation. To describe transitions between different spin states requires a more general formulation of the scattering problem as a multi-channel one. We consider the general theory of scattering between different channels, describe rates of inelastic processes, and show how Feshbach resonances arise in Sec. 5.4. In the final section, Sec. 5.5, we describe experimental techniques for investigating atom–atom interactions and summarize current knowledge of scattering lengths for hydrogen and for alkali atoms.

5.1 Interatomic potentials and the van der Waals interaction

Interactions between polarized alkali atoms are very different from those between unpolarized atoms. To understand why this is so, we recall that interactions between atoms with electrons outside closed shells have an attractive contribution because two electrons with opposite spin can occupy the same orbital. This is the effect responsible for covalent bonding. However, if two electrons are in the same spin state, they cannot have the same spatial wave function, and therefore the reduction in energy due to two electrons sharing the same orbital is absent. To illustrate this, we show in

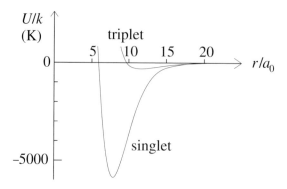

Fig. 5.1. Sketch of the interaction potentials $U(r)$ as functions of the atomic separation r for two ground-state rubidium atoms with electrons in singlet and triplet states.

Fig. 5.1 the interactions for two rubidium atoms in their ground state when the two valence electrons are in the singlet spin state and in the triplet one. For small separations the interactions are dominated by a strong repulsive core due to the overlapping of electron clouds, but at greater separations the attractive well is very much deeper for the singlet state than for the triplet state. The singlet potential has a minimum with a depth of nearly 6000 K in temperature units when the atoms are about $8a_0$ apart. By contrast, the depth of the minimum of the triplet potential that occurs for an atomic separation of about $12a_0$ is only a few hundred K. For large atomic separations there is an attraction due to the van der Waals interaction, but it is very weak compared with the attractive interactions due to covalent bonding.

While the van der Waals interaction is weak relative to covalent bonding, it is still strong in the sense that the triplet potential has many molecular bound states, as we shall see later from more detailed calculations. We remark that the electronic spin state for a pair of atoms in definite hyperfine states is generally a superposition of electronic triplet and singlet contributions, and consequently the interaction contains both triplet and singlet terms.

Two-body interactions at low energies are characterized by their scattering lengths, and it is remarkable that for polarized alkali atoms these are typically about two orders of magnitude greater than the size of an atom, $\sim a_0$. Before turning to detailed calculations we give qualitative arguments to show that the van der Waals interaction can give rise to such large scattering lengths. The van der Waals interaction is caused by the electric dipole–dipole interaction between the atoms, and it has the form $-\alpha/r^6$, where r is the atomic separation. The length scale r_0 in the Schrödinger

Table 5.1. *Calculated values of the van der Waals coefficient C_6. The value for Li is from Ref. [3] and those for other alkali atoms from Ref. [4].*

Element	C_6
H–H	6.5
Li–Li	1393
Na–Na	1556
K–K	3897
Rb–Rb	4691
Cs–Cs	6851

equation at zero energy, which sets the basic scale for the scattering length, may be estimated by dimensional arguments. The kinetic energy is of order $\hbar^2/mr_0^2$, where m is the atomic mass, and the van der Waals interaction is of order α/r_0^6. Equating these two energies gives the characteristic length r_0, which is of order $(\alpha m/\hbar^2)^{1/4}$. On dimensional grounds, the coefficient α must be of the form of a typical atomic energy, e_0^2/a_0, times the sixth power of the atomic length scale a_0, that is $\alpha = C_6 e_0^2 a_0^5$, where the dimensionless coefficient C_6 gives the strength of the van der Waals interaction in atomic units. Thus the length scale is given by

$$r_0 \approx (C_6 m/m_e)^{1/4} a_0. \tag{5.1}$$

This gives the general magnitude of scattering lengths but, as we shall see in Sec. 5.3, the sign and numerical value of the scattering length are determined by the short-range part of the interaction. The large scattering lengths for alkali atoms are thus a consequence of two effects. First, atomic masses are of order $10^3 A$ times the electron mass, A being the mass number, which gives more than a factor of 10 in the length scale. Second, van der Waals coefficients for alkali atoms lie between 10^3 and 10^4, as one can see from the theoretical values of C_6 listed in Table 5.1. A number of the measurements that we shall describe in Sec. 5.5 provide information about C_6 and, for example, a study of Feshbach resonances for ^{133}Cs gives $C_6 = 6890 \pm 35$ [5, 6]. The large values of C_6 give a further increase in the length scale by almost one order of magnitude, so typical scattering lengths are of order $10^2 a_0$.

The $1/r^6$ contribution to the potential is the leading term in an expansion of the long-range part of the two-body interaction $U(r)$ in inverse powers of r. If quantities are measured in atomic units the expansion is given more generally by

$$U(r) = -\frac{C_6}{r^6} - \frac{C_8}{r^8} - \frac{C_{10}}{r^{10}} + \cdots. \qquad (5.2)$$

The higher-order coefficients C_8 and C_{10} are typically $10^2 C_6$ and $10^4 C_6$, respectively, and therefore at distances of order r_0 a pure $1/r^6$ potential is a good first approximation.

Magnitude of the van der Waals interaction

The large van der Waals interactions for alkali atoms, like the large polarizabilities (see Sec. 3.3), are a consequence of the strong resonance lines in the optical spectrum of these atoms. To derive a microscopic expression for C_6 we recall that the van der Waals interaction is due to the electric dipole–dipole interaction between atoms, which has the form

$$U_{\text{ed}} = \frac{1}{4\pi\epsilon_0 r^3}[\mathbf{d}_1 \cdot \mathbf{d}_2 - 3(\mathbf{d}_1 \cdot \hat{\mathbf{r}})(\mathbf{d}_2 \cdot \hat{\mathbf{r}})], \qquad (5.3)$$

where $\hat{\mathbf{r}} = \mathbf{r}/r$, $\mathbf{r}$ being the vector separation of the two atoms. The ground states of atoms are to a very good approximation eigenstates of parity, and consequently diagonal matrix elements of the electric dipole operator vanish. The leading contribution to the interaction energy is of second order in the dipole–dipole interaction and has the form [7, §89]

$$U(r) = -\frac{6}{(4\pi\epsilon_0)^2 r^6} \sum_{n,n'} \frac{|\langle n|d_z|0\rangle|^2 |\langle n'|d_z|0\rangle|^2}{E_n + E_{n'} - 2E_0}, \qquad (5.4)$$

where the factor 6 comes from using the fact that, for atoms with $L=0$ in the ground state, the sum is independent of which Cartesian component of the dipole operator is chosen. Expressing the result in terms of the oscillator strength (see Eq. (3.38)) and measuring excitation energies in atomic units one finds

$$C_6 = \frac{3}{2} \sum_{n,n'} \frac{f_{n0}^z f_{n'0}^z}{(E_n - E_0)(E_{n'} - E_0)(E_n + E_{n'} - 2E_0)}. \qquad (5.5)$$

Just as for the polarizability, which we discussed in Sec. 3.3 (see Eq. (3.42)), the dominant contribution to the sum comes from the resonance line, and if we assume that the total oscillator strength for transitions from the ground state to all sublevels of the resonance doublet is unity and neglect all other transitions, one finds

$$C_6 \approx \frac{3}{4(\Delta E_{\text{res}})^3}. \qquad (5.6)$$

Here ΔE_{res} is the energy of the resonance line in atomic units. For alkali atoms this is less than 0.1 atomic units, and therefore the values of C_6

are more than 10^3, in agreement with more detailed calculations and with experiment. As an example consider sodium. The resonance line energy is 0.0773 atomic units, and therefore Eq. (5.6) yields $C_6 \approx 1620$, while detailed calculations give 1556, according to Table 5.1. For the heavier alkali atoms, electrons other than the valence one contribute significantly to the sum, and consequently C_6 is greater than the estimate (5.6). The simple formula (5.6) enables one to understand why the trend in the magnitude of C_6 for alkali atoms is similar to the variation of the atomic polarizabilities discussed in Sec. 3.3. For hydrogen the simple estimate evaluated using the Lyman-α energy is $C_6 = 128/9 \approx 14.2$, which is larger than the actual value ≈ 6.5 since the oscillator strength for the transition is significantly less than unity.

We note that to calculate the interatomic potential at large distances the dipolar interaction may not be treated as static, and retardation of the interaction must be taken into account. The interaction is then found to vary as $1/r^7$, rather than $1/r^6$. This effect becomes important when the separation between atoms is comparable to or larger than the wavelength of a photon with energy equal to that of the resonance transition. The latter is of order $10^3 a_0$ or more, which is larger than the distances of importance in determining the scattering length. Consequently retardation plays little role.

5.2 Basic scattering theory

Here we review aspects of scattering theory for a single channel and introduce the scattering length, which characterizes low-energy interactions between a pair of particles. More extensive treatments may be found in standard texts (see, e.g., [7, §123]). Consider the scattering of two particles of mass m_1 and m_2, which we assume for the moment to have no internal degrees of freedom. We shall further assume here that the two particles are distinguishable, and the effects of identity of the particles will be described later. As usual, we transform to centre-of-mass and relative coordinates. The wave function for the centre-of-mass motion is a plane wave, while that for the relative motion satisfies a Schrödinger equation with the mass equal to the reduced mass $m_r = m_1 m_2/(m_1 + m_2)$ of the two particles. To describe scattering, one writes the wave function for the relative motion as the sum of an incoming plane wave and a scattered wave,[2]

$$\psi = e^{ikz} + \psi_{\text{sc}}(\mathbf{r}), \tag{5.7}$$

---

[2] We shall use plane-wave states $e^{i\mathbf{k}\cdot\mathbf{r}}$ without an explicit factor of $1/V^{1/2}$.

where we have chosen the relative velocity in the incoming wave to be in the z direction. At large interatomic separations the scattered wave is an outgoing spherical wave $f(\mathbf{k})\exp(ikr)/r$, where $f(\mathbf{k})$ is the *scattering amplitude* and $\mathbf{k}$ specifies the wave vector of the scattered wave. We shall assume that the interaction between atoms is spherically symmetric, and the scattering amplitude $f(\theta)$ then depends on direction only through the scattering angle θ, which is the angle between the directions of the relative momentum of the atoms before and after scattering. The wave function for large r is thus

$$\psi = e^{ikz} + f(\theta)\frac{e^{ikr}}{r}. \tag{5.8}$$

The energy of the state is given by

$$E = \frac{\hbar^2 k^2}{2m_\mathrm{r}}. \tag{5.9}$$

At very low energies it is sufficient to consider s-wave scattering, as we shall argue below. In this limit the scattering amplitude $f(\theta)$ approaches a constant, which we denote by $-a$, and the wave function (5.8) becomes

$$\psi = 1 - \frac{a}{r}. \tag{5.10}$$

The constant a is called the *scattering length*. It gives the intercept of the asymptotic wave function (5.10) on the r axis.

In the following we discuss the connection between the scattering length and the phase shifts, which in general determine the scattering cross section. The differential cross section, that is the cross section per unit solid angle, is given by

$$\frac{d\sigma}{d\Omega} = |f(\theta)|^2. \tag{5.11}$$

For scattering through an angle between θ and $\theta + d\theta$, the element of solid angle is $d\Omega = 2\pi \sin\theta \, d\theta$. Since the potential is spherically symmetric, the solution of the Schrödinger equation has axial symmetry with respect to the direction of the incident particle. The wave function for the relative motion therefore may be expanded in terms of Legendre polynomials $P_l(\cos\theta)$,

$$\psi = \sum_{l=0}^{\infty} A_l P_l(\cos\theta) R_{kl}(r). \tag{5.12}$$

The radial wave function $R_{kl}(r)$ satisfies the equation

$$R''_{kl}(r) + \frac{2}{r}R'_{kl}(r) + \left[k^2 - \frac{l(l+1)}{r^2} - \frac{2m_\mathrm{r}}{\hbar^2}U(r)\right]R_{kl}(r) = 0, \tag{5.13}$$

where $U(r)$ is the potential, and the prime denotes a derivative with respect to r. For $r \to \infty$ the radial function is given in terms of the *phase shifts* δ_l according to the equation

$$R_{kl}(r) \simeq \frac{1}{kr} \sin(kr - \frac{\pi}{2}l + \delta_l). \tag{5.14}$$

By comparing (5.12) and (5.14) with (5.8) and expanding the plane wave $\exp(ikz)$ in Legendre polynomials one finds that $A_l = i^l(2l+1)e^{i\delta_l}$ and

$$f(\theta) = \frac{1}{2ik} \sum_{l=0}^{\infty} (2l+1)(e^{i2\delta_l} - 1) P_l(\cos\theta). \tag{5.15}$$

The total scattering cross section is obtained by integrating the differential cross section over all solid angles, and it is given by

$$\sigma = 2\pi \int_{-1}^{1} d(\cos\theta) |f(\theta)|^2. \tag{5.16}$$

When (5.15) is inserted in this expression, and use is made of the fact that the Legendre polynomials are orthogonal, one obtains the total cross section in terms of the phase shifts:

$$\sigma = \frac{4\pi}{k^2} \sum_{l=0}^{\infty} (2l+1) \sin^2 \delta_l. \tag{5.17}$$

For a finite-range potential the phase shifts vary as k^{2l+1} for small k. For a potential varying as r^{-n} at large distances, this result is true provided $l < (n-3)/2$, but for higher partial waves $\delta_l \propto k^{n-2}$ [7, §132]. Thus for potentials that behave as $1/r^6$ or $1/r^7$, all phase shifts become small as k approaches zero. The scattering cross section is thus dominated by the $l=0$ term (s-wave scattering), corresponding to a scattering amplitude $f = \delta_0/k$. If one writes the $l=0$ component of the asymptotic low-energy solution (5.14) at large distances as

$$R_0 \simeq c_1 \frac{\sin kr}{kr} + c_2 \frac{\cos kr}{r}, \tag{5.18}$$

where c_1 and c_2 are coefficients, the phase shift for $k \to 0$ is given by

$$\delta_0 = \frac{kc_2}{c_1}. \tag{5.19}$$

From the definition of the scattering length in terms of the wave function for $k \to 0$ given by (5.10), one finds that

$$\delta_0 = -ka, \tag{5.20}$$

which shows that a is determined by the coefficients in the asymptotic solution (5.18),

$$a = -\left.\frac{c_2}{c_1}\right|_{k\to 0}. \tag{5.21}$$

In this limit the total cross section, Eq. (5.17), is determined only by a. It is

$$\sigma = \frac{4\pi}{k^2}\delta_0^2 = 4\pi a^2. \tag{5.22}$$

Let us now consider scattering of identical particles in the same internal state. The wave function must be symmetric under interchange of the coordinates of the two particles if they are bosons, and antisymmetric if they are fermions. Interchange of the two particle coordinates corresponds to changing the sign of the relative coordinate, that is $\mathbf{r} \to -\mathbf{r}$, or $r \to r, \theta \to \pi - \theta$ and $\varphi \to \pi + \varphi$, where φ is the azimuthal angle. The symmetrized wave function corresponding to Eq. (5.8) is thus

$$\psi = e^{ikz} \pm e^{-ikz} + [f(\theta) \pm f(\pi - \theta)]\frac{e^{ikr}}{r}. \tag{5.23}$$

The amplitude for scattering a particle in the direction specified by the polar angle θ is therefore $f(\theta) \pm f(\pi - \theta)$, and the differential cross section is

$$\frac{d\sigma}{d\Omega} = |f(\theta) \pm f(\pi - \theta)|^2, \tag{5.24}$$

the plus sign applying to bosons and the minus sign to fermions.

The physical content of this equation is that the amplitude for a particle to be scattered into some direction is the sum or difference of the amplitude for one of the particles to be scattered through an angle θ and the amplitude for the other particle to be scattered through an angle $\pi - \theta$. The total cross section is obtained by integrating the differential cross section over all distinct final states. Because of the symmetry of the wave function, the state specified by angles θ, φ is identical with that for angles $\pi - \theta, \varphi + \pi$, and therefore to avoid double counting, one should integrate only over half of the total 4π solid angle, for example by integrating over the range $0 \le \theta \le \pi/2$ and $0 \le \varphi \le 2\pi$. Thus if scattering is purely s-wave, the total cross section is

$$\sigma = 8\pi a^2 \tag{5.25}$$

for identical bosons, and it vanishes for identical fermions.

In Sec. 5.3 we shall derive an explicit expression for the scattering length for a model of the interaction potential for alkali atoms. Before doing that we introduce the concept of an effective interaction and demonstrate how it is related to the scattering length.

5.2.1 Effective interactions and the scattering length

Interactions between atoms are strong, but they occur only when two atoms are close together. In dilute gases this is rather unlikely, since interactions are very small for typical atomic separations. For most configurations of the system, the many-body wave function varies slowly in space, but when two atoms approach each other, there are rapid spatial variations. To avoid having to calculate short-range correlations between atoms in detail, it is convenient to introduce the concept of an effective interaction. This describes interactions among long-wavelength, low-frequency degrees of freedom of a system when coupling of these degrees of freedom via interactions with those at shorter wavelengths has been taken into account. The short-wavelength degrees of freedom are said to have been 'integrated out'. In recent years this approach has found applications in numerous branches of physics, ranging from critical phenomena to elementary particle physics and nuclear physics.

To make these ideas quantitative, consider the problem of two-particle scattering again, this time in the momentum representation. The particles are assumed to have equal masses m, and therefore $m_r = m/2$. The wave function in coordinate space is given by (5.7), which in the momentum representation is[3]

$$\psi(\mathbf{k'}) = (2\pi)^3 \delta(\mathbf{k'} - \mathbf{k}) + \psi_{sc}(\mathbf{k'}), \qquad (5.26)$$

where the second term on the right hand side of this equation is the Fourier transform of the scattered wave in (5.7). The wave function (5.26) satisfies the Schrödinger equation, which is[4]

$$\left(\frac{\hbar^2 k^2}{m} - \frac{\hbar^2 k'^2}{m}\right)\psi_{sc}(\mathbf{k'}) = U(\mathbf{k'}, \mathbf{k}) + \frac{1}{V}\sum_{\mathbf{k''}} U(\mathbf{k'}, \mathbf{k''})\psi_{sc}(\mathbf{k''}), \qquad (5.27)$$

---

[3] A function $F(\mathbf{r})$ and its Fourier transform $F(\mathbf{q})$ are related by

$$F(\mathbf{r}) = \frac{1}{V}\sum_{\mathbf{q}} F(\mathbf{q})e^{i\mathbf{q}\cdot\mathbf{r}} = \int \frac{d\mathbf{q}}{(2\pi)^3} F(\mathbf{q})e^{i\mathbf{q}\cdot\mathbf{r}},$$

where V is the volume. Thus $F(\mathbf{q}) = \int d\mathbf{r} F(\mathbf{r})e^{-i\mathbf{q}\cdot\mathbf{r}}$.

[4] We caution the reader that there are different ways of normalizing states in the continuum. In some of the atomic physics literature it is common to use integrals over energy rather than sums over wave numbers and to work with states which differ from the ones used here by a factor of the square root of the density of states. Matrix elements of the potential are then dimensionless quantities, rather than ones with dimensions of energy times volume, as they are here.

where $\hbar^2 k^2/m\ (=E)$ is the energy eigenvalue and $U(\mathbf{k'}, \mathbf{k''}) = U(\mathbf{k'} - \mathbf{k''})$ is the Fourier transform of the bare atom–atom interaction. The scattered wave is thus given by

$$\psi_{\text{sc}}(\mathbf{k'}) = \left(\frac{\hbar^2 k^2}{m} - \frac{\hbar^2 k'^2}{m} + i\delta\right)^{-1} \left(U(\mathbf{k'}, \mathbf{k}) + \frac{1}{V}\sum_{\mathbf{k''}} U(\mathbf{k'}, \mathbf{k''})\psi_{\text{sc}}(\mathbf{k''})\right), \tag{5.28}$$

where we have in the standard way introduced the infinitesimal imaginary part δ to ensure that only outgoing waves are present in the scattered wave. This equation may be written in the form

$$\psi_{\text{sc}}(\mathbf{k'}) = \left(\frac{\hbar^2 k^2}{m} - \frac{\hbar^2 k'^2}{m} + i\delta\right)^{-1} T(\mathbf{k'}, \mathbf{k}; \hbar^2 k^2/m), \tag{5.29}$$

where the scattering matrix T satisfies the so-called Lippmann–Schwinger equation

$$T(\mathbf{k'}, \mathbf{k}; E) = U(\mathbf{k'}, \mathbf{k}) + \frac{1}{V}\sum_{\mathbf{k''}} U(\mathbf{k'}, \mathbf{k''}) \left(E - \frac{\hbar^2 k''^2}{m} + i\delta\right)^{-1} T(\mathbf{k''}, \mathbf{k}; E). \tag{5.30}$$

The scattered wave at large distances and for zero energy ($E = k = 0$) may be calculated from Eq. (5.29). Using the Fourier transform

$$\int \frac{d\mathbf{k'}}{(2\pi)^3} \frac{e^{i\mathbf{k'} \cdot \mathbf{r}}}{k'^2} = \frac{1}{4\pi r}, \tag{5.31}$$

we find

$$\psi_{\text{sc}}(r) = -\frac{mT(0, 0; 0)}{4\pi \hbar^2 r}. \tag{5.32}$$

We have replaced the argument k' in the T matrix by zero, since the values of k' of importance in the Fourier transform are of order $1/r$. The expression (5.32) may thus be identified with (5.10), which implies that the scattering matrix at zero energy and the scattering length are related by the expression

$$a = \frac{m}{4\pi \hbar^2} T(0, 0; 0) \tag{5.33}$$

or

$$T(0, 0; 0) = \frac{4\pi \hbar^2 a}{m}. \tag{5.34}$$

More generally the scattering amplitude and the T matrix are related by the equation

$$f(\mathbf{k}, \mathbf{k'}) = -\frac{m}{4\pi \hbar^2} T(\mathbf{k'}, \mathbf{k}; E = \hbar^2 k^2/m). \tag{5.35}$$

In the Born approximation, which is obtained by taking only the first term on the right hand side of the Lippmann–Schwinger equation, the scattering length is given by

$$a_{\text{Born}} = \frac{m}{4\pi\hbar^2}U(0) = \frac{m}{4\pi\hbar^2}\int d\mathbf{r}U(\mathbf{r}), \quad (5.36)$$

corresponding to $|\mathbf{k}-\mathbf{k}'|=0$. Thus the scattering matrix, T, may be regarded as an effective interaction, in the sense that when inserted into the Born-approximation expression for the scattered wave function it gives the exact result when the atoms are far apart. All the effects of short-wavelength components of the wave function that reflect the correlations between the two particles have been implicitly taken into account by replacing $U(0)$ by T.

To obtain further insight into effective interactions we now adopt another point of view. Let us divide the intermediate states in the Lippmann–Schwinger equation into two groups: those with energy greater than some cut-off value $\epsilon_c = \hbar^2 k_c^2/m$, and those with lower energy. We can perform the summation over intermediate states in (5.30) in two stages. First we sum over all intermediate states with energy in excess of ϵ_c, and then over the remaining states. The first stage leads to a quantity $\tilde{U}(\mathbf{k}',\mathbf{k};E)$ which satisfies the equation

$$\tilde{U}(\mathbf{k}',\mathbf{k};E) = U(\mathbf{k}',\mathbf{k}) \\ + \frac{1}{V}\sum_{\mathbf{k}'',k''>k_c} U(\mathbf{k}',\mathbf{k}'')\left(E - \frac{\hbar^2 k''^2}{m} + i\delta\right)^{-1}\tilde{U}(\mathbf{k}'',\mathbf{k};E), \quad (5.37)$$

and the second stage builds in the correlations associated with lower-energy states:

$$T(\mathbf{k}',\mathbf{k};E) = \tilde{U}(\mathbf{k}',\mathbf{k};E) \\ + \frac{1}{V}\sum_{\mathbf{k}'',k''<k_c} \tilde{U}(\mathbf{k}',\mathbf{k}'';E)\left(E - \frac{\hbar^2 k''^2}{m} + i\delta\right)^{-1} T(\mathbf{k}'',\mathbf{k};E). \quad (5.38)$$

The latter equation shows that if one uses $\tilde{U}$ as the interaction in a scattering problem in which intermediate states with energies in excess of ϵ_c do not appear explicitly, it produces the correct scattering matrix. In this sense it is an effective interaction, which describes interactions between a limited set of states. The difference between the effective interaction and the bare one is due to the influence of the high-momentum states. It is important to observe that the effective potential depends explicitly on the choice of the energy ϵ_c. However, the final result for the scattering amplitude is, of course, independent of this choice.

If one takes the limit $k_c \to 0$, the effective interaction reduces to the scattering matrix. For small k_c, that is for describing interactions between very-long-wavelength excitations and at low energies, the effective interaction becomes simply

$$\tilde{U}(0,0;0)|_{k_c \to 0} = \frac{4\pi\hbar^2 a}{m} \equiv U_0. \qquad (5.39)$$

We shall make extensive use of this effective interaction, which is also referred to as a pseudopotential, for calculating properties of dilute gases. The key point is that the effective interaction may be used to make precise calculations for dilute systems without the necessity of calculating short-range correlations. This implies that the Born approximation for scattering, and a mean-field approach such as the Hartree or Hartree–Fock ones for calculating energies give the correct results provided one uses the effective interaction rather than the bare one.

Later, in discussing the microscopic theory of the dilute Bose gas, we shall need an expression for the effective interaction for small but non-zero k_c. This may be found from (5.38) and is

$$\tilde{U}(\mathbf{k'},\mathbf{k};E) \simeq T(\mathbf{k'},\mathbf{k};E)$$

$$- \frac{1}{V} \sum_{\mathbf{k''}, k'' < k_c} T(\mathbf{k'},\mathbf{k''};E) \left(E - \frac{\hbar^2 k''^2}{m} + i\delta \right)^{-1} T(\mathbf{k''},\mathbf{k};E). \qquad (5.40)$$

In the opposite limit, $k_c \to \infty$, the effective interaction is the bare interaction, because no degrees of freedom are integrated out.

5.3 Scattering length for a model potential

As we have described earlier, the interaction between two alkali atoms at large separation is dominated by the van der Waals attraction. We now evaluate the scattering length for a model potential which has the van der Waals form $\sim 1/r^6$ at large distances, and which is cut off at short distances by an infinitely hard core of radius r_c [8, 9]:

$$U(r) = \infty \quad \text{for} \quad r \leq r_c, \qquad U(r) = -\frac{\alpha}{r^6} \quad \text{for} \quad r > r_c. \qquad (5.41)$$

This simplified model captures the essential aspects of the physics when the potential has many bound states, as it does for alkali atoms. The core radius is to be regarded as a way of parametrizing the short-distance behaviour of the potential, rather than as a realistic representation of it.

5.3 Scattering length for a model potential

The Schrödinger equation (5.13) for the relative motion is conveniently written in terms of the function $\chi = rR$,

$$\chi''(r) + \left[k^2 - \frac{l(l+1)}{r^2} - \frac{2m_r}{\hbar^2}U(r)\right]\chi(r) = 0. \tag{5.42}$$

Since we consider low-energy s-wave scattering, we set k and l equal to zero in (5.42), which results in

$$\chi''(r) + \frac{2m_r\alpha}{\hbar^2 r^6}\chi(r) = 0, \quad \text{for} \quad r > r_c. \tag{5.43}$$

Due to the presence of the hard core, χ must vanish at $r = r_c$.

Our strategy is to use (5.43) to determine χ at large r, where it has the form

$$\chi = c_1 r + c_2, \tag{5.44}$$

in terms of which the scattering length is given by $a = -c_2/c_1$ (compare (5.18) above). First we introduce the dimensionless variable $x = r/r_0$, where

$$r_0 = \left(\frac{2m_r\alpha}{\hbar^2}\right)^{1/4}. \tag{5.45}$$

This is the length scale we derived in Sec. 5.1 from qualitative arguments, except that here we have allowed for the possibility of the masses of the two atoms being different. The Schrödinger equation then becomes

$$\frac{d^2\chi(x)}{dx^2} + \frac{1}{x^6}\chi(x) = 0. \tag{5.46}$$

In order to turn the differential equation (5.46) into one whose solutions are known, we write

$$\chi = x^\beta f(x^\gamma) \tag{5.47}$$

and try to choose values of β and γ which result in a known differential equation for $f = f(y)$ with $y = x^\gamma$. We obtain

$$\frac{d^2\chi(x)}{dx^2} = \beta(\beta-1)y^{(\beta-2)/\gamma} f(y) + [\gamma(\gamma-1) + 2\beta\gamma]y^{1+(\beta-2)/\gamma}\frac{df(y)}{dy}$$
$$+ \gamma^2 y^{2+(\beta-2)/\gamma}\frac{d^2 f(y)}{dy^2} \tag{5.48}$$

and

$$\frac{1}{x^6}\chi(x) = y^{(\beta-6)/\gamma} f(y). \tag{5.49}$$

If we choose $\beta = 1/2$ and $\gamma = -2$, introduce the new variable $z = y/2$, and write $f(y) = g(z)$, we find

$$\frac{d^2g}{dz^2} + \frac{1}{z}\frac{dg}{dz} + (1 - \frac{1}{16z^2})g = 0, \tag{5.50}$$

which is Bessel's equation. The general solution of (5.50) may be written as a linear combination of the Bessel functions $J_{1/4}(z)$ and $J_{-1/4}(z)$,

$$g = AJ_{1/4}(z) + BJ_{-1/4}(z), \tag{5.51}$$

where

$$z = \frac{r_0^2}{2r^2}. \tag{5.52}$$

Since χ must vanish at $r = r_c$, the coefficients satisfy the condition

$$\frac{A}{B} = -\frac{J_{-1/4}(r_0^2/2r_c^2)}{J_{1/4}(r_0^2/2r_c^2)}. \tag{5.53}$$

In terms of the original variables, the radial function $\chi(r) = rR(r)$ is given by

$$\chi(r) = A(r/r_0)^{1/2}\left[J_{1/4}(r_0^2/2r^2) - \frac{J_{1/4}(r_0^2/2r_c^2)}{J_{-1/4}(r_0^2/2r_c^2)}J_{-1/4}(r_0^2/2r^2)\right]. \tag{5.54}$$

This function is plotted in Fig. 5.2 for different values of the parameter $r_0^2/2r_c^2$.

For alkali atoms, the interatomic potentials become repulsive for $r \lesssim 10a_0$, and therefore an appropriate choice of r_c is of this magnitude. Since $r_0 \sim 100a_0$ the condition

$$r_0 \gg r_c, \tag{5.55}$$

is satisfied, which implies that the potential has many bound states, as we shall argue below. We may then evaluate (5.53) by using the asymptotic expansion

$$J_p(z) \simeq \sqrt{\frac{2}{\pi z}}\cos\left[z - (p + \frac{1}{2})\frac{\pi}{2}\right] \tag{5.56}$$

valid for large z. We find from (5.53) that

$$\frac{A}{B} = -\frac{\cos(r_0^2/2r_c^2 - \pi/8)}{\cos(r_0^2/2r_c^2 - 3\pi/8)}. \tag{5.57}$$

To determine the scattering length we must examine the wave function χ at large distances $r \gg r_0$, that is for small values of z. In this limit the

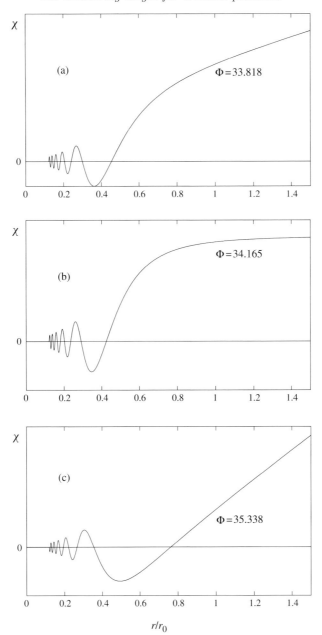

Fig. 5.2. The radial wave function χ as a function of r/r_0 for selected values of $\Phi = r_0^2/2r_c^2$. From top to bottom, the three curves correspond to core radii $r_c \approx 0.121\,59\,r_0, 0.120\,97\,r_0$, and $0.118\,95\,r_0$, and illustrate the sensitivity of the wave function to small changes in the short-range part of the potential. The normalization constant has been chosen to make the wave function at large r positive, and as a consequence the wave function for $r \to r_c$ has the opposite sign for case (c) compared with the other two cases.

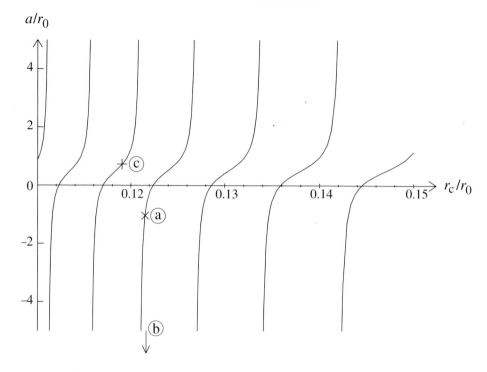

Fig. 5.3. The scattering length a as a function of the hard core radius r_c. Both are measured in units of r_0. ⓐ to ⓒ correspond to the three values of r_c used to generate curves (a) to (c), respectively, in Fig. 5.2; for case (b) a is $-186.0 r_0$.

leading term in the Bessel function is

$$J_p(z) \simeq \frac{z^p}{2^p \Gamma(1+p)}, \tag{5.58}$$

and therefore the radial wave function has the form

$$\chi \simeq A \frac{1}{\sqrt{2}\Gamma(5/4)} + B \frac{\sqrt{2}}{\Gamma(3/4)} \frac{r}{r_0}. \tag{5.59}$$

By comparing this result with the general expression (5.33) for the wave function at large distances we conclude that

$$a = -r_0 \frac{\Gamma(3/4)}{2\,\Gamma(5/4)} \frac{A}{B}. \tag{5.60}$$

The scattering length obtained from (5.53) and (5.60) is plotted in Fig. 5.3 as a function of the hard core radius. Inserting the ratio A/B from (5.57) we arrive at the expression

$$a = r_0 \frac{\Gamma(3/4)\cos(r_0^2/2r_c^2 - \pi/8)}{2\,\Gamma(5/4)\cos(r_0^2/2r_c^2 - 3\pi/8)}, \qquad (5.61)$$

which may alternatively be written as

$$a = r_0 \frac{\Gamma(3/4)}{2\sqrt{2}\,\Gamma(5/4)}[1 - \tan(\Phi - 3\pi/8)] \approx 0.478 r_0[1 - \tan(\Phi - 3\pi/8)], \quad (5.62)$$

where $\Phi = r_0^2/2r_c^2$.

From Eq. (5.62) we can draw a number of important conclusions. First, the scale of scattering lengths is set by the quantity r_0, in agreement with the dimensional arguments made at the beginning of this section. Second, the scattering length can be either positive (corresponding to a repulsive effective interaction) or negative (corresponding to an attractive one). Third, the scattering length depends on the details of the short-range part of the interaction (in this case the parameter r_c), and therefore it is impossible on the basis of this simple model to obtain realistic estimates of the scattering length. This is due to the fact that the sign of the effective interaction at zero energy depends on the energy of the highest bound state. One may estimate the number of bound states by imagining slowly increasing the strength of the potential from zero to its physical value. A bound state appears whenever the scattering length tends to minus infinity and therefore the number N_b of bound states is given by the integer part of $(\Phi/\pi - 3/8)$ or

$$N_b \approx \frac{r_0^2}{2\pi r_c^2} = \frac{\Phi}{\pi}, \qquad (5.63)$$

if Φ is large.

One can make statistical arguments about the relative likelihood of attractive and repulsive interactions by assuming that all values of Φ in a range much greater than π are equally probable. According to (5.62) the scattering length a is positive unless $\Phi - 3\pi/8 - \nu\pi$ lies in the interval between $\pi/4$ and $\pi/2$, with ν being an integer. Since the length of this interval is $\pi/4$, there is a 'probability' $1/4$ of the scattering length being negative and a 'probability' $3/4$ of it being positive. Thus, on average, repulsive interactions should be three times more common than attractive ones.

The van der Waals potential plays such an important role in low-energy scattering of alkali atoms because, in the range of atomic separations for which it dominates the interaction, it is so strong that it can cause many spatial oscillations of the zero-energy wave function. The qualitative conclusions we have arrived at using the simple model apply for more general

5.4 Scattering between different internal states

In treating atom–atom scattering we have so far neglected the internal degrees of freedom of the atoms due to the nuclear and electronic spins. These give rise to the hyperfine and Zeeman splittings, as discussed in Chapter 3. For an alkali atom in its ground state, the electronic spin is 1/2 and the total number of nuclear spin states is $2I+1$, where I is the nuclear spin. For two alkali atoms in their ground states the total number of hyperfine states is thus $[2(2I+1)]^2$, and we shall label them by indices α and β which specify the hyperfine states of each of the atoms. Two atoms initially in the state $|\alpha\beta\rangle$ may be scattered by atom–atom interactions to the state $|\alpha'\beta'\rangle$ and, as a consequence, scattering becomes a multi-channel problem. In this section we indicate how to generalize the theory of scattering to this situation.

In the absence of interactions between atoms, the Hamiltonian for two atoms consists of the kinetic energy associated with the centre-of-mass motion, the kinetic energy of the relative motion, and the hyperfine and Zeeman energies, Eq. (3.8). As in the single-channel problem, the centre-of-mass motion is simple, since the corresponding momentum is conserved. We can therefore confine our attention to the relative motion, the Hamiltonian for which is

$$H_{\rm rel} = H_0 + U, \tag{5.64}$$

where

$$H_0 = \frac{\hat{\mathbf{p}}^2}{2m_{\rm r}} + H_{\rm spin}(1) + H_{\rm spin}(2). \tag{5.65}$$

Here the first term in H_0 is the kinetic energy operator for the relative motion, $\hat{\mathbf{p}}$ being the operator for the relative momentum, $H_{\rm spin}$ is the Hamiltonian (3.8), the labels 1 and 2 refer to the two atoms, and U is the atom–atom interaction. The eigenstates of H_0 may be denoted by $|\alpha\beta, \mathbf{k}\rangle$, where $\hbar\mathbf{k}$ is the relative momentum. If the eigenvalues of the spin Hamiltonian are given by

$$H_{\rm spin}|\alpha\rangle = \epsilon_\alpha|\alpha\rangle, \tag{5.66}$$

the energies of the eigenstates of H_0 are

$$E_{\alpha\beta}(k_{\alpha\beta}) = \frac{\hbar^2 k_{\alpha\beta}^2}{2m_{\rm r}} + \epsilon_\alpha + \epsilon_\beta. \tag{5.67}$$

5.4 Scattering between different internal states

The scattering amplitude is now introduced by generalizing (5.8) to allow for the internal states. The asymptotic form of the wave function corresponding to (5.8) is thus

$$\psi = e^{i\mathbf{k}_{\alpha\beta}\cdot\mathbf{r}}|\alpha\beta\rangle + \sum_{\alpha'\beta'} f_{\alpha\beta}^{\alpha'\beta'}(\mathbf{k}_{\alpha\beta}, \mathbf{k}'_{\alpha'\beta'}) \frac{e^{ik'_{\alpha'\beta'}r}}{r}|\alpha'\beta'\rangle, \tag{5.68}$$

where $\hbar\mathbf{k}_{\alpha\beta}$ is the relative momentum in the incoming state, which is referred to as the entrance channel. The scattered wave has components in different internal states $\alpha'\beta'$ which are referred to as the exit channels.

It is important to note that if the entrance and exit channels are different, their hyperfine and Zeeman energies are generally different, and therefore the magnitude of the relative momentum $\hbar\mathbf{k}'_{\alpha'\beta'}$ in the exit channel $\alpha'\beta'$ is different from that in the entrance channel. The two relative momenta are related by the requirement that the total energy E be conserved, $E = E_{\alpha\beta}(k_{\alpha\beta}) = E_{\alpha'\beta'}(k'_{\alpha'\beta'})$, and thus the condition on the wave numbers for the relative motion is

$$\frac{\hbar^2 k'^2_{\alpha'\beta'}}{2m_r} = \frac{\hbar^2 k^2_{\alpha\beta}}{2m_r} + \epsilon_\alpha + \epsilon_\beta - \epsilon_{\alpha'} - \epsilon_{\beta'}. \tag{5.69}$$

If $k'^2_{\alpha'\beta'} \leq 0$, the channel is said to be closed, since there is insufficient energy for the pair of atoms to be at rest far from each other, and the corresponding term should not be included in the sum. Another way of expressing this is that a channel $\alpha'\beta'$ is closed if the energy in the relative motion as given by the Hamiltonian Eq. (5.65) is less than a *threshold energy* $E_{\text{th}}(\alpha'\beta')$ given by

$$E_{\text{th}}(\alpha'\beta') = \epsilon_{\alpha'} + \epsilon_{\beta'}. \tag{5.70}$$

For many purposes it is convenient to work in terms of the T matrix rather than the scattering amplitude. This is defined by a Lippmann–Schwinger equation analogous to that for the single-channel problem, Eq. (5.30), which we write formally as

$$T = U + UG_0T, \tag{5.71}$$

where the propagator for the pair of atoms in the absence of interactions between them is given by

$$G_0 = (E - H_0 + i\delta)^{-1}. \tag{5.72}$$

The scattering amplitude is related to the matrix element of the T matrix by the same factor as for a single channel. This is given in Eq. (5.35) for two particles of equal mass, and the generalization to unequal masses is straightforward, amounting to the replacement of m by $2m_\text{r}$. We therefore get

$$f_{\alpha\beta}^{\alpha'\beta'}(\mathbf{k}_{\alpha\beta}, \mathbf{k}'_{\alpha'\beta'}) = -\frac{m_\text{r}}{2\pi\hbar^2}\langle\alpha'\beta'|T(\mathbf{k}'_{\alpha'\beta'},\mathbf{k}_{\alpha\beta};E)|\alpha\beta\rangle. \tag{5.73}$$

The central part of the interaction

To estimate rates of processes we need to invoke the specific properties of the interatomic interaction. The largest contribution is the central part U^c, which we discussed in Sec. 5.1. This depends on the separation of the atoms, and on the electronic spin state of the two atoms. For alkali atoms and hydrogen, the electronic spin is $1/2$, and therefore the electronic spin state of a pair of atoms can be either a singlet or a triplet. One therefore writes the interaction in terms of the electron spin operators $\mathbf{S}_1$ and $\mathbf{S}_2$. By expressing the scalar product $\mathbf{S}_1\cdot\mathbf{S}_2$ in terms of the total spin by analogy with (3.3) one sees that $\mathbf{S}_1\cdot\mathbf{S}_2$ has eigenvalues $1/4$ (for triplet states) and $-3/4$ (for singlet states). Consequently the interaction may be written in the form

$$U^\text{c} = U_\text{s}\mathcal{P}_0 + U_\text{t}\mathcal{P}_1 = \frac{U_\text{s}+3U_\text{t}}{4} + (U_\text{t} - U_\text{s})\mathbf{S}_1\cdot\mathbf{S}_2, \tag{5.74}$$

where $\mathcal{P}_0 = 1/4 - \mathbf{S}_1\cdot\mathbf{S}_2$ and $\mathcal{P}_1 = 3/4 + \mathbf{S}_1\cdot\mathbf{S}_2$ are projection operators for the two-electron singlet and triplet states, respectively. At large distances the singlet and triplet potentials are dominated by the van der Waals interaction. The interaction (5.74) is invariant under rotations in coordinate space, and therefore it conserves orbital angular momentum. However, it can exchange the spins of the atoms, flipping one from up to down, and the other from down to up, for example. Because it is invariant under rotations in spin space, it conserves the total electronic spin angular momentum.

The magnetic dipole–dipole interaction

Some transitions are forbidden for the central part of the interaction, and under such circumstances the magnetic dipole–dipole interaction between electronic spins can be important. This has a form analogous to the electric dipole–dipole interaction (5.3) and it may be written as

$$U_\text{md} = \frac{\mu_0(2\mu_\text{B})^2}{4\pi r^3}[\mathbf{S}_1\cdot\mathbf{S}_2 - 3(\mathbf{S}_1\cdot\hat{\mathbf{r}})(\mathbf{S}_2\cdot\hat{\mathbf{r}})]. \tag{5.75}$$

This is independent of nuclear spins and is invariant under simultaneous rotations in coordinate space and electron spin space, and therefore it conserves the total angular momentum. However, it does not conserve separately orbital angular momentum and electronic spin angular momentum.

The interaction transforms as spherical tensors of rank 2 in both coordinate space and spin space, and it may be written in the form

$$U_{\mathrm{md}} = -\left(\frac{24}{5\pi}\right)^{1/2} \frac{\mu_0 \mu_B^2}{r^3} \sum_{\mu=-2}^{2} Y_{2\mu}^*(\hat{\mathbf{r}}) \Sigma_{2,\mu}, \qquad (5.76)$$

where $Y_{lm}(\hat{\mathbf{r}})$ is a spherical harmonic and $\Sigma_{2,\mu}$ is a spherical tensor of rank 2 made up from the two spin operators. Its components are

$$\Sigma_{2,0} = -\sqrt{\frac{3}{2}}(S_{1z}S_{2z} - \mathbf{S}_1 \cdot \mathbf{S}_2/3),$$

$$\Sigma_{2,\pm 1} = \pm\frac{1}{2}(S_{1z}S_{2\pm} + S_{1\pm}S_{2z}),$$

and

$$\Sigma_{2,\pm 2} = -\frac{1}{2}S_{1\pm}S_{2\pm}. \qquad (5.77)$$

This interaction can induce transitions in which the orbital angular momentum quantum number l changes by -2, 0, or $+2$. Two incoming atoms in an s-wave state can thus be scattered to a d-wave state, the angular momentum being taken from the electronic spins. As we shall show, typical non-vanishing matrix elements of the dipole–dipole interaction are roughly one or two orders of magnitude less than those for spin exchange, which involve the $\mathbf{S}_1 \cdot \mathbf{S}_2$ term in (5.74).

Low-energy collisions

The scattering amplitude is determined by solving the Schrödinger equation in essentially the same way as was done for a single channel. It is convenient to work in a basis of angular-momentum eigenstates of the form $|lm\alpha\beta\rangle$. The quantum number l specifies the total orbital angular momentum due to the relative motion of the atoms, and m its projection on some axis. Expressing the state in terms of this basis corresponds to the partial wave expansion in the single-channel problem supplemented by the channel label $\alpha\beta$ specifying the electronic and nuclear spin state of the two atoms. The result is a set of coupled second-order differential equations for the different channels specified by the quantum numbers l, m, α, and β. The picture of a low-energy collision that emerges from such calculations is that the two incoming atoms are initially in single-atom states obtained by diagonalizing the hyperfine and Zeeman terms in Eq. (5.65). If we forget for the moment about the dipole–dipole interaction, the interaction at large separations is dominated by the van der Waals term, which does not mix different hyperfine states. However, for smaller separations the central part of the interaction is different for triplet and singlet electronic spin states. As a simple model, let us assume that for separations greater than some value R_0 the interaction may be taken to be independent of hyperfine state and therefore there is no mixing of the different hyperfine states. At smaller separations, where the dependence of the interaction on the electronic spin state becomes important, we shall neglect the hyperfine interaction, and therefore the interaction depends only on the electronic spin state. The wave function for the incoming particles at R_0 may be expressed in terms of electronic singlet and triplet states, and then for $r < R_0$ the singlet and triplet states propagate in different ways. Finally, the outgoing wave at $r = R_0$ must be re-expressed in terms of the hyperfine states for two atoms. Because the potential for $r < R_0$ depends on the electronic spin state this will generally give rise to an outgoing wave function with a hyperfine composition different from that of the incoming state. In practice, a multi-channel calculation is necessary to obtain reliable quantitative results because the transition between the outer and inner regions is gradual, not sharp. For a more detailed account, we refer to Ref. [2].

A convenient way of summarizing data on low-energy scattering is in terms of ficticious singlet and triplet scattering lengths, which are the scattering lengths a_s for the singlet part of the central potential and a_t for the triplet when the hyperfine splittings and the dipole–dipole interaction are neglected. Since the long-range part of the interaction is well-characterized, from these quantities it is possible to calculate scattering lengths for arbitrary combinations of hyperfine states and the rates of inelastic collisions.

If both atoms occupy the same doubly polarized state $F = F_{\max} = I + 1/2$ and $m_F = \pm F_{\max}$, the electron spins are either both up or both down. Consequently the electronic spin state of the two atoms is a triplet, and only the triplet part of the interaction contributes. For pairs of atoms in other hyperfine states the electronic spin state is a superposition of singlet and triplet components, and therefore both triplet and singlet parts of the interaction play a role, and in particular they will mix different channels.

The coupling of channels has two effects. First, atoms can be scattered between different magnetic substates. Since the trap potential depends on the magnetic substate, this generally leads to loss of atoms from the trap, as described in Sec. 4.6. Second, the elastic scattering amplitude and the effective interaction are altered by the coupling between channels. We now discuss these two topics in greater detail.

5.4.1 Inelastic processes

Rates of processes may be calculated in terms of the scattering amplitude. The differential cross section is the current per unit solid angle in the final state divided by the flux in the initial state. With the wave function (5.68) the flux in the entrance channel is the relative velocity $v_{\alpha\beta} = \hbar k_{\alpha\beta}/m_r$, and the current in the exit channel is $|f_{\alpha\beta}^{\alpha'\beta'}(\mathbf{k}_{\alpha\beta}, \mathbf{k}'_{\alpha'\beta'})|^2 v'_{\alpha'\beta'}$ per unit solid angle. The differential cross section is therefore

$$\frac{d\sigma_{\alpha\beta}^{\alpha'\beta'}}{d\Omega} = |f_{\alpha\beta}^{\alpha'\beta'}(\mathbf{k}_{\alpha\beta}, \mathbf{k}'_{\alpha'\beta'})|^2 \frac{v'_{\alpha'\beta'}}{v_{\alpha\beta}} = |f_{\alpha\beta}^{\alpha'\beta'}(\mathbf{k}_{\alpha\beta}, \mathbf{k}'_{\alpha'\beta'})|^2 \frac{k'_{\alpha'\beta'}}{k_{\alpha\beta}}. \quad (5.78)$$

The rate at which two atoms in states α and β and contained within a volume V are scattered to states α' and β' is given by $\mathcal{K}_{\alpha\beta}^{\alpha'\beta'}/V$, where

$$\begin{aligned}\mathcal{K}_{\alpha\beta}^{\alpha'\beta'} &= v_{\alpha\beta} \int d\Omega \frac{d\sigma_{\alpha\beta}^{\alpha'\beta'}}{d\Omega} \\ &= \frac{2\pi}{\hbar} N_{\alpha'\beta'}(E) \int \frac{d\Omega}{4\pi} |\langle \alpha'\beta'|T(\mathbf{k}'_{\alpha'\beta'}, \mathbf{k}_{\alpha\beta}; E)|\alpha\beta\rangle|^2. \end{aligned} \quad (5.79)$$

Here

$$N_{\alpha'\beta'}(E) = \frac{m_r^2 v'_{\alpha'\beta'}}{2\pi^2 \hbar^3} \quad (5.80)$$

is the density of final states of the relative motion per unit energy and per unit volume. Equation (5.80) is equivalent to Eq. (2.5) for a single particle, but with the mass of an atom replaced by the reduced mass and the particle momentum by the relative momentum $m_r v'_{\alpha'\beta'}$. The relative velocity in the final state is given by

$$v'_{\alpha'\beta'} = \frac{\hbar k'_{\alpha'\beta'}}{m_r} = \left[\frac{2(E - \epsilon_{\alpha'} - \epsilon_{\beta'})}{m_r}\right]^{1/2}. \quad (5.81)$$

In the Born approximation one replaces the T matrix by the potential itself, and the result (5.79) then reduces to Fermi's Golden Rule for the rate of transitions. If either the incoming atoms are in the same internal state, or the outgoing ones are in the same state, the T matrix must be symmetrized or antisymmetrized according to the statistics of the atoms, as we described earlier for elastic scattering (see Eq. (5.24)).

In a gas at temperatures high enough that the effects of quantum degeneracy can be neglected, the total rate of a process is obtained by multiplying the rate (5.79) for a pair of particles by the distribution functions for atoms in internal states α and β and then integrating over the momenta of the particles. This leads to equations for the rate of change of the densities of atoms:

$$\frac{dn_\alpha}{dt} = \frac{dn_\beta}{dt} = -\frac{dn_{\alpha'}}{dt} = -\frac{dn_{\beta'}}{dt} = -K^{\alpha'\beta'}_{\alpha\beta} n_\alpha n_\beta. \quad (5.82)$$

When the effects of degeneracy become important, statistical factors must be included for the final states to take into account induced emission for bosons and Pauli blocking for fermions.

The rate coefficients K are temperature dependent and are defined by

$$K^{\alpha'\beta'}_{\alpha\beta} = \overline{\mathcal{K}^{\alpha'\beta'}_{\alpha\beta}} = \overline{\frac{2\pi}{\hbar}|\langle\alpha'\beta'|T|\alpha\beta\rangle|^2 N_{\alpha'\beta'}(E)} = \overline{\sigma^{\alpha'\beta'}_{\alpha\beta}(E)v_{\alpha\beta}}, \quad (5.83)$$

where the bar indicates an average over the distribution of the relative velocities v of the colliding atoms and over angles for the final relative momentum. These coefficients have the dimensions of volume divided by time, and a typical lifetime for an atom in state α to be lost by this process is given by $\tau_\alpha = 1/K^{\alpha'\beta'}_{\alpha\beta} n_\beta$.

Spin-exchange processes

To estimate the rate of processes that can proceed via the central part of the interaction, we use the fact that the interaction may be written in the form (5.74). If we adopt the simplified picture of a collision given earlier and also neglect the hyperfine and magnetic dipole–dipole interactions during the scattering process we may write the effective interaction for the spin degrees of freedom as

$$U_{\text{ex}}(\mathbf{r}) = \frac{4\pi\hbar^2(a_\text{t} - a_\text{s})}{m} \mathbf{S}_1 \cdot \mathbf{S}_2 \, \delta(\mathbf{r}), \quad (5.84)$$

which has the usual form of an exchange interaction. Here we have replaced $U_\text{t} - U_\text{s}$ in (5.74) by the difference between the corresponding pseudopotentials, following the same line of argument that led to Eq. (5.39). The matrix elements of the T matrix are therefore given by

$$\langle\alpha'\beta'|T|\alpha\beta\rangle = \frac{4\pi\hbar^2(a_\text{t} - a_\text{s})}{m}\langle\alpha'\beta'|\mathbf{S}_1\cdot\mathbf{S}_2|\alpha\beta\rangle. \quad (5.85)$$

For incoming atoms with zero kinetic energy the rate coefficient (5.83) is therefore

$$K^{\alpha'\beta'}_{\alpha\beta} = 4\pi(a_\text{t} - a_\text{s})^2 v'_{\alpha'\beta'}|\langle\alpha'\beta'|\mathbf{S}_1\cdot\mathbf{S}_2|\alpha\beta\rangle|^2. \quad (5.86)$$

For hydrogen, the difference of the scattering lengths is of order a_0, and therefore for atoms in the upper hyperfine state, the rate coefficient is of order 10^{-13} cm^3 s^{-1} if the spin matrix element is of order unity. For alkali atoms the corresponding values are of order 10^{-11} cm^3 s^{-1} because of the larger scattering lengths. Since typical densities in experiments are of order 10^{13} cm^{-3} or more, the estimates for rate coefficients indicate that atoms will be lost by exchange collisions in a fraction of a second. For atoms in the lower hyperfine multiplet in low magnetic fields the rate coefficients are generally smaller because of the reduced amount of phase space available to the outgoing atoms.

The fact that rates of spin-exchange processes vary as $(a_\text{t} - a_\text{s})^2$ makes it possible to deduce information about scattering lengths from data on the lifetimes of atoms in traps. As we shall describe in Sec. 5.5, the lifetime of a double condensate of ^{87}Rb atoms was found to be unexpectedly long, and this provided evidence for the triplet and singlet scattering lengths being roughly equal.

Dipolar processes

Some transitions cannot occur via the central part of the interaction because of angular momentum selection rules. For example, two atoms which are both in the doubly polarized state $|F = I + 1/2, m_F = F\rangle$ cannot scatter to any other channel because there are no other states with the same projection of the total spin angular momentum. Consider next two atoms which are both in the maximally stretched state $|F = I - 1/2, m_F = -F\rangle$ in a low magnetic field. This state is partly electronic spin triplet and partly singlet. Consequently, the central part of the interaction does have matrix elements to states other than the original one. However if the nuclear magnetic moment is positive, as it is for the bosonic alkali isotopes we consider, these other states have $F = I + 1/2$, and therefore they lie above the original states by an energy equal to the hyperfine splitting. Therefore, for temperatures at which the thermal energy kT is less than the hyperfine splitting, transitions to the upper hyperfine state are suppressed. Clouds of atoms containing either the doubly polarized state or the maximally stretched state can, however, decay via the magnetic dipole–dipole interaction, but rates for dipolar processes are typically 10^{-2}–10^{-4} times those for spin-exchange processes, as we shall explain below. Similar arguments apply for the states $|F = I + 1/2, m_F = -F\rangle$ and $|F = I - 1/2, m_F = F\rangle$. The doubly polarized state $|F = I + 1/2, m_F = F\rangle$ and the maximally stretched state $|F = I - 1/2, m_F = -F\rangle$ have low inelastic scattering rates and, since they can be trapped magnetically, these states are favoured for experiments on dilute gases.

Polarized gases in which all atoms are either in the doubly polarized state or the maximally stretched state with $F = I - 1/2$ can decay via the magnetic dipole–dipole interaction. We shall estimate rate coefficients for dipole–dipole relaxation of two incoming atoms with zero relative momentum using the Born approximation, in which one replaces the T matrix in Eq. (5.83) by the matrix element of the magnetic dipole–dipole interaction (5.75) between an incoming state with relative momentum zero and an outgoing one with relative momentum $\mathbf{k}'_{\alpha'\beta'}$ whose magnitude is defined by Eq. (5.81) for $E = \epsilon_\alpha + \epsilon_\beta$.

If the relative momentum of the incoming atoms is zero, the result is

$$K^{\alpha'\beta'}_{\alpha\beta} = \frac{m^2}{4\pi\hbar^4} \overline{|\langle\alpha'\beta'|U_{\mathrm{md}}(\mathbf{k}'_{\alpha'\beta'}, 0)|\alpha\beta\rangle|^2}\, v'_{\alpha'\beta'}, \tag{5.87}$$

where the bar denotes an average over directions of the final relative momentum. The wave function for the initial state, which has zero momentum, is unity, and that for the final state is $\exp i\mathbf{k}'_{\alpha'\beta'}\cdot\mathbf{r}$, and therefore the matrix element of the magnetic dipole–dipole interaction (5.75) is

$$U_{\mathrm{md}}(\mathbf{k}'_{\alpha'\beta'}, 0) = \int d\mathbf{r}\, U_{\mathrm{md}}(\mathbf{r}) e^{-i\mathbf{k}'_{\alpha'\beta'}\cdot\mathbf{r}}, \tag{5.88}$$

and it is an operator in the spin variables. A plane wave is given in terms of spherical waves by [7, §34]

$$e^{i\mathbf{k}\cdot\mathbf{r}} = 4\pi\sum_{lm} i^l j_l(kr) Y^*_{lm}(\hat{\mathbf{k}}) Y_{lm}(\hat{\mathbf{r}}), \tag{5.89}$$

where j_l is the spherical Bessel function and $\hat{\mathbf{k}}$ denotes a unit vector in the direction of $\mathbf{k}$. The expression (5.76) for the magnetic dipole–dipole interaction shows that it has only contributions corresponding to an orbital angular momentum $l = 2$, and consequently the matrix element reduces to

$$U_{\mathrm{md}}(\mathbf{k}'_{\alpha'\beta'}, 0) = \left(\frac{24}{5\pi}\right)^{1/2} \mu_0 \mu_B^2 \left(\int d\mathbf{r}\, \frac{j_2(k'_{\alpha'\beta'} r)}{r^3}\right) \sum_\mu Y^*_{2\mu}(\hat{\mathbf{k}}'_{\alpha'\beta'}) \Sigma_{2,\mu}. \tag{5.90}$$

The integral containing the spherical Bessel function may easily be evaluated using the fact that

$$j_l(x) = (-1)^l x^l \left(\frac{1}{x}\frac{d}{dx}\right)^l \frac{\sin x}{x}, \tag{5.91}$$

and it is equal to $4\pi/3$. The matrix element is thus

$$U_{\mathrm{md}}(\mathbf{k}'_{\alpha'\beta'}, 0) = 4\pi^2 \left(\frac{8}{15\pi}\right)^{1/2} \frac{\hbar^2 r_e}{m_e} \sum_\mu Y^*_{2\mu}(\hat{\mathbf{k}}'_{\alpha'\beta'}) \Sigma_{2,\mu}, \tag{5.92}$$

since $\mu_0\mu_B^2 = \pi\hbar^2 r_e/m_e$, where $r_e = e_0^2/m_e c^2 = \alpha_{\mathrm{fs}}^2 a_0$ is the classical electron radius. We have introduced the length r_e so that the expression for the matrix element resembles the effective interaction $4\pi\hbar^2 a/m$ for the central part of the potential.

It is striking that this result is independent of the magnitude of the relative momentum in the final state. This is because of the slow fall-off of the dipole–dipole interaction at large distances. By contrast the matrix element for a short-range interaction having the same tensor structure as the magnetic dipole–dipole interaction depends on k' and is proportional to k'^2 for small k'. The suppression of the matrix element in that case is due to the reduction by the centrifugal barrier of the wave function at atomic separations less than the wavelength $\lambda \sim 1/k'$ corresponding to the relative momentum of the outgoing atoms. The matrix element for the dipole–dipole interaction remains constant at small relative momenta because the main contribution to the integral comes from separations of order $\lambda \sim 1/k'$ where the magnitude of the Bessel function has its first (and largest) maximum. While the value of the interaction is of order $\mu_0 \mu_B^2 \lambda^{-3} \propto k'^3$ and therefore tends to zero as $k' \to 0$, the volume of space in which the integrand has this magnitude is of order λ^3, and the integral is therefore of order $\mu_0 \mu_B^2 \sim \hbar^2 r_e/m_e$, which is independent of k'. The matrix element for spin-exchange processes is of order $4\pi\hbar^2 (a_t - a_s)/m$, and therefore the ratio of the magnitude of the dipole matrix element to that for spin exchange is

$$\left| \frac{U_{\text{md}}}{U_{\text{ex}}} \right| \sim A \frac{m_p}{m_e} \alpha_{\text{fs}}^2 \frac{a_0}{|a_t - a_s|} \approx \frac{A}{10} \frac{a_0}{|a_t - a_s|}, \quad (5.93)$$

if one neglects differences between the matrix elements of the spin operators. This ratio is of order 10^{-1} for hydrogen, and for the alkalis it lies between 10^{-1} and 10^{-2} since scattering lengths for alkali atoms are of order $100\, a_0$.

The rate coefficient for dipolar transitions is obtained by inserting the matrix element (5.92) into the expression (5.79), and, because of the orthogonality of the spherical harmonics, for incoming particles with zero relative momentum it is given by

$$\begin{aligned} K^{\alpha'\beta'}_{\alpha\beta} &= \frac{m^2 v'_{\alpha'\beta'}}{4\pi\hbar^4} \int \frac{d\Omega}{4\pi} |\langle \alpha'\beta' | U_{\text{md}}(\mathbf{k}'_{\alpha'\beta'}, 0) | \alpha\beta \rangle|^2 \\ &= \frac{32\pi^2}{15} \left(\frac{m r_e}{m_e} \right)^2 v'_{\alpha'\beta'} \sum_\mu |\langle \alpha'\beta' | \Sigma_{2,\mu} | \alpha\beta \rangle|^2. \end{aligned} \quad (5.94)$$

The states of greatest interest experimentally are the doubly polarized state and, for atoms with a positive nuclear magnetic moment, the maximally stretched lower hyperfine state, since these cannot decay via the relatively rapid spin-exchange processes. Detailed calculations for hydrogen give dipolar rate coefficients of order 10^{-15} cm^3 s^{-1} or less depending on the particular transition and the magnetic field strength [10]. Estimates for alkali atoms are comparable for the doubly polarized state [11]. Dipolar rates can be much greater or much less than the typical values cited above because coupling between channels can give resonances, and also because of phase-space limitations. The latter are important for atoms initially in the lower hyperfine multiplet, for example a maximally stretched state. The phase space available is reflected in the factor $v'_{\alpha'\beta'}$ in Eq. (5.94). In this case the relative velocity in the final state is due solely to the Zeeman splitting, and not the hyperfine interaction as it is for collisions between two doubly polarized atoms. Since $v'_{\alpha'\beta'} \propto B^{1/2}$ for small fields, the dipole rate vanishes in the limit of low magnetic fields. Consequently, these states are attractive ones for experimental study. In experiments on alkali atoms, dipolar processes do not generally limit the lifetime. However, for hydrogen the densities achieved are so high that the characteristic time for the dipolar process is of order one second, and this process is the dominant loss mechanism, as we mentioned in Sec. 4.7. For the hydrogen atom, which has nuclear spin 1/2, there is only one lower hyperfine state ($F = 0$) and it is a high-field seeker. Consequently, for hydrogen, dipolar losses in magnetic traps cannot be avoided by working with a state in the lower hyperfine multiplet.

In the heavier alkalis Rb and Cs, relativistic effects are important and they provide another mechanism for losses [12]. This is the spin–orbit interaction, which acting in second order gives rise to transitions like those that can occur via the magnetic dipole–dipole interaction. The sign of the interaction is opposite that of the dipolar interaction. In Rb, the second-order spin–orbit matrix element is smaller in magnitude than that for the dipole process, and the result is that the inelastic rate coefficient for the doubly polarized and maximally stretched states are *decreased* compared with the result for the dipolar interaction alone. For Cs, with its higher atomic number, spin–orbit effects are larger and, while still of opposite sign compared with the dipolar interaction, they are so large that they overwhelm the dipolar contribution and give an *increased* inelastic scattering rate. To date, these losses have thwarted attempts to observe Bose–Einstein condensation of cesium atoms.

Three-body processes

Thus far in this section we have only considered two-body reactions, but as we indicated in Chapter 4, three-body recombination puts stringent limits on the densities that can be achieved in traps. The rate of this process is given by the equation

$$\frac{dn}{dt} = -Ln^3, \tag{5.95}$$

where L is the rate coefficient and n is the density of the atomic species. The rate is proportional to the third power of the density because the process is a three-body one, and the probability of three atoms being close to each other varies as n^3. The rate coefficients L for alkali atoms have been calculated by Moerdijk et al. [13], who find them to be 2.6×10^{-28} cm^6 s^{-1} for ^{7}Li, 2.0×10^{-28} cm^6 s^{-1} for ^{23}Na and 4×10^{-30} cm^6 s^{-1} for ^{87}Rb. The magnitude of the three-body rates for the alkalis may be crudely understood by noting that for a three-body process to occur, three atoms must be within a volume r_0^3, where r_0 is the distance out to which the two-body interaction is significant. The probability of an atom being within a volume r_0^3 is of order nr_0^3, and thus the rate of three-body processes compared with that of two-body ones is roughly given by the same factor, if we forget about differences between phase space for the two processes. Therefore the rate coefficient for three-body processes is of order r_0^3 compared with the rate coefficient for two-body processes, that is $L \sim Kr_0^3$, assuming that all processes are allowed to occur via the central part of the interaction. In Sec. 5.1 we showed for the alkalis that $r_0 \sim 10^2 a_0$, which, together with $K \sim 10^{-11}$ cm^3 s^{-1} (see below (5.86)), gives a crude estimate of the three-body rate coefficient of order 10^{-30} cm^6 s^{-1}. For hydrogen, the rate is very much smaller, of order 10^{-38} cm^6 s^{-1}, because of the much weaker interatomic potential. Consequently, three-body processes can be important in experiments with alkali atoms, but not with hydrogen. Rates of three-body interactions, like those of two-body ones, are affected by the identity of the atoms participating in the process. In Sec. 13.2 we shall demonstrate how the rate of three-body processes is reduced by the appearance of a condensate.

5.4.2 Elastic scattering and Feshbach resonances

A qualitative difference compared with the single-channel problem is that the scattering amplitude is generally not real at $k = 0$ if there are other open channels. However, this effect is rather small under typical experimental conditions for alkali atoms at low temperatures because the phase space available for real transitions is small, and to a very good approximation elastic scattering can still be described in terms of a real scattering length, as in the single-channel theory. Another effect is that elastic scattering in one channel can be altered dramatically if there is a low-energy bound state in a second channel which is closed. This phenomenon, which is referred to as a

Feshbach resonance, was first investigated in the context of nuclear physics [14]. These resonances have become an important tool in investigations of the basic atomic physics of cold atoms, and, because they make it possible to tune scattering lengths and other quantities by adjusting an external parameter such as the magnetic field, they have applications in experiments on trapped gases.

Feshbach resonances appear when the total energy in an open channel matches the energy of a bound state in a closed channel, as illustrated in Fig. 5.4. To first order in the coupling between open and closed channels the scattering is unaltered, because there are, by definition, no continuum states in the closed channels. However, two particles in an open channel can scatter to an intermediate state in a closed channel, which subsequently decays to give two particles in one of the open channels. Such second-order processes are familiar from our earlier treatments of the atomic polarizability in Sec. 3.3, and therefore from perturbation theory one would expect there to be a contribution to the scattering length having the form of a sum of terms of the type

$$a \sim \frac{C}{E - E_{\text{res}}}, \qquad (5.96)$$

where E is the energy of the particles in the open channel and E_{res} is the energy of a state in the closed channels. Consequently there will be large effects if the energy E of the two particles in the entrance channel is close to the energy of a bound state in a closed channel. As one would expect from second-order perturbation theory for energy shifts, coupling between channels gives rise to a repulsive interaction if the energy of the scattering particles is greater than that of the bound state, and an attractive one if it is less. The closer the energy of the bound state is to the energy of the incoming particles in the open channels, the larger the effect on the scattering. Since the energies of states depend on external parameters, such as the magnetic field, these resonances make it possible to tune the effective interaction between atoms.

Basic formalism

We now describe the general formalism for treating Feshbach resonances. The space of states describing the spatial and spin degrees of freedom may be divided into two subspaces, P, which contains the open channels, and Q, which contains the closed ones. We write the state vector $|\Psi\rangle$ as the sum of its projections onto the two subspaces,

$$\Psi = |\Psi_P\rangle + |\Psi_Q\rangle, \qquad (5.97)$$

5.4 Scattering between different internal states

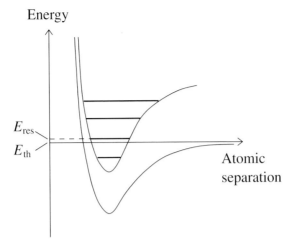

Fig. 5.4. Schematic plot of potential energy curves for two different channels that illustrate the formation of Feshbach resonances. E_{th} is the threshold energy, Eq. (5.70), for the entrance channel, and E_{res} is the energy of a state in a closed channel.

where $|\Psi_P\rangle = \mathcal{P}|\Psi\rangle$ and $|\Psi_Q\rangle = \mathcal{Q}|\Psi\rangle$. Here $\mathcal{P}$ and $\mathcal{Q}$ are projection operators for the two subspaces, and they satisfy the conditions $\mathcal{P} + \mathcal{Q} = 1$ and $\mathcal{P}\mathcal{Q} = 0$.

By multiplying the Schrödinger equation $H|\Psi\rangle = E|\Psi\rangle$ on the left by $\mathcal{P}$ and $\mathcal{Q}$ we obtain two coupled equations for the projections of the state vector onto the two subspaces:

$$(E - H_{PP})|\Psi_P\rangle = H_{PQ}|\Psi_Q\rangle \qquad (5.98)$$

and

$$(E - H_{QQ})|\Psi_Q\rangle = H_{QP}|\Psi_P\rangle, \qquad (5.99)$$

where $H_{PP} = \mathcal{P}H\mathcal{P}$, $H_{QQ} = \mathcal{Q}H\mathcal{Q}$, $H_{PQ} = \mathcal{P}H\mathcal{Q}$ and $H_{QP} = \mathcal{Q}H\mathcal{P}$. The operator H_{PP} is the Hamiltonian in the P subspace, H_{QQ} that in the Q subspace, and H_{PQ} and H_{QP} represent the coupling between the two subspaces. The formal solution of (5.99) is

$$|\Psi_Q\rangle = (E - H_{QQ} + i\delta)^{-1} H_{QP}|\Psi_P\rangle, \qquad (5.100)$$

where we have added a positive infinitesimal imaginary part δ in the denominator to ensure that the scattered wave has only outgoing terms. When Eq. (5.100) is inserted into (5.98), the resulting equation for $|\Psi_P\rangle$ becomes

$$\left(E - H_{PP} - H'_{PP}\right)|\Psi_P\rangle = 0. \qquad (5.101)$$

Here

$$H'_{PP} = H_{PQ}(E - H_{QQ} + i\delta)^{-1} H_{QP} \qquad (5.102)$$

is the term that describes Feshbach resonances. It represents an effective interaction in the P subspace due to transitions from that subspace to the Q subspace and back again to the P subspace. In agreement with our earlier qualitative arguments, it has a form similar to the energy shift in second-order perturbation theory, and corresponds to a non-local potential in the open channels. Due to the energy dependence of the interaction, it is also retarded in time.

It is convenient to divide the diagonal parts $H_{PP}+H_{QQ}$ of the Hamiltonian into a term H_0 independent of the separation of the two atoms, and an interaction contribution. Here H_0 is the sum of the kinetic energy of the relative motion and the hyperfine and Zeeman terms, Eq. (3.8). We write

$$H_{PP} = H_0 + U_1, \qquad (5.103)$$

where U_1 is the interaction term for the P subspace. Equation (5.101) may be rewritten as

$$(E - H_0 - U)|\Psi_P\rangle = 0, \qquad (5.104)$$

where the total effective atom–atom interaction in the subspace of open channels is given by

$$U = U_1 + U_2 \qquad (5.105)$$

with

$$U_2 = H'_{PP}. \qquad (5.106)$$

A simple example

For the purpose of illustration, let us begin by considering the T matrix when there is no interaction in the open channels if the open and closed channels are not coupled. We also treat U_2 in the first Born approximation, which corresponds to calculating the matrix element of U_2 for plane-wave states. The diagonal element of the T matrix in the channel $\alpha\beta$ may be evaluated by using the identity $1 = \sum_n |n\rangle\langle n|$, where the states $|n\rangle$ form a complete set.[5] If one takes for the states the energy eigenstates in the absence of coupling between channels, the result for two atoms with zero

---

[5] In the sum there are both bound states and continuum states. The normalization condition for the bound states is $\int d\mathbf{r}|\psi_n(\mathbf{r})|^2 = 1$, and the continuum states are normalized as usual.

relative momentum in the incoming state is

$$\langle \alpha\beta, 0|T|\alpha\beta, 0\rangle = \sum_n \frac{|\langle n|H_{QP}|\alpha\beta, 0\rangle|^2}{E_{\text{th}}(\alpha\beta) - E_n + i\delta}. \tag{5.107}$$

Here $E_{\text{th}}(\alpha\beta)$ is the threshold energy (5.70), and we denote the eigenstates of the Hamiltonian H_{QQ} by E_n and the state vectors by $|n\rangle$. This result has the form anticipated, and it shows how the scattering length can be large if the energy of the two scattering particles is close to the energy of a state in a closed channel.

General solution

Before deriving more general results, we describe their key features. The expression for the scattering length retains essentially the same form when the interaction U_1 is included, the only difference being that the open-channel plane-wave state $|\alpha\beta, 0\rangle$ is replaced by a scattering state which is an energy eigenstate of the Hamiltonian including the potential U_1. When higher-order terms in U_2 are taken into account, the energy of the resonant state is shifted, and the state may acquire a non-zero width due to decay to states in the open channels.

To solve the problem more generally we calculate the T matrix corresponding to the interaction given in Eq. (5.101). As before, Eqs. (5.71) and (5.72), we write the Lippmann–Schwinger equation (5.30) as an operator equation

$$T = U + UG_0T. \tag{5.108}$$

The quantity G_0 is given by

$$G_0 = (E - H_0 + i\delta)^{-1}, \tag{5.109}$$

where H_0 is the Hamiltonian (5.65). It represents the free propagation of atoms, and is the Green function for the Schrödinger equation. The formal solution of (5.108) is

$$T = (1 - UG_0)^{-1}U = U(1 - G_0U)^{-1}. \tag{5.110}$$

The second equality in (5.110) may be proved by multiplying $(1-UG_0)^{-1}U$ on the right by $G_0G_0^{-1}$ and using the fact that UG_0 commutes with $(1 - UG_0)^{-1}$, together with the identities

$$UG_0(1 - UG_0)^{-1}G_0^{-1} = U[G_0(1 - UG_0)G_0^{-1}]^{-1} = U(1 - G_0U)^{-1}. \tag{5.111}$$

By inserting (5.109) into the first equality in (5.110) we obtain

$$T = (E + i\delta - H_0)(E + i\delta - H_0 - U)^{-1}U. \tag{5.112}$$

Putting $A = E + i\delta - H_0 - U$ and $B = U_2$ in the matrix identity

$$(A - B)^{-1} = A^{-1}(1 + B(A - B)^{-1}) \tag{5.113}$$

one finds

$$(E+i\delta-H_0-U)^{-1} = (E+i\delta-H_0-U_1)^{-1}[1+U_2(E+i\delta-H_0-U)^{-1}], \tag{5.114}$$

which when inserted into Eq. (5.112) gives the useful result

$$T = T_1 + (1 - U_1 G_0)^{-1} U_2 (1 - G_0 U)^{-1}. \tag{5.115}$$

Here T_1, which satisfies the equation $T_1 = U_1 + U_1 G_0 T_1$, is the T matrix in the P subspace if transitions to the Q subspace are neglected.

Let us now interpret the result (5.115) by considering its matrix elements between plane-wave states with relative momenta $\mathbf{k}$ and $\mathbf{k}'$:

$$\langle \mathbf{k}'|T|\mathbf{k}\rangle = \langle \mathbf{k}'|T_1|\mathbf{k}\rangle + \langle \mathbf{k}'|(1 - U_1 G_0)^{-1} U_2 (1 - G_0 U)^{-1}|\mathbf{k}\rangle. \tag{5.116}$$

The scattering amplitude is generally a matrix labelled by the quantum numbers of the entrance and exit channels, but for simplicity we suppress the indices. Acting on a plane-wave state for the relative motion $|\mathbf{k}\rangle$ the operator $\Omega_U = (1 - G_0 U)^{-1}$ generates an eigenstate of the Hamiltonian $H_0 + U$, as one may verify by operating on the state with the Hamiltonian. This result is equivalent to Eqs. (5.26) and (5.27) for the single-channel problem. At large separations the state $\Omega_U|\mathbf{k}\rangle$ consists of a plane wave and a spherical wave, which is outgoing because of the $i\delta$ in G_0. We denote this state by $|\mathbf{k}; U, +\rangle$, the plus sign indicating that the state has outgoing spherical waves. The operators U_1 and H_0 are Hermitian and therefore $(1 - U_1 G_0)^\dagger = 1 - G_0^- U_1$ for E real, where $G_0^- = (E - H_0 - i\delta)^{-1}$ is identical with G_0 defined in Eq. (5.109) except that the sign of the infinitesimal imaginary part is opposite. We may thus write

$$\langle \mathbf{k}'|(1 - U_1 G_0)^{-1} = [(1 - G_0^- U_1)^{-1}|\mathbf{k}'\rangle]^\dagger \equiv [|\mathbf{k}'; U_1, -\rangle]^\dagger. \tag{5.117}$$

As a consequence of the difference of the sign of δ, the state contains *incoming* spherical waves at large distances, and this is indicated by the minus sign in the notation for the eigenstates.[6] The scattering amplitude (5.116) may therefore be written as

$$\langle \mathbf{k}'|T|\mathbf{k}\rangle = \langle \mathbf{k}'|T_1|\mathbf{k}\rangle + \langle \mathbf{k}'; U_1, -|U_2|\mathbf{k}; U, +\rangle. \tag{5.118}$$

This is the general expression for the scattering amplitude in the P subspace.

---

[6] If the relative velocity of the incoming particles is zero and provided real scattering between different open channels may be neglected, scattering states with 'outgoing' spherical waves are the same as those with 'incoming' ones since there are no phase factors $\sim e^{\pm ikr}$, and therefore both functions behave as $1 - a/r$ at large distances.

Tuning effective interactions

We now apply the above results, and we begin by considering the contribution of first order in U_2. This is equivalent to replacing U by U_1 in Eq. (5.116), which gives

$$T \simeq T_1 + (1 - U_1 G_0)^{-1} U_2 (1 - G_0 U_1)^{-1}. \tag{5.119}$$

The matrix elements between plane-wave states are given by

$$\langle \mathbf{k}'|T|\mathbf{k}\rangle = \langle \mathbf{k}'|T_1|\mathbf{k}\rangle + \langle \mathbf{k}'; U_1, -|U_2|\mathbf{k}; U_1, +\rangle. \tag{5.120}$$

This result differs from the simple example with $U_1 = 0$ that we considered above in two respects: the interaction in the P subspace gives a contribution T_1 to the T matrix, and the contribution due to the U_2 term is to be evaluated using scattering states that take into account the potential U_1, not plane waves.

Let us now neglect coupling between the open channels. The scattering in a particular channel for particles with zero relative velocity is then specified by the scattering length, which is related to the T matrix by the usual factor. As explained in footnote 5, we may then neglect the difference between the scattering states with incoming and outgoing spherical waves and we shall denote the state simply by $|\psi_0\rangle$. We again evaluate the matrix element of U_2, defined in Eqs. (5.102) and (5.106), by inserting the unit operator and find

$$\frac{4\pi\hbar^2}{m} a = \frac{4\pi\hbar^2}{m} a_P + \sum_n \frac{|\langle \psi_n | H_{QP} | \psi_0 \rangle|^2}{E_{\text{th}} - E_n}, \tag{5.121}$$

where the sum is over all states in the Q subspace and a_P is the scattering length when coupling between open and closed channels is neglected.

If the energy E_{th} is close to the energy E_{res} of one particular bound state, contributions from all other states will vary slowly with energy, and they, together with the contribution from the potential U_1, may be represented by a non-resonant scattering length a_{nr} whose energy dependence may be neglected. However in the resonance term the energy dependence must be retained. The scattering length may then be written as

$$\frac{4\pi\hbar^2}{m} a = \frac{4\pi\hbar^2}{m} a_{\text{nr}} + \frac{|\langle \psi_{\text{res}} | H_{QP} | \psi_0 \rangle|^2}{E_{\text{th}} - E_{\text{res}}}. \tag{5.122}$$

This displays the energy dependence characteristic of a Feshbach resonance.

Atomic interactions may be tuned by exploiting the fact that the energies of states depend on external parameters, among which are the strengths of magnetic and electric fields. For definiteness, let us consider an external magnetic field. We imagine that the energy denominator in Eq. (5.122) vanishes for a particular value of the magnetic field, $B = B_0$. Expanding the energy denominator about this value of the magnetic field we find

$$E_{\text{th}} - E_{\text{res}} \approx (\mu_{\text{res}} - \mu_\alpha - \mu_\beta)(B - B_0), \tag{5.123}$$

where

$$\mu_\alpha = -\frac{\partial \epsilon_\alpha}{\partial B} \quad \text{and} \quad \mu_\beta = -\frac{\partial \epsilon_\beta}{\partial B} \tag{5.124}$$

are the magnetic moments of the two atoms in the open channel, and

$$\mu_{\text{res}} = -\frac{\partial E_{\text{res}}}{\partial B} \tag{5.125}$$

is the magnetic moment of the molecular bound state. The scattering length is therefore given by

$$a = a_{\text{nr}}\left(1 + \frac{\Delta B}{B - B_0}\right) \tag{5.126}$$

where the width parameter ΔB is given by

$$\Delta B = \frac{m}{4\pi\hbar^2 a_{\text{nr}}} \frac{|\langle\psi_{\text{res}}|H_{QP}|\psi_0\rangle|^2}{(\mu_{\text{res}} - \mu_\alpha - \mu_\beta)}. \tag{5.127}$$

This shows that the characteristic range of magnetic fields over which the resonance significantly affects the scattering length depends on the magnetic moments of the states, and on the coupling between channels. Consequently, measurement of Feshbach resonances provides a means of obtaining information about interactions between atoms. Equation (5.126) shows that in this approximation the scattering length diverges to $\pm\infty$ as B approaches B_0. Because of the dependence on $1/(B - B_0)$, large changes in the scattering length can be produced by small changes in the field. It is especially significant that the sign of the interaction can be changed by a small change in the field.

Higher-order contributions in U_2 may be included, and the exact solution (5.116) may be expressed in the form

$$T = T_1 + (1 - U_1 G_0)^{-1} H_{PQ}(E - H_{QQ} - H'_{QQ})^{-1} H_{QP}(1 - U_1 G_0)^{-1}, \tag{5.128}$$

where

$$H'_{QQ} = H_{QP}(E - H_{PP})^{-1} H_{PQ}. \tag{5.129}$$

By comparing this result with the earlier one (5.119) we see that the only effect of higher-order terms in U_2 is to introduce an extra contribution H'_{QQ} in the Hamiltonian acting in the Q subspace. This effective interaction in the Q subspace takes into account the influence of coupling to the open channels, and it is therefore the analogue of the interaction H'_{PP} in the open channels. Physically it has two effects. One is to shift the energies of the bound states, and the second is to give them a non-zero width $\hbar\Gamma_{\text{res}}$ if decay into open channels is possible. The width leads to a scattering amplitude of the Breit–Wigner form $\sim 1/(E_{\text{th}} - E_{\text{res}} + i\hbar\Gamma_{\text{res}}/2)$. For $|E_{\text{th}} - E_{\text{res}}| \gg \hbar\Gamma_{\text{res}}$ the scattering amplitude shows the $1/(E_{\text{th}} - E_{\text{res}})$ behaviour predicted by the simpler calculations. For $|E_{\text{th}} - E_{\text{res}}| \lesssim \hbar\Gamma_{\text{res}}$ the divergence is cut off and the scattering amplitude behaves smoothly. However, the width of resonant states close to the threshold energy in the open channel is generally very small because of the small density of final states, and Feshbach resonances for cold atoms are consequently very sharp.

Detailed calculations of Feshbach resonances have been made for lithium and sodium, using the methods outlined above, by Moerdijk et al. [15]. Feshbach resonances have been investigated experimentally for ^{23}Na [16], ^{85}Rb [17, 18] and ^{133}Cs [5]. Level shifts due to the interaction of atoms with a laser field can also induce Feshbach resonances, as has recently been observed in sodium [19]. As we shall describe in Sec. 5.5, Feshbach resonances are becoming a key tool in extracting information about interactions between atoms. In addition they provide a means of varying the scattering length almost at will, and this allows one to explore properties of condensates under novel conditions. For example, in the experiment by Cornish et al. on ^{85}Rb described in Ref. [18] the effective interaction was changed from positive to negative by varying the magnetic field, and this caused the cloud to collapse.

5.5 Determination of scattering lengths

In this section we survey the methods used to determine scattering lengths of alkali atoms. Most experiments do not give directly the value of the scattering length, but they give information about atom–atom interactions, from which scattering lengths may be calculated. The general framework in which this is done was described immediately preceding Sec. 5.4.1.

Measurement of the collision cross section

For identical bosons, the cross section for elastic collisions at low energy is given by $8\pi a^2$, and therefore a measurement of the elastic cross section gives the magnitude of the scattering length, but not its sign. In Sec. 11.3 we shall describe how the scattering length is related to the damping rate

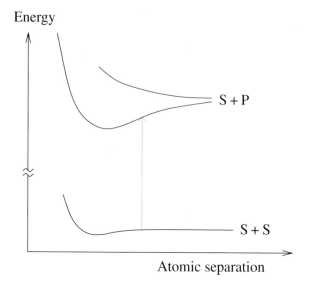

Fig. 5.5. Schematic representation of potential energy curves to illustrate the method of photoassociative spectroscopy.

for collective modes and temperature anisotropies. One difficulty of this technique is that it requires knowledge of the particle density.

Molecular spectra

Measurements of the frequencies of transitions between molecular bound states of two polarized atoms give information about the interaction potential. However, this method is not particularly sensitive to the long-range part of the interaction, which is crucial for determining the scattering length.

Photoassociative spectroscopy

The understanding of interactions between cold atoms has advanced rapidly over the past decade due to the use of *photoassociative spectroscopy*, in which one measures the rate at which two interacting ground-state atoms in an unbound state are excited by means of a laser to a molecular state in which one of the atoms is in a P state. The principle of the method is illustrated in Fig. 5.5. For two atoms in the ground state, the potential is of the van der Waals form ($\sim r^{-6}$) at large distances, as described in Sec. 5.1. This is represented by the lower curve, labelled "S + S". When one atom is excited to a P state, the electric dipole interaction, Eq. (5.3), gives a first-order contribution to the energy because the state with one atom excited and the other in the ground state is degenerate with the state in which the two atoms are interchanged, and this varies as r^{-3} at large distances, as does the dipole interaction itself. Its sign can be either positive or negative, depending on the symmetry of the excited state of the two atoms. The potentials for this case are labelled in the figure by "S + P".

The long-range part of the potential when one atom is excited is very much stronger than the van der Waals interaction and, consequently, the excited-state potential, when attractive, has many bound states, some of which have sizes much greater than those of the bound states for the ground-state potential. The transition rate to an excited molecular state is given in terms of the matrix element of the perturbation due to the laser, between the state with two ground-state atoms and the excited state. The spatial variation of the ground-state wave function is generally slow compared with that of the excited molecular state, since the excited-state potential is much stronger. Consequently, the largest contribution to the matrix element comes from the region where the relative momentum of the two atoms in the excited state is small, that is near the classical turning points of the motion. In regions where the relative momentum is larger, the product of the two wave functions oscillates rapidly in space, and, when integrated over space, gives little contribution to the matrix element. Thus the transition probability depends sensitively on the magnitude of the ground-state wave function near the turning point of the motion in the excited molecular state. The position of the turning point depends on the molecular state under consideration, and therefore the variation of the strengths of transitions to different molecular states, which have different energies and are therefore resonant at different laser frequencies, shows strong features due the oscillations in the ground-state wave function for two atoms. From a knowledge of the excited-state potential and the transition rates for a number of molecular excited states it is possible to deduce properties of the ground-state wave function and scattering length.

Excited molecular states may decay to give atoms which have kinetic energies high enough that they can overcome the trapping potential. Therefore, one way of detecting transitions to excited molecular states is by measuring trap loss. Alternatively, a second light pulse may be used to excite the atom further, possibly to an ionized state which is easily detectable. Extensive discussions of the method and developments of it are given in Refs. [1] and [2].

Other methods

Other measurements give information about atomic interactions. Since rates of inelastic processes are sensitive to the atomic interaction, especially to differences between singlet and triplet scattering lengths, measurement of them can give information about the potential. This has been particularly important for ^{87}Rb. Surprisingly, it was found experimentally that a double condensate of atoms in the $F = 2, m_F = 2$ and $F = 1, m_F = -1$ states lived much longer than suggested by an order-of-magnitude estimate of the lifetime due to exchange collisions based on Eq. (5.86) with the assumption that $|a_t - a_s|$ is comparable to $|a_t|$. The explanation for this is that the triplet and singlet scattering lengths are nearly equal. This means physically that the singlet and triplet components of the wave function are essentially in phase again when the two particles have been scattered, and consequently they reconstruct the initial hyperfine state, with only a small admixture of other hyperfine states.

As explained in Sec. 5.4.2, the positions of Feshbach resonances and the width ΔB, Eq. (5.127), are sensitive to the interatomic interaction. Therefore by studying these resonances it is possible to deduce properties of interatomic potentials. Techniques that make use of the properties of Feshbach resonances are becoming increasingly important in extracting information about potentials, as is exemplified by the discussion of results for Rb isotopes and ^{133}Cs given below.

5.5.1 Scattering lengths for alkali atoms and hydrogen

Below we list current values of the scattering lengths for hydrogen and the members of the alkali series. The reader is referred to Ref. [2] and the original literature for further discussion. The triplet scattering length is denoted by a_t, the singlet one by a_s, and that for two atoms in the maximally stretched lower hyperfine state $F = I - 1/2, m_F = -(I - 1/2)$ by a_{ms}. All scattering lengths are given in atomic units (1 a.u. = $a_0 = 4\pi\epsilon_0\hbar^2/m_e e^2 \approx 0.0529$ nm).

Hydrogen

The hydrogen atom is sufficiently simple that scattering lengths may be calculated reliably from first principles. Recent calculations give $a_t = 1.2$ [20] and $a_s = 0.41$ [21].

Lithium

Scattering lengths for the lithium isotopes have been reported by Abraham et al. [22], who employed photoassociative spectroscopy.

^{6}Li For this fermion they obtained $a_s = 45.5 \pm 2.5$ and $a_t = -2160 \pm 250$, the latter exceeding by more than one order of magnitude the estimate based on the dimensional arguments of Sec. 5.1.

^{7}Li   They found $a_s = 33 \pm 2$ and $a_t = -27.6 \pm 0.5$. As we shall see in Chapter 6, the negative value of the triplet scattering length for ^{7}Li prevents the formation of a condensate with more than a few thousand atoms.

Sodium

^{23}Na   Scattering lengths have been derived by Tiesinga *et al.* [23] from photoassociative spectroscopy data. They found $a_t = 85 \pm 3$ and $a_{\rm ms} = 52 \pm 5$. A more recent analysis of different types of data gives $a_t = 65.3 \pm 0.9$, $a_s = 19.1 \pm 2.1$, and $a_{\rm ms} = 55.4 \pm 1.2$ [24].

Potassium

Scattering lengths for potassium isotopes have been derived from photoassociative spectroscopy by Bohn *et al.* [25]. Their results are:

^{39}K $a_t = -17 \pm 25$, $a_s = 140^{+3}_{-6}$, and $a_{\rm ms} = -20^{+42}_{-64}$.

^{40}K $a_t = 194^{+114}_{-35}$ and $a_s = 105^{+2}_{-3}$.

^{41}K $a_t = 65^{+13}_{-8}$, $a_s = 85 \pm 2$, and $a_{\rm ms} = 69^{+14}_{-9}$.

The results are somewhat different from earlier ones obtained from molecular spectroscopy [26, 27].

Rubidium

^{85}Rb   From photoassociative spectroscopy and study of a Feshbach resonance, Roberts *et al.* found $a_t = -369 \pm 16$ and $a_s = 2400^{+600}_{-350}$ [18]. Vogels *et al.* found $a_{\rm ms} = -450 \pm 140$ from photoassociative spectroscopy [28].

^{87}Rb   All the scattering lengths are closely equal. Roberts *et al.* found $a_t = 106 \pm 4$ and $a_s = 90 \pm 1$ from the same kinds of data as they used for the lighter isotope [18]. From inelastic scattering data obtained with the double condensate and from photoassociative spectroscopy Julienne *et al.* found $a_{\rm ms} = 103 \pm 5$ [29].

Cesium

^{133}Cs   The cross section for the scattering of ^{133}Cs atoms in the doubly polarized state, $F = 4, m_F = 4$, was measured by Arndt *et al.* [30] in the temperature range from 5 to 60 µK. It was found to be inversely proportional to the temperature, characteristic of resonant scattering ($|a| \to \infty$). Since the cross section did not saturate at the lowest temperatures, the measurements yield only a lower bound on the magnitude of the scattering length, $|a_t| \geq 260$. Chin *et al.* [5] have made high-resolution measurements of a large number of Feshbach resonances, from which Leo et al. [6] deduced the values $a_\mathrm{t} = 2400 \pm 100$ and $a_\mathrm{s} = 280 \pm 10$. The calculations of Ref. [6] indicate that for some choices of the magnetic field, the ratio of elastic scattering cross sections to inelastic ones can be large enough that the possibility of Bose–Einstein condensing Cs atoms by evaporative cooling cannot be ruled out.

Problems

PROBLEM 5.1 Use the model of Sec. 5.3 to determine the value of r_c/r_0 at which the first bound state appears. In the light of this calculation, discuss why two spin-polarized hydrogen atoms in a triplet electronic state do not have a bound state. (The triplet-state potential for hydrogen is positive for distances less than approximately $7a_0$.)

PROBLEM 5.2 Calculate the rate of the process $d + d \to a + a$ for hydrogen atoms at zero temperature. This process was discussed in Sec. 4.7, and expressions for the states are given in Eqs. (3.24)–(3.27). Use the Born approximation and take into account only the magnetic dipole–dipole interaction. Give limiting results for low magnetic fields ($B \ll \Delta E_\mathrm{hf}/\mu_\mathrm{B}$) and high magnetic fields.

PROBLEM 5.3 Make numerical estimates of rates of elastic scattering, and inelastic two- and three-body processes for hydrogen and alkali atoms at low temperatures under typical experimental conditions.

References

[1] J. Weiner, V. S. Bagnato, S. Zilio, and P. S. Julienne, *Rev. Mod. Phys.* **71**, 1 (1999).

[2] D. J. Heinzen, in *Bose–Einstein Condensation in Atomic Gases*, Proceedings of the Enrico Fermi International School of Physics, Vol. CXL, ed. M. Inguscio, S. Stringari, and C. E. Wieman, (IOS Press, Amsterdam, 1999), p. 351.

[3] Z.-C. Yan, J. F. Babb, A. Dalgarno, and G. W. F. Drake, *Phys. Rev. A* **54**, 2824 (1996).

[4] A. Derevianko, W. R. Johnson, M. S. Safronova, and J. F. Babb, *Phys. Rev. Lett.* **82**, 3589 (1999).

[5] C. Chin, V. Vuletić, A. J. Kerman, and S. Chu, *Phys. Rev. Lett.* **85**, 2717 (2000).

[6] P. J. Leo, C. J. Williams, and P. S. Julienne, *Phys. Rev. Lett.* **85**, 2721 (2000).

[7] L. D. Landau and E. M. Lifshitz, *Quantum Mechanics*, Third edition, (Pergamon, New York, 1977).

[8] G. F. Gribakin and V. V. Flambaum, *Phys. Rev. A* **48**, 546 (1993).

[9] G. V. Shlyapnikov, J. T. M. Walraven, and E. L. Surkov, *Hyp. Int.* **76**, 31 (1993).

[10] H. T. C. Stoof, J. M. V. A. Koelman, and B. J. Verhaar, *Phys. Rev. B* **38**, 4688 (1988).

[11] A. J. Moerdijk and B. J. Verhaar, *Phys. Rev. A* **53**, 19 (1996).

[12] F. H. Mies, C. J. Williams, P. S. Julienne, and M. Krauss, *J. Res. Natl. Inst. Stand. Technol.* **101**, 521 (1996).

[13] A. J. Moerdijk, H. M. J. M. Boesten, and B. J. Verhaar, *Phys. Rev. A* **53**, 916 (1996).

[14] H. Feshbach, *Ann. Phys.* **19**, 287 (1962).

[15] A. J. Moerdijk, B. J. Verhaar, and A. Axelsson, *Phys. Rev. A* **51**, 4852 (1995).

[16] S. Inouye, M. R. Andrews, J. Stenger, H.-J. Miesner, D. M. Stamper-Kurn, and W. Ketterle, *Nature* **392**, 151 (1998).

[17] P. Courteille, R. S. Freeland, D. J. Heinzen, F. A. van Abeelen, and B. J. Verhaar, *Phys. Rev. Lett.* **81,** 69 (1998).

[18] J. L. Roberts, N. R. Claussen, J. P. Burke, C. H. Greene, E. A. Cornell, and C. A. Wieman, *Phys. Rev. Lett.* **81**, 5109 (1998); S. L. Cornish, N. R. Claussen, J. L. Roberts, E. A. Cornell, and C. A. Wieman, *Phys. Rev. Lett.* **85**, 1795 (2000).

[19] F. K. Fatemi, K. M. Jones, and P. D. Lett, *Phys. Rev. Lett.* **85**, 4462 (2000).

[20] M. J. Jamieson, A. Dalgarno, and M. Kimura, *Phys. Rev. A* **51**, 2626 (1995).

[21] M. J. Jamieson, A. Dalgarno, and J. N. Yukich, *Phys. Rev. A* **46**, 6956 (1992).

[22] E. R. I. Abraham, W. I. McAlexander, J. M. Gerton, R. G. Hulet, R. Côté, and A. Dalgarno, *Phys. Rev. A* **55**, 3299 (1997).

[23] E. Tiesinga, C. J. Williams, P. S. Julienne, K. M. Jones, P. D. Lett, and W. D. Phillips, *J. Res. Nat. Inst. Stand. Technol.* **101**, 505 (1996).

[24] F. A. van Abeelen and B. J. Verhaar, *Phys. Rev. A* **59**, 578 (1999).

[25] J. L. Bohn, J. P. Burke, C. H. Greene, H. Wang, P. L. Gould, and W. C. Stwalley, *Phys. Rev. A* **59**, 3660 (1999).

[26] H. M. J. M. Boesten, J. M. Vogels, J. G. C. Tempelaars, and B. J. Verhaar, *Phys. Rev. A* **54**, 3726 (1996).

[27] R. Côté, A. Dalgarno, H. Wang, and W. C. Stwalley, *Phys. Rev. A* **57**, 4118 (1998).

[28] J. M. Vogels, C. C. Tsai, R. S. Freeland, S. J. J. M. F. Kokkelmans, B. J. Verhaar, and D. J. Heinzen, *Phys. Rev. A* **56**, 1067 (1997).

[29] P. S. Julienne, F. H. Mies, E. Tiesinga, and C. J. Williams, *Phys. Rev. Lett.* **78**, 1880 (1997).

[30] M. Arndt, M. Ben Dahan, D. Guéry-Odelin, M. W. Reynolds, and J. Dalibard, *Phys. Rev. Lett.* **79**, 625 (1997).

6
Theory of the condensed state

In the present chapter we consider the structure of the Bose–Einstein condensed state in the presence of interactions. Our discussion is based on the Gross–Pitaevskii equation [1], which describes the zero-temperature properties of the non-uniform Bose gas when the scattering length a is much less than the mean interparticle spacing. We shall first derive the Gross–Pitaevskii equation at zero temperature by treating the interaction between particles in a mean-field approximation (Sec. 6.1). Following that, in Sec. 6.2 we discuss the ground state of atomic clouds in a harmonic-oscillator potential. We compare results obtained by variational methods with those derived in the Thomas–Fermi approximation, in which the kinetic energy operator is neglected in the Gross–Pitaevskii equation. The Thomas–Fermi approximation fails near the surface of a cloud, and in Sec. 6.3 we calculate the surface structure using the Gross–Pitaevskii equation. Finally, in Sec. 6.4 we determine how the condensate wave function 'heals' when subjected to a localized disturbance.

6.1 The Gross–Pitaevskii equation

In the previous chapter we have shown that the effective interaction between two particles at low energies is a constant in the momentum representation, $U_0 = 4\pi\hbar^2 a/m$. In coordinate space this corresponds to a contact interaction $U_0 \delta(\mathbf{r} - \mathbf{r}')$, where $\mathbf{r}$ and $\mathbf{r}'$ are the positions of the two particles. To investigate the energy of many-body states we adopt a Hartree or mean-field approach, and assume that the wave function is a symmetrized product of single-particle wave functions. In the fully condensed state, all bosons are in the same single-particle state, $\phi(\mathbf{r})$, and therefore we may write the wave function of the N-particle system as

$$\Psi(\mathbf{r}_1, \mathbf{r}_2, \ldots, \mathbf{r}_N) = \prod_{i=1}^{N} \phi(\mathbf{r}_i). \tag{6.1}$$

The single-particle wave function $\phi(\mathbf{r}_i)$ is normalized in the usual way,

$$\int d\mathbf{r} |\phi(\mathbf{r})|^2 = 1. \tag{6.2}$$

This wave function does not contain the correlations produced by the interaction when two atoms are close to each other. These effects are taken into account by using the effective interaction $U_0 \delta(\mathbf{r} - \mathbf{r}')$, which includes the influence of short-wavelength degrees of freedom that have been eliminated, or integrated out, as described in Sec. 5.2.1. In the mean-field treatment, we shall not take into account explicitly interactions between degrees of freedom corresponding to length scales less than the interparticle spacing, and therefore we can effectively set the cut-off wave number k_c to zero. The effective interaction is thus equal to U_0, the T matrix at zero energy, and the effective Hamiltonian may be written

$$H = \sum_{i=1}^{N} \left[\frac{\mathbf{p}_i^2}{2m} + V(\mathbf{r}_i) \right] + U_0 \sum_{i<j} \delta(\mathbf{r}_i - \mathbf{r}_j), \tag{6.3}$$

$V(\mathbf{r})$ being the external potential. The energy of the state (6.1) is given by

$$E = N \int d\mathbf{r} \left[\frac{\hbar^2}{2m} |\nabla \phi(\mathbf{r})|^2 + V(\mathbf{r}) |\phi(\mathbf{r})|^2 + \frac{(N-1)}{2} U_0 |\phi(\mathbf{r})|^4 \right]. \tag{6.4}$$

In the Hartree approximation, all atoms are in the state whose wave function we denote by ϕ. In the true wave function, some atoms will be in states with more rapid spatial variation, due to the correlations at small atomic separations, and therefore the total number of atoms in the state ϕ will be less than N. However, as we shall demonstrate in Sec. 8.1 from microscopic theory for the uniform Bose gas, the relative reduction of the number of particles in the condensate, the so-called *depletion* of the condensate due to interactions, is of order $(na^3)^{1/2}$, where n is the particle density. As a measure of the particle separation we introduce the radius r_s of a sphere having a volume equal to the average volume per particle. This is related to the density by the equation

$$n = \frac{1}{(4\pi/3)r_s^3}. \tag{6.5}$$

The depletion is thus of order $(a/r_s)^{3/2}$ which is typically of order one per cent or less in experiments performed to date, and therefore depletion of the condensate due to interactions may be neglected under most circumstances.

We begin by considering the uniform Bose gas. In a uniform system of volume V, the wave function of a particle in the ground state is $1/V^{1/2}$, and therefore the interaction energy of a pair of particles is U_0/V. The energy of a state with N bosons all in the same state is this quantity multiplied by the number of possible ways of making pairs of bosons, $N(N-1)/2$. In this approximation, the energy is

$$E = \frac{N(N-1)}{2V} U_0 \approx \frac{1}{2} V n^2 U_0, \tag{6.6}$$

where $n = N/V$. In writing the last expression we have assumed that $N \gg 1$.

It is convenient to introduce the concept of the wave function of the condensed state,

$$\psi(\mathbf{r}) = N^{1/2} \phi(\mathbf{r}). \tag{6.7}$$

The density of particles is given by

$$n(\mathbf{r}) = |\psi(\mathbf{r})|^2, \tag{6.8}$$

and, with the neglect of terms of order $1/N$, the energy of the system may therefore be written as

$$E(\psi) = \int d\mathbf{r} \left[\frac{\hbar^2}{2m} |\boldsymbol{\nabla} \psi(\mathbf{r})|^2 + V(\mathbf{r})|\psi(\mathbf{r})|^2 + \frac{1}{2} U_0 |\psi(\mathbf{r})|^4 \right]. \tag{6.9}$$

To find the optimal form for ψ, we minimize the energy (6.9) with respect to independent variations[1] of $\psi(\mathbf{r})$ and its complex conjugate $\psi^*(\mathbf{r})$ subject to the condition that the total number of particles

$$N = \int d\mathbf{r} |\psi(\mathbf{r})|^2 \tag{6.10}$$

be constant. The constraint is conveniently taken care of by the method of Lagrange multipliers. One writes $\delta E - \mu \delta N = 0$, where the chemical potential μ is the Lagrange multiplier that ensures constancy of the particle number and the variations of ψ and ψ^* may thus be taken to be arbitrary. This procedure is equivalent to minimizing the quantity $E - \mu N$ at fixed μ. Equating to zero the variation of $E - \mu N$ with respect to $\psi^*(\mathbf{r})$ gives

$$-\frac{\hbar^2}{2m} \nabla^2 \psi(\mathbf{r}) + V(\mathbf{r}) \psi(\mathbf{r}) + U_0 |\psi(\mathbf{r})|^2 \psi(\mathbf{r}) = \mu \psi(\mathbf{r}), \tag{6.11}$$

---

[1] ψ is given in terms of two real functions, its real and imaginary parts. In carrying out the variations, the real and imaginary parts should be considered to be independent. This is equivalent to regarding ψ and ψ^* as independent quantities.

which is the time-independent Gross–Pitaevskii equation. This has the form of a Schrödinger equation in which the potential acting on particles is the sum of the external potential V and a non-linear term $U_0|\psi(\mathbf{r})|^2$ that takes into account the mean field produced by the other bosons. Note that the eigenvalue is the chemical potential, not the energy per particle as it is for the usual (linear) Schrödinger equation. For non-interacting particles all in the same state the chemical potential is equal to the energy per particle, but for interacting particles it is not.

For a uniform Bose gas, the Gross–Pitaevskii equation (6.11) is

$$\mu = U_0|\psi(\mathbf{r})|^2 = U_0 n, \qquad (6.12)$$

which agrees with the result of using the thermodynamic relation $\mu = \partial E/\partial N$ to calculate the chemical potential from the energy of the uniform state, Eq. (6.6).

6.2 The ground state for trapped bosons

We now examine the solution of the Gross–Pitaevskii equation for bosons in a trap [2]. For definiteness, and because of their experimental relevance, we shall consider harmonic traps, but the formalism may easily be applied to more general traps.

Before embarking on detailed calculations let us consider qualitative properties of the solution. For simplicity, we neglect the anisotropy of the oscillator potential, and take it to be of the form $V = m\omega_0^2 r^2/2$. If the spatial extent of the cloud is $\sim R$, the potential energy of a particle in the oscillator potential is $\sim m\omega_0^2 R^2/2$, and the kinetic energy is of order $\hbar^2/2mR^2$ per particle, since a typical particle momentum is of order $\hbar/R$ from Heisenberg's uncertainty principle. Thus in the absence of interactions, the total energy varies as $1/R^2$ for small R and as R^2 for large R, and it has a minimum when the kinetic and potential energies are equal. The corresponding value of the radius of the cloud is of order

$$a_{\mathrm{osc}} = \left(\frac{\hbar}{m\omega_0}\right)^{1/2}, \qquad (6.13)$$

which is the characteristic quantum-mechanical length scale for the harmonic oscillator. This result is what one would anticipate, because we have made what amounts to a variational calculation of the ground state of a single particle in an oscillator potential.

We now consider the effect of interactions. A typical particle density is $n \sim N/R^3$, and the interaction energy of a particle is therefore of order $nU_0 \sim U_0 N/R^3$. For repulsive interactions, the effect of an additional contribution to the energy varying as R^{-3} shifts the minimum of the total energy to larger values of R, and consequently, for increasing values of Na, the kinetic energy term becomes less important. It is instructive to investigate a strong-coupling limit, in which the kinetic energy may be neglected. The equilibrium size is found by minimizing the sum of the potential and interaction energies, and this occurs when the two contributions to the energy are of the same order of magnitude. By equating the two energies, one finds the equilibrium radius to be given by

$$R \sim a_{\rm osc} \left(\frac{Na}{a_{\rm osc}} \right)^{1/5}, \qquad (6.14)$$

and the energy per particle is

$$\frac{E}{N} \sim \hbar \omega_0 \left(\frac{Na}{a_{\rm osc}} \right)^{2/5}. \qquad (6.15)$$

The quantity $Na/a_{\rm osc}$ is a dimensionless measure of the strength of the interaction, and in most experiments performed to date for atoms with repulsive interactions it is much larger than unity, so the radius R is somewhat larger than $a_{\rm osc}$. For $|a| \sim 10$ nm and $a_{\rm osc} \sim 1$ μm (see Eq. (2.35)), with N between 10^4 and 10^6, the ratio $R/a_{\rm osc}$ is seen to range from 2.5 to 6. In equilibrium, the oscillator and interaction energies are both proportional to R^2 and therefore the ratio between the kinetic energy, which is proportional to R^{-2}, and the potential (or interaction) energy is proportional to $(a_{\rm osc}/Na)^{4/5}$. This confirms that the kinetic energy is indeed negligible for clouds containing a sufficiently large number of particles.[2]

Let us now turn to attractive interactions. For a small number of particles, the total energy as a function of R is similar to that for non-interacting particles, except that at very small R the energy diverges to $-\infty$ as $-1/R^3$. Consequently, for a sufficiently small number of particles the energy has a local minimum near that for non-interacting particles, but at a smaller radius. This state is metastable, since for small departures from the minimum, the energy increases, but for small R the energy eventually varies as $-1/R^3$ and becomes less than that at the local minimum. With increasing particle number, the local minimum becomes shallower, and at a critical particle number $N_{\rm c}$ it disappears. For larger numbers of particles there is no metastable state. As one might expect, the critical number is determined by

---

[2] As we shall see from the detailed calculations in Sec. 6.3, the leading term is of order $(a_{\rm osc}/Na)^{4/5} \ln(Na/a_{\rm osc})$.

the condition that the dimensionless coupling parameter be of order -1, that is $N_c \sim a_{\rm osc}/|a|$. For ^{7}Li the (triplet) scattering length is $-27.6a_0 = -1.46$ nm, and therefore in traps with frequencies of order 100 Hz, corresponding to $a_{\rm osc}$ of order microns, the critical number is of order 10^3, which is what is found experimentally [3].

We now consider the problem quantitatively. We shall determine the ground-state energy for a gas trapped in an anisotropic three-dimensional harmonic-oscillator potential V given by

$$V(x,y,z) = \frac{1}{2}m(\omega_1^2 x^2 + \omega_2^2 y^2 + \omega_3^2 z^2), \tag{6.16}$$

where the three oscillator frequencies ω_i ($i=1,2,3$) may differ from each other. Many traps used in experiments have an axis of symmetry, so that two of the frequencies are equal, but we shall consider the general case. The Gross–Pitaevskii equation (6.11) may be solved directly by numerical integration, but it is instructive to derive some analytical results. We begin with a variational calculation based on a Gaussian trial function and then go on to the Thomas–Fermi approximation.

6.2.1 A variational calculation

In the absence of interparticle interactions the lowest single-particle state has the familiar wave function,

$$\phi_0(\mathbf{r}) = \frac{1}{\pi^{3/4}(a_1 a_2 a_3)^{1/2}} e^{-x^2/2a_1^2} e^{-y^2/2a_2^2} e^{-z^2/2a_3^2}, \tag{6.17}$$

where the oscillator lengths a_i ($i=1,2,3$) are given by $a_i^2 = \hbar/m\omega_i$ according to Eq. (2.34). The density distribution $n(\mathbf{r}) = N\phi_0(\mathbf{r})^2$ is thus Gaussian. Interatomic interactions change the dimensions of the cloud, and we adopt as our trial function for ψ the same form as (6.17),

$$\psi(\mathbf{r}) = \frac{N^{1/2}}{\pi^{3/4}(b_1 b_2 b_3)^{1/2}} e^{-x^2/2b_1^2} e^{-y^2/2b_2^2} e^{-z^2/2b_3^2}, \tag{6.18}$$

where the lengths b_i are variational parameters. The trial function satisfies the normalization condition (6.10). Substitution of (6.18) into (6.9) yields the energy expression

$$E(b_1,b_2,b_3) = N \sum_i \hbar\omega_i \left(\frac{a_i^2}{4b_i^2} + \frac{b_i^2}{4a_i^2}\right) + \frac{N^2 U_0}{2(2\pi)^{3/2} b_1 b_2 b_3}. \tag{6.19}$$

If we evaluate (6.19), putting the b_i equal to their values a_i in the absence of interaction, one finds

$$\begin{aligned} E &\approx N \sum_i \frac{\hbar \omega_i}{2} + \frac{N^2 U_0}{2(2\pi)^{3/2} a_1 a_2 a_3} \\ &= N \sum_i \frac{\hbar \omega_i}{2} + \frac{N^2}{2} \langle 00|v|00\rangle, \end{aligned} \quad (6.20)$$

where

$$\langle 00|v|00\rangle = \frac{4\pi \hbar^2 a}{m} \int d\mathbf{r} |\phi_0(\mathbf{r})|^4 \quad (6.21)$$

is the interaction energy for two particles in the ground state of the oscillator. The result (6.20) corresponds to a perturbation theory estimate, and it is a good approximation as long as the interaction energy per particle is small compared with any of the zero-point energies. If the magnitudes of the three oscillator frequencies are comparable ($\omega_i \sim \omega_0$), the ratio of the interaction energy to the zero-point energy of the oscillator is of order Na/a_{osc}, which, as we argued above, is a dimensionless measure of the strength of the interaction. It gives the ratio of the interaction energy to the oscillator energy $\hbar\omega_0$ when the wave function is that for the particles in the ground state of the oscillator. The condition $Na/a_{\text{osc}} \sim 1$ marks the crossover between the perturbative regime and one where equilibrium is determined by competition between the interaction energy and the potential energy due to the trap.

As the interaction becomes stronger, the cloud expands, and the optimal wave function becomes more extended, corresponding to larger values of the lengths b_i. It is convenient to introduce dimensionless lengths x_i defined by

$$x_i = \frac{b_i}{a_i}. \quad (6.22)$$

Minimizing E with respect to the variational parameters x_i ($i = 1, 2, 3$) then yields the three equations

$$\frac{1}{2}\hbar\omega_i(x_i^2 - \frac{1}{x_i^2}) - \frac{1}{2(2\pi)^{3/2}} \frac{NU_0}{\bar{a}^3} \frac{1}{x_1 x_2 x_3} = 0. \quad (6.23)$$

Here we have introduced the characteristic length

$$\bar{a} = \sqrt{\frac{\hbar}{m\bar\omega}} \quad (6.24)$$

for an oscillator of frequency

$$\bar\omega = (\omega_1 \omega_2 \omega_3)^{1/3}, \quad (6.25)$$

the geometric mean of the oscillator frequencies for the three directions. In the general case we obtain the optimal parameters for the trial function by solving this set of coupled equations. Let us here, however, consider the simpler situation when the number of particles is sufficiently large that the interaction energy per particle is large compared with $\hbar\omega_i$ for all ω_i. Then it is permissible to neglect the kinetic energy terms (proportional to $1/x_i^2$) in (6.23). By solving for x_i we find

$$x_i^5 = \left(\frac{2}{\pi}\right)^{1/2} \frac{Na}{\bar{a}} \left(\frac{\bar{\omega}}{\omega_i}\right)^{5/2}, \quad (6.26)$$

or

$$b_i = \left(\frac{2}{\pi}\right)^{1/10} \left(\frac{Na}{\bar{a}}\right)^{1/5} \frac{\bar{\omega}}{\omega_i} \bar{a}, \quad (6.27)$$

and the leading contribution to the energy per particle is given by

$$\frac{E}{N} = \frac{5}{4} \left(\frac{2}{\pi}\right)^{1/5} \left(\frac{Na}{\bar{a}}\right)^{2/5} \hbar\bar{\omega}. \quad (6.28)$$

According to the variational estimate (6.28) the energy per particle is proportional to $N^{2/5}$ in the limit when the kinetic energy is neglected, and is of order $(Na/\bar{a})^{2/5}$ times greater than the energy in the absence of interactions. As we shall see below, this is also true in the Thomas–Fermi approximation, which is exact when the particle number is large.

In Fig. 6.1 we illustrate for an isotropic oscillator the dependence of E/N on the variational parameter b ($= b_1 = b_2 = b_3$) for different values of the dimensionless parameter Na/a_{osc}. We have included examples of attractive interactions, corresponding to negative values of the scattering length a. As shown in the figure, a local minimum exists for negative a provided N is less than some value N_c, but for larger values of N the cloud will collapse. The critical particle number is found from the condition that the first and second derivatives of E/N with respect to b are both equal to zero, which gives [4]

$$\frac{N_c|a|}{a_{\text{osc}}} = \frac{2(2\pi)^{1/2}}{5^{5/4}} \approx 0.67. \quad (6.29)$$

The minimum energy per particle (in units of $\hbar\omega_0$) is plotted in Fig. 6.2 as a function of Na/a_{osc} within the range of stability $-0.67 < Na/a_{\text{osc}} < \infty$. For comparison, we also exhibit the result of the Thomas–Fermi approximation discussed in the following subsection. A numerical integration of the Gross–Pitaevskii equation gives $N_c|a|/a_{\text{osc}} \approx 0.57$ [5].

If the kinetic energy is included as a perturbation, the total energy per

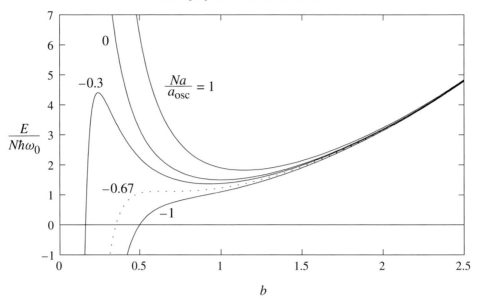

Fig. 6.1. Variational expression for the energy per particle for an isotropic harmonic trap as a function of the variational parameter b, for different values of the dimensionless parameter $Na/a_{\rm osc}$. The dotted curve corresponds to the critical value, approximately -0.67, at which the cloud becomes unstable.

particle contains an additional term proportional to $(Na/a_{\rm osc})^{-2/5}$ (Problem 6.1). In Sec. 6.3 we shall calculate the kinetic energy more accurately for large $Na/a_{\rm osc}$ and show that it is proportional to $(Na/a_{\rm osc})^{-2/5}\ln(Na/a_{\rm osc})$.

6.2.2 The Thomas–Fermi approximation

For sufficiently large clouds, an accurate expression for the ground-state energy may be obtained by neglecting the kinetic energy term in the Gross–Pitaevskii equation. As we have seen for a harmonic trap in the preceding subsection, when the number of atoms is large and interactions are repulsive, the ratio of kinetic to potential (or interaction) energy is small. A better approximation for the condensate wave function for large numbers of atoms may be obtained by solving the Gross–Pitaevskii equation, neglecting the kinetic energy term from the start. Thus from Eq. (6.11) one finds

$$\left[V(\mathbf{r}) + U_0|\psi(\mathbf{r})|^2\right]\psi(\mathbf{r}) = \mu\psi(\mathbf{r}), \tag{6.30}$$

where μ is the chemical potential. This has the solution

$$n(\mathbf{r}) = |\psi(\mathbf{r})|^2 = [\mu - V(\mathbf{r})]/U_0 \tag{6.31}$$

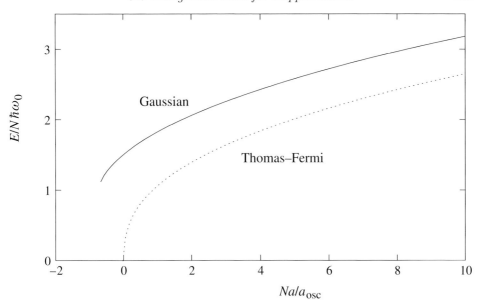

Fig. 6.2. Variational estimate of the energy per particle for an isotropic harmonic trap as a function of the dimensionless parameter Na/a_{osc}. The dotted line is the result in the Thomas–Fermi approximation.

in the region where the right hand side is positive, while $\psi = 0$ outside this region. The boundary of the cloud is therefore given by

$$V(\mathbf{r}) = \mu. \tag{6.32}$$

The physical content of this approximation is that the energy to add a particle at any point in the cloud is the same everywhere. This energy is given by the sum of the external potential $V(\mathbf{r})$ and an interaction contribution $n(\mathbf{r})U_0$ which is the chemical potential of a uniform gas having a density equal to the local density $n(\mathbf{r})$. Since this approximation is reminiscent of the Thomas–Fermi approximation in the theory of atoms, it is generally referred to by the same name. For atoms, the total electrostatic potential takes the place of the trapping potential, and the local Fermi energy that of the mean-field energy $U_0|\psi|^2 = U_0 n$.

In the Thomas–Fermi approximation the extension of the cloud in the three directions is given by the three semi-axes R_i obtained by inserting (6.16) into (6.32),

$$R_i^2 = \frac{2\mu}{m\omega_i^2}, \quad i = 1, 2, 3. \tag{6.33}$$

The lengths R_i may be evaluated in terms of trap parameters once the chemical potential has been determined. The normalization condition on ψ, Eq. (6.10), yields a relation between the chemical potential μ and the total number of particles N. For a harmonic trap with a potential given by Eq. (6.16) one finds

$$N = \frac{8\pi}{15}\left(\frac{2\mu}{m\bar{\omega}^2}\right)^{3/2}\frac{\mu}{U_0}, \tag{6.34}$$

as may be seen by scaling each spatial coordinate by $(2\mu/m\omega_i^2)^{1/2}$ and integrating over the interior of the unit sphere. Solving (6.34) for μ we obtain the following relation between μ and $\hbar\bar{\omega}$:

$$\mu = \frac{15^{2/5}}{2}\left(\frac{Na}{\bar{a}}\right)^{2/5}\hbar\bar{\omega}. \tag{6.35}$$

The quantity $\bar{R} = (R_1 R_2 R_3)^{1/3}$ is a convenient measure of the spatial extent of the cloud. By combining (6.33) and (6.35) we obtain

$$\bar{R} = 15^{1/5}\left(\frac{Na}{\bar{a}}\right)^{1/5}\bar{a} \approx 1.719\left(\frac{Na}{\bar{a}}\right)^{1/5}\bar{a}, \tag{6.36}$$

which implies that $\bar{R}$ is somewhat greater than $\bar{a}$ under typical experimental conditions. In Fig. 6.3 we compare wave functions for the variational calculation described in the previous section and the Thomas–Fermi approximation.

Since $\mu = \partial E/\partial N$ and $\mu \propto N^{2/5}$ according to Eq. (6.35) the energy per particle is

$$\frac{E}{N} = \frac{5}{7}\mu. \tag{6.37}$$

This is the exact result for the leading contribution to the energy at large N, and it is smaller than the variational estimate (6.28) by a numerical factor $(3600\pi)^{1/5}/7 \approx 0.92$. The central density of the cloud is $n(0) = \mu/U_0$ within the Thomas–Fermi approximation.

In order to see how the total energy is distributed between potential and interaction energies we insert the Thomas–Fermi solution given by (6.31) into (6.9) and evaluate the last two terms, neglecting the kinetic energy. The calculation is carried out most easily by scaling the spatial coordinates so that the potential $V(\mathbf{r})$ and the Thomas–Fermi solution both become spherically symmetric. The ratio between the interaction energy E_{int} and

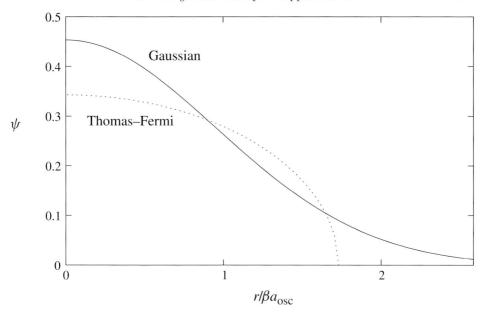

Fig. 6.3. The ground-state wave function in the Gaussian variational approximation (full line) and in the Thomas–Fermi approximation (dotted line) for an isotropic harmonic-oscillator potential. The wave functions are given in units of $N^{1/2}/(\beta a_{\rm osc})^{3/2}$, and $\beta = (Na/a_{\rm osc})^{1/5}$.

the potential energy $E_{\rm pot}$ then becomes

$$\frac{E_{\rm int}}{E_{\rm pot}} = \frac{\int_0^1 dr\, r^2(1-r^2)^2/2}{\int_0^1 dr\, r^4(1-r^2)} = \frac{2}{3}. \quad (6.38)$$

This result is an expression of the virial theorem, see Problem 6.2. The interaction energy in the Thomas–Fermi approximation is thus equal to $2/5$ times the total energy. Since the total energy per particle is $5\mu/7$ we conclude that the interaction energy per particle and the chemical potential are related by

$$\frac{E_{\rm int}}{N} = \frac{2}{7}\mu. \quad (6.39)$$

We shall return to this result in Chapter 11 when calculating the effect of interactions on properties of clouds at non-zero temperature.

The Thomas–Fermi approximation gives an excellent account of the gross properties of clouds when $Na/\bar{a}$ is large compared with unity. However, in a number of important problems of physical interest, the kinetic energy plays a crucial role. In the next section we consider the surface structure of clouds, and in Sec. 9.2 vortex states.

6.3 Surface structure of clouds

The Thomas–Fermi approach is applicable provided the order parameter varies sufficiently slowly in space. It fails near the edge of the cloud, as one may see by estimating the contributions to the energy functional for the Thomas–Fermi wave function. The density profile is $n(\mathbf{r}) = [\mu - V(\mathbf{r})]/U_0$, and if we expand the external potential about a point $\mathbf{r}_0$ in the surface, which is given by $V(\mathbf{r}_0) = \mu$, this becomes $n(\mathbf{r}) = \mathbf{F} \cdot (\mathbf{r} - \mathbf{r}_0)/U_0$, where

$$\mathbf{F} = -\boldsymbol{\nabla} V(\mathbf{r}_0) \tag{6.40}$$

is the force that the external potential exerts on a particle at the surface. The condensate wave function is given by

$$\psi(r) = \left[\frac{\mathbf{F} \cdot (\mathbf{r} - \mathbf{r}_0)}{U_0}\right]^{1/2}. \tag{6.41}$$

If we denote the coordinate in the direction of $\boldsymbol{\nabla} V(\mathbf{r}_0)$ by x, and denote the position of the surface by $x = x_0$, the interior of the cloud corresponds to $x \leq x_0$. The Thomas–Fermi wave function for the cloud varies as $(x_0 - x)^{1/2}$ for $x \leq x_0$, and therefore its derivative with respect to x is proportional to $(x_0 - x)^{-1/2}$. Consequently, the kinetic energy term in the energy functional behaves as $1/(x_0 - x)$, and the total kinetic energy per unit area of the surface, which is obtained by integrating this expression over x, diverges as $-\ln(x_0 - x)$, as x approaches x_0 from below. To estimate the distance from the surface at which the kinetic energy term becomes important, we observe that the kinetic energy contribution to the energy functional is of order

$$\frac{\hbar^2 |d\psi/dx|^2}{2m|\psi|^2} \sim \frac{\hbar^2}{2m(x_0 - x)^2}, \tag{6.42}$$

per atom. The difference between the chemical potential and the external potential is

$$\mu - V(\mathbf{r}) \simeq F(x_0 - x), \tag{6.43}$$

where F is the magnitude of the trapping force acting on a particle at the surface of the cloud. Thus the kinetic energy term dominates for $x_0 - x \lesssim \delta$, where

$$\delta = (\hbar^2/2mF)^{1/3}, \tag{6.44}$$

which is the same length scale as occurs in the quantum mechanics of a free particle in a linear potential. For an isotropic harmonic potential $V =$

$m\omega_0^2 r^2/2$, F is $m\omega_0^2 R$, where R is the radius of the cloud, and therefore

$$\delta = \left(\frac{a_{\text{osc}}^4}{2R}\right)^{1/3} = \left(\frac{\hbar\omega_0}{\mu}\right)^{2/3}\frac{R}{2}, \tag{6.45}$$

where $a_{\text{osc}} = (\hbar/m\omega_0)^{1/2}$ and in writing the second expression we have used the fact that $\mu = m\omega_0^2 R^2/2$. Consequently, the fraction of the volume of the cloud where the Thomas–Fermi approximation is poor is proportional to $(a_{\text{osc}}/R)^{4/3}$, which is small for a large enough number of particles.

We now study the surface region starting from the Gross–Pitaevskii equation. If the external potential varies slowly on the length scale δ, we may expand the potential about the position of the surface as we did above, and the problem becomes essentially one-dimensional [6, 7]. In terms of the coordinate x introduced above, and with the origin chosen to be at the position of the surface, the Gross–Pitaevskii equation is

$$\left[-\frac{\hbar^2}{2m}\frac{d^2}{dx^2} + Fx + U_0|\psi(x)|^2\right]\psi(x) = 0. \tag{6.46}$$

In the discussion above we identified the length scale δ associated with the surface structure and, as one would expect, the Gross–Pitaevskii equation simplifies if one measures lengths in units of δ. In addition, it is convenient to measure the wave function of the condensate in terms of its value $b = (F\delta/U_0)^{1/2}$ in the Thomas–Fermi theory at a distance δ from the edge of the cloud. After introducing a scaled length variable $y = x/\delta$ and a scaled wave function given by $\Psi = \psi/b$ we obtain the equation

$$\Psi'' = y\Psi + \Psi^3, \tag{6.47}$$

where the prime denotes differentiation with respect to y. The solution in the Thomas–Fermi approximation is

$$\Psi = \sqrt{-y} \text{ for } y \leq 0, \quad \Psi = 0 \text{ for } y > 0. \tag{6.48}$$

First let us consider the behaviour for $x \gg \delta$, corresponding to $y \gg 1$. Since the condensate wave function is small we may neglect the cubic term. The resulting equation is that for the Airy function, and its asymptotic solution is

$$\Psi \simeq \frac{C}{y^{1/4}}e^{-2y^{3/2}/3}. \tag{6.49}$$

Deep inside the cloud, corresponding to $y \ll -1$, the Thomas–Fermi solution $\Psi \simeq \sqrt{-y}$ is approximately valid. To determine the leading correction to the wave function, we write $\Psi = \Psi_0 + \Psi_1$ and linearize (6.47), thereby finding

$$-\Psi_1'' + y\Psi_1 + 3\Psi_0^2\Psi_1 = \Psi_0''. \tag{6.50}$$

Using $\Psi_0 = (-y)^{1/2}$ from (6.48) and neglecting Ψ_1'' in (6.50) since it contributes to terms of higher order in $1/y$, we arrive at the result

$$\Psi_1 \simeq -\frac{1}{8y^2\sqrt{-y}}. \tag{6.51}$$

The asymptotic solution is thus

$$\Psi = \sqrt{-y}\left(1 + \frac{1}{8y^3}\right). \tag{6.52}$$

Equation (6.47) may be solved numerically and this enables one to evaluate the coefficient C in Eq. (6.49), which is found to be approximately 0.3971 [7].

We now evaluate the kinetic energy per unit area perpendicular to the x axis,

$$\frac{<p^2>}{2m} = \frac{\hbar^2}{2m}\int dx|\boldsymbol{\nabla}\psi|^2. \tag{6.53}$$

Let us first use the Thomas–Fermi wave function (6.48) which we expect to be valid only in the region $x \ll -\delta$. Since the integral diverges for $x \to 0$, as we discussed at the beginning of this section, we evaluate the integral for x less than some cut-off value $-l$. We take the lower limit of the integration to be $-L$, where L is large compared with δ. We expect that the kinetic energy will be given approximately by the Thomas–Fermi result if the cut-off distance is chosen to be $\sim -\delta$, the distance at which the Thomas–Fermi approximation fails,

$$\frac{<p^2>}{2m} = \frac{\hbar^2}{2m}\int_{-L}^{-l} dx(\psi')^2 \simeq \frac{\hbar^2}{8m}\frac{F}{U_0}\ln\frac{L}{l}. \tag{6.54}$$

We now compare this result with the kinetic energy per unit area calculated numerically using the true wave function,

$$\frac{<p^2>}{2m} = \frac{\hbar^2}{2m}\int_{-L}^{\infty} dx(\psi')^2 \approx \frac{\hbar^2}{8m}\frac{F}{U_0}\ln\frac{4.160L}{\delta}, \tag{6.55}$$

which is valid for large values of $\ln(L/\delta)$.

We conclude that one obtains the correct asymptotic behaviour of the kinetic energy if one uses the Thomas–Fermi approach and cuts the integral off at $x = -l$, where

$$l = 0.240\delta. \tag{6.56}$$

As we shall demonstrate below, the same effective cut-off may be used for calculating the kinetic energy in more general situations.

We now turn to the system of physical interest, a cloud of N atoms trapped in a three-dimensional harmonic-oscillator potential. For simplicity, we consider only the isotropic case, where the potential is $V(r) = m\omega_0^2 r^2/2$. The Gross–Pitaevskii equation for the ground-state wave function is

$$\left[-\frac{\hbar^2}{2mr^2}\frac{d}{dr}(r^2\frac{d}{dr}) + \frac{1}{2}m\omega_0^2 r^2 + \frac{4\pi\hbar^2 a}{m}|\psi(r)|^2\right]\psi(r) = \mu\psi(r). \quad (6.57)$$

By the substitution $\chi = r\psi$ we obtain

$$-\frac{\hbar^2}{2m}\frac{d^2\chi}{dr^2} + \frac{1}{2}m\omega_0^2(r^2 - R^2)\chi(r) + \frac{4\pi\hbar^2 a}{mr^2}|\chi(r)|^2\chi(r) = 0, \quad (6.58)$$

since $\mu = m\omega_0^2 R^2/2$. The Thomas–Fermi solution is

$$\chi_{TF} = r\left(\frac{R^2 - r^2}{8\pi a a_{osc}^4}\right)^{1/2}. \quad (6.59)$$

By expanding about $r = R$ in Eq. (6.58) we arrive at an equation of the form (6.47), with the length scale δ given by Eq. (6.45).

To calculate the kinetic energy, we use the Thomas–Fermi wave function and cut the integral off at a radius $R - l$, where l is given by the calculation for the linear ramp, Eq. (6.56). The result is

$$\frac{E_{kin}}{N} = \frac{\hbar^2}{2m}\frac{\int_0^{R-l} dr\, r^2 (d\psi/dr)^2}{\int_0^R dr\, r^2 \psi^2} \simeq \frac{\hbar^2}{2mR^2}\left(\frac{15}{4}\ln\frac{2R}{l} - \frac{5}{2}\right). \quad (6.60)$$

This expression agrees well with the numerical result [7] for $\ln(R/\delta)$ greater than 3, the relative difference being less than 2.5%.

6.4 Healing of the condensate wave function

In the previous section we considered the condensate wave function for an external potential that varied relatively smoothly in space. It is instructive to investigate the opposite extreme, a condensate confined by a box with infinitely hard walls. At the wall, the wave function must vanish, and in the interior of the box the condensate density approaches its bulk value. The distance over which the wave function rises from zero at the wall to close to its bulk value may be estimated from the Gross–Pitaevskii equation, since away from the wall the wave function is governed by competition between the interaction energy term $\sim nU_0$ and the kinetic energy one. If one denotes the spatial scale of variations by ξ, the kinetic energy per particle is of order $\hbar^2/2m\xi^2$ and the two energies are equal when

$$\frac{\hbar^2}{2m\xi^2} = nU_0, \qquad (6.61)$$

or

$$\xi^2 = \frac{\hbar^2}{2mnU_0} = \frac{1}{8\pi na} = \frac{r_s^3}{6a}, \qquad (6.62)$$

where the particle separation r_s is defined in Eq. (6.5). Since in experiments the distance between atoms is typically much larger than the scattering length, the coherence length is larger than the atomic separation. The length ξ is referred to in the condensed matter literature as the *coherence length*, where the meaning of the word 'coherence' is different from that in optics. Since it describes the distance over which the wave function tends to its bulk value when subjected to a localized perturbation, it is also referred to as the *healing length*.

To investigate the behaviour of the condensate wave function quantitatively, we begin with the Gross–Pitaevskii equation (6.11), and assume that the potential vanishes for $x \geq 0$, and is infinite for $x < 0$. The ground-state wave function is uniform in the y and z directions, and therefore the Gross–Pitaevskii equation is

$$-\frac{\hbar^2}{2m}\frac{d^2\psi(x)}{dx^2} + U_0|\psi(x)|^2\psi(x) = \mu\psi(x). \qquad (6.63)$$

For bulk uniform matter, the chemical potential is given by Eq. (6.12), and thus we may write $\mu = U_0|\psi_0|^2$ where ψ_0 is the wave function far from the wall, where the kinetic energy term becomes negligible. The equation then becomes

$$\frac{\hbar^2}{2m}\frac{d^2\psi(x)}{dx^2} = -U_0(|\psi_0|^2 - |\psi(x)|^2)\psi(x). \qquad (6.64)$$

When ψ is real, one may regard ψ as being a spatial coordinate and x as being the 'time'. Then (6.64) has the same form as the classical equation of motion of a particle in a potential $\sim \psi_0^2\psi^2 - \psi^4/2$. The equation may be solved analytically, subject to the boundary conditions that $\psi(0) = 0$ and $\psi(\infty) = \psi_0$, with the result

$$\psi(x) = \psi_0 \tanh(x/\sqrt{2}\xi). \qquad (6.65)$$

This confirms that the wave function approaches its bulk value over a distance $\sim \xi$, in agreement with the qualitative arguments above.

Problems

PROBLEM 6.1 Consider N bosons interacting via repulsive interactions in an isotropic harmonic trap. Use the Gaussian trial function (6.18) to calculate the kinetic energy per particle of a cloud in its ground state when $Na/a_{\rm osc}$ is large.

PROBLEM 6.2 Consider a condensate which is trapped by a potential which is a homogeneous function of degree ν of the radial coordinate but with arbitrary dependence on the angular coordinates ($V(\lambda \mathbf{r}) = \lambda^\nu V(\mathbf{r})$). In equilibrium, the energy of the condensate must be unchanged by a small change in the wave function from its value in the ground state, subject to the number of particles being constant. By considering a change of spatial scale of the wave function, with its form being unaltered, show that the kinetic, trap and interaction energies, which are given by $E_{\rm kin} = (\hbar^2/2m)\int d\mathbf{r} |\boldsymbol{\nabla}\psi(\mathbf{r})|^2$, $E_{\rm trap} = \int d\mathbf{r} V(\mathbf{r})|\psi(\mathbf{r})|^2$, and $E_{\rm int} = \frac{1}{2}U_0 \int d\mathbf{r}|\psi(\mathbf{r})|^4$ satisfy the condition

$$2E_{\rm kin} - \alpha E_{\rm trap} + 3E_{\rm int} = 0,$$

which is a statement of the virial theorem for this problem. Show in addition that the chemical potential is given by

$$\mu N = E_{\rm kin} + E_{\rm trap} + 2E_{\rm int},$$

and determine the ratio between the chemical potential and the total energy per particle in the limit when the kinetic energy may be neglected.

PROBLEM 6.3 Consider a cloud of 10^5 atoms of ^{87}Rb in an isotropic harmonic-oscillator potential with the oscillation frequency ω_0 given by $\omega_0/2\pi = 200$ Hz. Take the scattering length a to be $100a_0$ and calculate the total energy, the chemical potential μ, the radius R, the coherence length ξ at the centre of the cloud, and the length δ giving the scale of surface structure.

References

[1] L. P. Pitaevskii, *Zh. Eksp. Teor. Fiz.* **40**, 646 (1961) [*Sov. Phys.–JETP* **13**, 451 (1961)]; E. P. Gross, *Nuovo Cimento* **20**, 454 (1961); *J. Math. Phys.* **4**, 195 (1963).

[2] Our discussion is based on that of G. Baym and C. J. Pethick, *Phys. Rev. Lett.* **76**, 6 (1996). Some of the results were obtained earlier by R. V. E. Lovelace and T. J. Tommila, *Phys. Rev. A* **35**, 3597 (1987).

[3] C. C. Bradley, C. A. Sackett, J. J. Tollett, and R. G. Hulet, *Phys. Rev. Lett.* **75**, 1687 (1995); C. C. Bradley, C. A. Sackett, and R. G. Hulet, *Phys. Rev. Lett.* **78**, 985 (1997).

[4] A. L. Fetter, cond-mat/9510037.
[5] P. A. Ruprecht, M. J. Holland, K. Burnett, and M. Edwards, *Phys. Rev. A* **51**, 4704 (1995).
[6] F. Dalfovo, L. P. Pitaevskii, and S. Stringari, *Phys. Rev. A* **54**, 4213 (1996).
[7] E. Lundh, C. J. Pethick, and H. Smith, *Phys. Rev. A* **55**, 2126 (1997).

7
Dynamics of the condensate

The time-dependent behaviour of Bose–Einstein condensed clouds, such as collective modes and the expansion of a cloud when released from a trap, is an important source of information about the physical nature of the condensate. In addition, the spectrum of elementary excitations of the condensate is an essential ingredient in calculations of thermodynamic properties. In this chapter we treat the dynamics of a condensate at zero temperature starting from a time-dependent generalization of the Gross–Pitaevskii equation used in Chapter 6 to describe static properties. From this equation one may derive equations very similar to those of classical hydrodynamics, which we shall use to calculate properties of collective modes.

We begin in Sec. 7.1 by describing the time-dependent Gross–Pitaevskii equation and deriving the hydrodynamic equations. We then use the hydrodynamic equations to determine the excitation spectrum of a homogeneous Bose gas (Sec. 7.2). Subsequently, we consider modes in trapped clouds (Sec. 7.3) within the hydrodynamic approach, and also describe the method of collective coordinates and the related variational method. In Sec. 7.4 we consider surface modes of oscillation, which resemble gravity waves on a liquid surface. The variational approach is used in Sec. 7.5 to treat the free expansion of a condensate upon release from a trap. Finally, in Sec. 7.6 we discuss solitons, which are exact one-dimensional solutions of the time-dependent Gross–Pitaevskii equation.

7.1 General formulation

In the previous chapter we saw that the equilibrium structure of the condensate is described by a time-independent Schrödinger equation with a non-linear contribution to the potential to take into account interactions between particles. To treat dynamical problems it is natural to use a time-dependent generalization of this Schrödinger equation, with the same non-linear interaction term. This equation is the time-dependent Gross–Pitaevskii equation,

$$-\frac{\hbar^2}{2m}\nabla^2\psi(\mathbf{r},t) + V(\mathbf{r})\psi(\mathbf{r},t) + U_0|\psi(\mathbf{r},t)|^2\psi(\mathbf{r},t) = i\hbar\frac{\partial\psi(\mathbf{r},t)}{\partial t}, \quad (7.1)$$

and it is the basis for our discussion of the dynamics of the condensate.

The time-independent Gross–Pitaevskii equation, Eq. (6.11), is a non-linear Schrödinger equation with the chemical potential replacing the energy eigenvalue in the time-independent Schrödinger equation. To ensure consistency between the time-dependent Gross–Pitaevskii equation and the time-independent one, under stationary conditions $\psi(\mathbf{r},t)$ must develop in time as $\exp(-i\mu t/\hbar)$. The phase factor reflects the fact that microscopically ψ is equal to the matrix element of the annihilation operator $\hat{\psi}$ between the ground state with N particles and that with $N-1$ particles,

$$\psi(\mathbf{r},t) = \langle N-1|\hat{\psi}(\mathbf{r})|N\rangle \propto \exp[-i(E_N - E_{N-1})t/\hbar], \quad (7.2)$$

since the states $|N\rangle$ and $|N-1\rangle$ develop in time as $\exp(-iE_N t/\hbar)$ and $\exp(-iE_{N-1}t/\hbar)$. For large N the difference in ground-state energies $E_N - E_{N-1}$ is equal to $\partial E/\partial N$, which is the chemical potential. Therefore this result is basically Josephson's relation for the development of the phase ϕ of the condensate wave function

$$\frac{d\phi}{dt} = -\frac{\mu}{\hbar}. \quad (7.3)$$

Both for formal reasons as well as for applications a variational formulation analogous to that for static problems is useful. The time-dependent Gross–Pitaevskii equation (7.1) may be derived from the action principle

$$\delta\int_{t_1}^{t_2} L\,dt = 0, \quad (7.4)$$

where the Lagrangian L is given by

$$\begin{aligned}L &= \int d\mathbf{r}\frac{i\hbar}{2}\left(\psi^*\frac{\partial\psi}{\partial t} - \psi\frac{\partial\psi^*}{\partial t}\right) - E \\ &= \int d\mathbf{r}\left[\frac{i\hbar}{2}\left(\psi^*\frac{\partial\psi}{\partial t} - \psi\frac{\partial\psi^*}{\partial t}\right) - \mathcal{E}\right].\end{aligned} \quad (7.5)$$

Here E is the energy, Eq. (6.9), and the energy density is given by

$$\mathcal{E} = \frac{\hbar^2}{2m}|\boldsymbol{\nabla}\psi|^2 + V(\mathbf{r})|\psi|^2 + \frac{U_0}{2}|\psi|^4. \quad (7.6)$$

In the variational principle (7.4) the variations of ψ (or ψ^*) are arbitrary, apart from the requirement that they vanish at $t = t_1$, $t = t_2$, and on any spatial boundaries for all t. With a physically motivated choice of trial function for ψ, this variational principle provides the foundation for approximate solutions of dynamical problems, as we shall illustrate in Sec. 7.3.3.

The physical content of the Gross–Pitaevskii equation (7.1) may be revealed by reformulating it as a pair of hydrodynamic equations, which we now derive.

7.1.1 The hydrodynamic equations

Under general, time-dependent conditions we may use instead of (7.1) an equivalent set of equations for the density, which is given by $|\psi|^2$, and the gradient of its phase, which is proportional to the local velocity of the condensate.

To understand the nature of the velocity of the condensate, we derive the continuity equation. If one multiplies the time-dependent Gross–Pitaevskii equation (7.1) by $\psi^*(\mathbf{r}, t)$ and subtracts the complex conjugate of the resulting equation, one arrives at the equation

$$\frac{\partial |\psi|^2}{\partial t} + \boldsymbol{\nabla} \cdot \left[\frac{\hbar}{2mi} (\psi^* \boldsymbol{\nabla} \psi - \psi \boldsymbol{\nabla} \psi^*) \right] = 0. \tag{7.7}$$

This is the same as one obtains from the usual (linear) Schrödinger equation, since the non-linear potential in the Gross–Pitaevskii equation is real. Equation (7.7) has the form of a continuity equation for the particle density, $n = |\psi|^2$, and it may be written as

$$\frac{\partial n}{\partial t} + \boldsymbol{\nabla} \cdot (n\mathbf{v}) = 0, \tag{7.8}$$

where the velocity of the condensate is defined by

$$\mathbf{v} = \frac{\hbar}{2mi} \frac{(\psi^* \boldsymbol{\nabla} \psi - \psi \boldsymbol{\nabla} \psi^*)}{|\psi|^2}. \tag{7.9}$$

The momentum density $\mathbf{j}$ is given by

$$\mathbf{j} = \frac{\hbar}{2i} (\psi^* \boldsymbol{\nabla} \psi - \psi \boldsymbol{\nabla} \psi^*), \tag{7.10}$$

and therefore the relation (7.9) is equivalent to the result

$$\mathbf{j} = mn\mathbf{v}, \tag{7.11}$$

which states that the momentum density is equal to the particle mass times the particle current density.

Simple expressions for the density and velocity may be obtained if we write ψ in terms of its amplitude f and phase ϕ,

$$\psi = fe^{i\phi}, \tag{7.12}$$

from which it follows that

$$n = f^2, \tag{7.13}$$

and the velocity $\mathbf{v}$ is

$$\mathbf{v} = \frac{\hbar}{m}\boldsymbol{\nabla}\phi. \tag{7.14}$$

From Eq. (7.14) we conclude that the motion of the condensate corresponds to potential flow, since the velocity is the gradient of a scalar quantity, which is referred to as the velocity potential. For a condensate, Eq. (7.14) shows that the velocity potential is $\hbar\phi/m$. Provided that ϕ is not singular, we can immediately conclude that the motion of the condensate must be irrotational, that is[1]

$$\boldsymbol{\nabla}\times\mathbf{v} = \frac{\hbar}{m}\boldsymbol{\nabla}\times\boldsymbol{\nabla}\phi = 0. \tag{7.15}$$

The possible motions of a condensate are thus much more restricted than those of a classical fluid.

The equations of motion for f and ϕ may be found by inserting (7.12) into (7.1) and separating real and imaginary parts. Since

$$i\frac{\partial \psi}{\partial t} = i\frac{\partial f}{\partial t}e^{i\phi} - \frac{\partial \phi}{\partial t}fe^{i\phi} \tag{7.16}$$

and

$$-\nabla^2\psi = [-\nabla^2 f + (\boldsymbol{\nabla}\phi)^2 f - i\nabla^2\phi f - 2i\boldsymbol{\nabla}\phi\cdot\boldsymbol{\nabla}f]e^{i\phi}, \tag{7.17}$$

we obtain the two equations

$$\frac{\partial(f^2)}{\partial t} = -\frac{\hbar}{m}\boldsymbol{\nabla}\cdot(f^2\boldsymbol{\nabla}\phi) \tag{7.18}$$

and

$$-\hbar\frac{\partial \phi}{\partial t} = -\frac{\hbar^2}{2mf}\nabla^2 f + \frac{1}{2}mv^2 + V(\mathbf{r}) + U_0 f^2. \tag{7.19}$$

Equation (7.18) is the continuity equation (7.8) expressed in the new

---

[1] Note that this result applies only if ϕ is not singular. This condition is satisfied in the examples we consider in this chapter, but it fails at, e.g., the core of a vortex line. The properties of vortices will be treated in Chapter 9.

variables. To find the equation of motion for the velocity, given by Eq. (7.14), we take the gradient of Eq. (7.19), and the resulting equation is

$$m\frac{\partial \mathbf{v}}{\partial t} = -\boldsymbol{\nabla}(\tilde{\mu} + \frac{1}{2}mv^2), \qquad (7.20)$$

where

$$\tilde{\mu} = V + nU_0 - \frac{\hbar^2}{2m\sqrt{n}}\nabla^2\sqrt{n}. \qquad (7.21)$$

Equation (7.19) may be expressed in terms of the functional derivative[2] $\delta E/\delta n$,

$$\frac{\partial \phi(\mathbf{r},t)}{\partial t} = -\frac{1}{\hbar}\frac{\delta E}{\delta n(\mathbf{r})}. \qquad (7.22)$$

The quantity $\delta E/\delta n(\mathbf{r})$ is the energy required to add a particle at point $\mathbf{r}$, and therefore this result is the generalization of the Josephson relation (7.3) to systems not in their ground states. Under stationary conditions $\tilde{\mu} + \frac{1}{2}mv^2$ is a constant, and if in addition the velocity is zero, that is ϕ is independent of position, $\tilde{\mu}$ is a constant, which is precisely the time-independent Gross–Pitaevskii equation (6.11).

The quantity nU_0 in Eq. (7.21) is the expression for the chemical potential of a uniform Bose gas, omitting contributions from the external potential. At zero temperature, changes in the chemical potential for a bulk system are related to changes in the pressure p by the Gibbs–Duhem relation $dp = nd\mu$, a result easily confirmed for the uniform dilute Bose gas, since $\mu = nU_0$ and $p = -\partial E/\partial V = n^2 U_0/2$ (see Eqs. (6.12) and (6.6), respectively). Equation (7.20) may therefore be rewritten in the form

$$\frac{\partial \mathbf{v}}{\partial t} = -\frac{1}{mn}\boldsymbol{\nabla} p - \boldsymbol{\nabla}\left(\frac{v^2}{2}\right) + \frac{1}{m}\boldsymbol{\nabla}\left(\frac{\hbar^2}{2m\sqrt{n}}\nabla^2\sqrt{n}\right) - \frac{1}{m}\boldsymbol{\nabla} V. \qquad (7.23)$$

Equations (7.8) and (7.23) are very similar to the hydrodynamic equations for a perfect fluid. If we denote the velocity of the fluid by $\mathbf{v}$, the continuity equation (7.8) has precisely the same form as for a perfect fluid, while the analogue of Eq. (7.23) is the Euler equation

$$\frac{\partial \mathbf{v}}{\partial t} + (\mathbf{v}\cdot\boldsymbol{\nabla})\mathbf{v} + \frac{1}{mn}\boldsymbol{\nabla} p = -\frac{1}{m}\boldsymbol{\nabla} V, \qquad (7.24)$$

---

[2] The functional derivative $\delta E/\delta n(\mathbf{r})$ of the energy is defined according to the equation $\delta E = \int d\mathbf{r}[\delta E/\delta n(\mathbf{r})]\delta n(\mathbf{r})$, and it is a function of $\mathbf{r}$ with the dimension of energy. It is given in terms of the energy density (7.6), which is a function of n and $\boldsymbol{\nabla} n$, by

$$\frac{\delta E}{\delta n(\mathbf{r})} = \frac{\delta \mathcal{E}}{\delta n} = \frac{\partial \mathcal{E}}{\partial n} - \sum_i \frac{\partial}{\partial x_i}\frac{\partial \mathcal{E}}{\partial(\partial n/\partial x_i)},$$

where the sum is over the three spatial coordinates.

or

$$\frac{\partial \mathbf{v}}{\partial t} - \mathbf{v}\times(\boldsymbol{\nabla}\times\mathbf{v}) = -\frac{1}{mn}\boldsymbol{\nabla}p - \boldsymbol{\nabla}\left(\frac{v^2}{2}\right) - \frac{1}{m}\boldsymbol{\nabla}V. \quad (7.25)$$

Here the pressure p is that of the fluid, which generally has a form different from that of the condensate.

There are two differences between equations (7.23) and (7.25). The first is that the Euler equation contains the term $\mathbf{v}\times(\boldsymbol{\nabla}\times\mathbf{v})$. However, since the velocity field of the superfluid corresponds to potential flow, $\boldsymbol{\nabla}\times\mathbf{v}=0$, the term $\mathbf{v}\times(\boldsymbol{\nabla}\times\mathbf{v})$ for such a flow would not contribute in the Euler equation. We shall comment further on this term in the context of vortex motion at the end of Sec. 9.4. The only difference between the two equations for potential flow is therefore the third term on the right hand side of Eq. (7.23), which is referred to as the *quantum pressure term*. This describes forces due to spatial variations in the magnitude of the wave function for the condensed state. Like the term $\boldsymbol{\nabla}v^2/2$, its origin is the kinetic energy term $\hbar^2|\boldsymbol{\nabla}\psi|^2/2m = mnv^2/2 + \hbar^2(\boldsymbol{\nabla}f)^2/2m$ in the energy density, but the two contributions correspond to different physical effects: the first is the kinetic energy of motion of particles, while the latter corresponds to 'zero-point motion', which does not give rise to particle currents. If the spatial scale of variations of the condensate wave function is l, the pressure term in Eq. (7.23) is of order nU_0/ml, while the quantum pressure term is of order $\hbar^2/m^2 l^3$. Thus the quantum pressure term dominates the usual pressure term if spatial variations of the density occur on length scales l less than or of order the coherence length $\xi \sim \hbar/(mnU_0)^{1/2}$ (see Eq. (6.62)), and it becomes less important on larger length scales.

As we have seen, motions of the condensate may be specified in terms of a local density and a local velocity. The reason for this is that the only degrees of freedom are those of the condensate wave function, which has a magnitude and a phase. Ordinary liquids and gases have many more degrees of freedom and, as a consequence, it is in general necessary to employ a microscopic description, e.g., in terms of the distribution function for the particles. However, a hydrodynamic description is possible for ordinary gases and liquids if collisions between particles are sufficiently frequent that thermodynamic equilibrium is established locally. The state of the fluid may then be specified completely in terms of the local particle density (or equivalently the mass density), the local velocity, and the local temperature. At zero temperature, the temperature is not a relevant variable, and the motion may be described in terms of the local density and the local fluid velocity, just as for a condensate. The reason that the equations of motion for a condensate and for a perfect fluid are so similar is that they are expressions of the conservation laws for particle number and for total momentum. However, the physical reasons for a description in terms of a local density and a local velocity being possible are quite different for the two situations.

7.2 Elementary excitations

The properties of elementary excitations may be investigated by considering small deviations of the state of the gas from equilibrium and finding periodic solutions of the time-dependent Gross–Pitaevskii equation. An equivalent approach is to use the hydrodynamic formulation given above, and we begin by describing this. In Chapter 8 we shall consider the problem on the basis of microscopic theory. We write the density as $n = n_\mathrm{eq} + \delta n$, where n_eq is the equilibrium density and δn the departure of the density from its equilibrium value. On linearizing Eqs. (7.8), (7.20), and (7.21) by treating the velocity $\mathbf{v}$ and δn as small quantities, one finds

$$\frac{\partial \delta n}{\partial t} = -\boldsymbol{\nabla} \cdot (n_\mathrm{eq} \mathbf{v}) \qquad (7.26)$$

and

$$m \frac{\partial \mathbf{v}}{\partial t} = -\boldsymbol{\nabla} \delta \tilde{\mu}, \qquad (7.27)$$

where $\delta \tilde{\mu}$ is obtained by linearizing (7.21). Taking the time derivative of (7.26) and eliminating the velocity by means of (7.27) results in the equation of motion

$$m \frac{\partial^2 \delta n}{\partial t^2} = \boldsymbol{\nabla} \cdot (n_\mathrm{eq} \boldsymbol{\nabla} \delta \tilde{\mu}). \qquad (7.28)$$

This equation describes the excitations of a Bose gas in an arbitrary potential. To keep the notation simple, we shall henceforth in this chapter denote the equilibrium density by n. Note that n is the density of the condensate, since we neglect the zero-temperature depletion of the condensate. In Chapter 10, which treats the dynamics at finite temperature, we shall denote the condensate density by n_0, in order to distinguish it from the total density, which includes a contribution from thermal excitations.

A uniform gas

As a first example we investigate the spectrum for a homogeneous gas, where the external potential V is constant. In the undisturbed state the density n is the same everywhere and it may therefore be taken outside the spatial derivatives. We look for travelling-wave solutions, proportional to $\exp(i\mathbf{q}\cdot\mathbf{r} - i\omega t)$, where $\mathbf{q}$ is the wave vector and ω the frequency. From Eq. (7.21) the change in $\tilde{\mu}$ is seen to be equal to

$$\delta\tilde{\mu} = \left(U_0 + \frac{\hbar^2 q^2}{4mn}\right)\delta n \qquad (7.29)$$

and the equation of motion becomes

$$m\omega^2 \delta n = \left(nU_0 q^2 + \frac{\hbar^2 q^4}{4m}\right)\delta n. \qquad (7.30)$$

To make contact with the microscopic calculations to be described later, it is convenient to work with the energy of an excitation, ϵ_q, rather than the frequency. Non-vanishing solutions of (7.30) are possible only if the frequency is given by $\omega = \pm \epsilon_q/\hbar$, where

$$\epsilon_q = \sqrt{2nU_0 \epsilon_q^0 + (\epsilon_q^0)^2}. \qquad (7.31)$$

Here

$$\epsilon_q^0 = \frac{\hbar^2 q^2}{2m} \qquad (7.32)$$

is the free-particle energy. This spectrum was first derived by Bogoliubov from microscopic theory [1]. In the following discussion we shall adopt the convention that the branch of the square root to be used is the positive one.

The excitation spectrum (7.31) is plotted in Fig. 7.1. For small q, ϵ_q is a linear function of q,

$$\epsilon_q \simeq s\hbar q, \qquad (7.33)$$

and the spectrum is sound-like. The velocity s is seen to be

$$s = \sqrt{nU_0/m}. \qquad (7.34)$$

This result agrees with the expression for the sound velocity calculated from the hydrodynamic result $s^2 = dp/d\rho = (n/m)d\mu/dn$, where $\rho = nm$ is the mass density. The repulsive interaction has thus turned the energy spectrum at long wavelengths, which is quadratic in q for free particles, into a linear one, in agreement with what is observed experimentally in liquid ^{4}He. As we shall see in Chapter 10, the linear spectrum at long wavelengths provides the key to superfluid behaviour, and it was one of the triumphs of Bogoliubov's pioneering calculation. In the hydrodynamic description the result is almost 'obvious', since sound waves are well-established excitations of hydrodynamic systems. What is perhaps surprising is that at short wavelengths the leading contributions to the spectrum are

$$\epsilon_q \simeq \epsilon_q^0 + nU_0, \qquad (7.35)$$

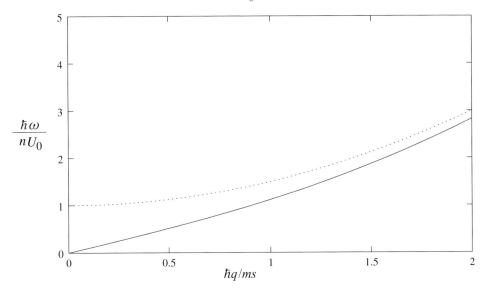

Fig. 7.1. Excitation spectrum of a homogeneous Bose gas (full line) plotted as a function of the wave number expressed as the dimensionless variable $\hbar q/ms$, where the sound velocity s is given by Eq. (7.34). The expansion (7.35) for high wave number is shown as a dotted line.

which is also shown in Fig. 7.1. This is the free-particle spectrum plus a mean-field contribution. The transition between the linear spectrum and the quadratic one occurs when the kinetic energy, $\hbar^2 q^2/2m$, becomes large compared with the potential energy of a particle $\sim nU_0$, or in other words the 'quantum pressure' term dominates the usual pressure term. This occurs at a wave number $\sim (2mnU_0)^{1/2}/\hbar$, which is the inverse of the coherence length, ξ, Eq. (6.62). The coherence length is related to the sound velocity, Eq. (7.34) by $\xi = \hbar/\sqrt{2}ms$. On length scales longer than ξ, atoms move collectively, while on shorter length scales, they behave as free particles. The spectrum of elementary excitations in superfluid liquid ^{4}He differs from that for a dilute gas because of the strong short-range correlations. The first satisfactory account of the roton part of the spectrum was given by Feynman [2].

As a generalization of the above approach one may calculate the response of the condensate to a space- and time-dependent external potential $V(\mathbf{r},t) = V_\mathbf{q}\exp(i\mathbf{q}\cdot\mathbf{r}-i\omega t)$. There is then an additional term $V_\mathbf{q}$ in the equation for $\delta\tilde{\mu}$, and one finds

$$m\left(\omega^2 - \frac{\epsilon_q^2}{\hbar^2}\right)\delta n = nq^2 V_\mathbf{q}, \tag{7.36}$$

or
$$\delta n = \chi(q,\omega) V_{\mathbf{q}}, \qquad (7.37)$$

where
$$\chi(q,\omega) = \frac{nq^2}{m(\omega^2 - \epsilon_q^2/\hbar^2)} \qquad (7.38)$$

is the density–density response function for the condensate. Thus the response diverges if the frequency of the external potential is equal to the frequency of an elementary excitation of the condensate. In the final chapter we shall use the expression for the response function to calculate how the interaction between two fermions is affected by the presence of a condensate of bosons.

The Bogoliubov equations

An alternative route to calculating the excitation spectrum is to start from the Gross–Pitaevskii equation directly, without introducing the hydrodynamic variables. This approach complements the hydrodynamic one since it emphasizes single-particle behaviour and shows how the collective effects at long wavelengths come about. Let us denote the change in ψ by $\delta\psi$. Linearizing the Gross–Pitaevskii equation (7.1), one finds

$$-\frac{\hbar^2}{2m}\nabla^2 \delta\psi(\mathbf{r},t) + V(\mathbf{r})\delta\psi(\mathbf{r},t) + U_0[2|\psi(\mathbf{r},t)|^2 \delta\psi(\mathbf{r},t) + \psi(\mathbf{r},t)^2 \delta\psi^*(\mathbf{r},t)]$$
$$= i\hbar \frac{\partial \delta\psi(\mathbf{r},t)}{\partial t} \qquad (7.39)$$

and

$$-\frac{\hbar^2}{2m}\nabla^2 \delta\psi^*(\mathbf{r},t) + V(\mathbf{r})\delta\psi^*(\mathbf{r},t) + U_0[2|\psi(\mathbf{r},t)|^2 \delta\psi^*(\mathbf{r},t) + \psi^*(\mathbf{r},t)^2 \delta\psi(\mathbf{r},t)]$$
$$= -i\hbar \frac{\partial \delta\psi^*(\mathbf{r},t)}{\partial t}. \qquad (7.40)$$

Here $\psi(\mathbf{r},t)$ is understood to be the condensate wave function in the unperturbed state, which we may write as $\psi = \sqrt{n(\mathbf{r})}e^{-i\mu t/\hbar}$, where $n(\mathbf{r})$ is the equilibrium density of particles and μ is the chemical potential of the unperturbed system. To avoid carrying an arbitrary phase factor along in our calculations we have taken the phase of the condensate wave function at $t=0$ to be zero. We wish to find solutions of these equations which are periodic in time, apart from the overall phase factor $e^{-i\mu t/\hbar}$ present for the unperturbed state. We therefore search for solutions of the form

$$\delta\psi(\mathbf{r},t) = e^{-i\mu t/\hbar}\left[u(\mathbf{r})e^{-i\omega t} - v^*(\mathbf{r})e^{i\omega t}\right], \qquad (7.41)$$

where $u(\mathbf{r})$ and $v(\mathbf{r})$ are functions to be determined. The overall phase factor $e^{-i\mu t/\hbar}$ is necessary to cancel the effects of the phases of $\psi(\mathbf{r},t)^2$ and $\psi^*(\mathbf{r},t)^2$ in Eqs. (7.39) and (7.40), and thereby ensure that the equations can be satisfied for all time. The choice of the sign in front of v is a matter of convention, and we take it to be negative so that u and v will have the same sign. Since the equations couple $\delta\psi$ and $\delta\psi^*$, they cannot be satisfied unless both positive and negative frequency components are allowed for. By inserting the ansatz (7.41) into the two equations (7.39) and (7.40) we obtain the following pair of coupled equations for $u(\mathbf{r})$ and $v(\mathbf{r})$:

$$\left[-\frac{\hbar^2}{2m}\nabla^2 + V(\mathbf{r}) + 2n(\mathbf{r})U_0 - \mu - \hbar\omega\right]u(\mathbf{r}) - n(\mathbf{r})U_0 v(\mathbf{r}) = 0 \quad (7.42)$$

and

$$\left[-\frac{\hbar^2}{2m}\nabla^2 + V(\mathbf{r}) + 2n(\mathbf{r})U_0 - \mu + \hbar\omega\right]v(\mathbf{r}) - n(\mathbf{r})U_0 u(\mathbf{r}) = 0, \quad (7.43)$$

which are referred to as the Bogoliubov equations.

We now apply this formalism to the uniform Bose gas, $V(\mathbf{r}) = 0$. Because of the translational invariance the solutions may be chosen to be of the form

$$u(\mathbf{r}) = u_q \frac{e^{i\mathbf{q}\cdot\mathbf{r}}}{V^{1/2}} \quad \text{and} \quad v(\mathbf{r}) = v_q \frac{e^{i\mathbf{q}\cdot\mathbf{r}}}{V^{1/2}}, \quad (7.44)$$

where we have introduced the conventional normalization factor $1/V^{1/2}$ explicitly, V being the volume of the system.

The chemical potential for the uniform system is given by nU_0 (Eq. (6.12)), and thus the Bogoliubov equations are

$$\left(\frac{\hbar^2 q^2}{2m} + nU_0 - \hbar\omega\right)u_q - nU_0 v_q = 0 \quad (7.45)$$

and

$$\left(\frac{\hbar^2 q^2}{2m} + nU_0 + \hbar\omega\right)v_q - nU_0 u_q = 0. \quad (7.46)$$

The two equations are consistent only if the determinant of the coefficients vanishes. With the definition (7.32) this leads to the condition

$$(\epsilon_q^0 + nU_0 + \hbar\omega)(\epsilon_q^0 + nU_0 - \hbar\omega) - n^2 U_0^2 = 0, \quad (7.47)$$

or

$$(\hbar\omega)^2 = (\epsilon_q^0 + nU_0)^2 - (nU_0)^2 = \epsilon_q^0(\epsilon_q^0 + 2nU_0), \quad (7.48)$$

which agrees with the spectrum (7.31) obtained earlier from the hydrodynamic approach.

The nature of the excitations may be elucidated by investigating the behaviour of the coefficients u_q and v_q. Here we shall consider the case of repulsive interactions. For the positive energy solutions, one has

$$v_q = \frac{nU_0}{\epsilon_q + \xi_q} u_q, \tag{7.49}$$

where

$$\xi_q = \epsilon_q^0 + nU_0 \tag{7.50}$$

is the energy of an excitation if one neglects coupling between u_q and v_q. The normalization of u_q and v_q is arbitrary, but as we shall see from the quantum-mechanical treatment in Sec. 8.1, a convenient one is

$$|u_q|^2 - |v_q|^2 = 1, \tag{7.51}$$

since this ensures that the new operators introduced there satisfy Bose commutation relations. The Bogoliubov equations are unaltered if u_q and v_q are multiplied by an arbitrary phase factor. Therefore, without loss of generality we may take u_q and v_q to be real. With this choice one finds that

$$u_q^2 = \frac{1}{2}\left(\frac{\xi_q}{\epsilon_q} + 1\right) \tag{7.52}$$

and

$$v_q^2 = \frac{1}{2}\left(\frac{\xi_q}{\epsilon_q} - 1\right). \tag{7.53}$$

In terms of ξ_q, the excitation energy is given by

$$\epsilon_q = \sqrt{\xi_q^2 - (nU_0)^2}. \tag{7.54}$$

The coefficients u_q and v_q are exhibited as functions of the dimensionless variable $\hbar q/ms$ in Fig. 7.2.

For the positive energy solution, v_q tends to zero as $1/q^2$ for large q, and in this limit $\delta\psi = e^{i(\mathbf{q}\cdot\mathbf{r}-\omega_q t)}/V^{1/2}$, with $\omega_q = \epsilon_q/\hbar$. This corresponds to addition of a single particle with momentum $\hbar\mathbf{q}$, and the removal of a particle in the zero-momentum state, as will be made explicit when the quantum-mechanical theory is presented in Chapter 8. At smaller momenta, excitations are linear superpositions of the state in which a particle with momentum $\hbar\mathbf{q}$ is added (and a particle in the condensate is removed) and the state in which a particle with momentum $-\hbar\mathbf{q}$ is removed (and a particle added to the condensate). At long wavelengths u_q and v_q diverge as $1/q^{1/2}$, and the two components of the wave function are essentially equal in magnitude.

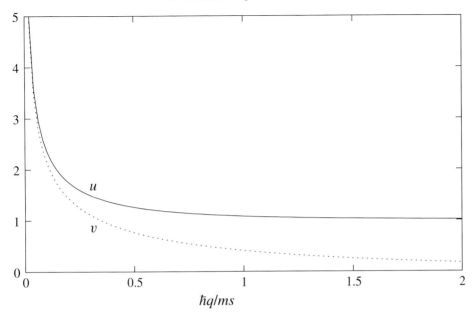

Fig. 7.2. The coefficients u_q and v_q given by Eqs. (7.52)–(7.53) as functions of the wave number, expressed as the dimensionless variable $\hbar q/ms$.

The algebraic expressions for the excitation spectrum and the factors u_q and v_q are completely analogous to those for a superconductor in the Bardeen–Cooper–Schrieffer (BCS) theory, apart from some sign changes due to the fact that we are here dealing with bosons rather than fermions. In the BCS theory, which we shall describe in Sec. 14.3 in the context of the transition to a superfluid state for dilute Fermi gases, the dispersion relation for an elementary excitation is $\epsilon_q = \sqrt{\xi_q^2 + \Delta^2}$, where ξ_q is the normal-state energy of a particle measured with respect to the chemical potential, as for the boson problem we consider, and Δ is the superconducting energy gap. Thus one sees that for bosons the excitation energy is obtained by replacing Δ^2 in the BCS expression by $-(nU_0)^2$.

Attractive interactions

If the interaction is attractive, the sound speed is imaginary, which indicates that long-wavelength modes grow or decay exponentially in time, rather than oscillate. This signals an instability of the system due to the attractive interaction tending to make atoms clump together. However at shorter wavelengths modes are stable, since the free-particle kinetic energy term dominates in the dispersion relation. The lowest wave number q_c for which the mode is stable is given by the condition that its frequency vanish. Thus from Eq. (7.31)

$$\epsilon_{q_c}^0 + 2nU_0 = 0 \qquad (7.55)$$

or

$$q_c^2 = -\frac{4mnU_0}{\hbar^2} = 16\pi n|a|, \qquad (7.56)$$

where we have used Eq. (5.39) to express U_0 in terms of the scattering length a. This shows that the spatial scale of unstable modes is greater than or of order the coherence length, Eq. (6.62), evaluated using the absolute magnitude of the scattering length.

It is instructive to relate these ideas for bulk matter to a cloud in a trap. For simplicity, we consider the trap to be a spherical container with radius R_0. The lowest mode has a wave number of order $1/R_0$, and the density is $n \sim N/R_0^3$. Thus according to (7.56) the lowest mode is stable if the number of particles is less than a critical value N_c given by $1/R_0^2 \sim N_c|a|/R_0^3$ or

$$N_c \sim \frac{R_0}{|a|}, \qquad (7.57)$$

and unstable for larger numbers of particles. The physics of the instability is precisely the same as that considered in Chapter 6 in connection with the energy of a cloud: for sufficiently large numbers of particles, the zero-point energy of atoms is too small to overcome the attraction between them. In the present formulation, the zero-point energy is the kinetic energy of the lowest mode in the well. To make contact with the calculations for a harmonic-oscillator trap in Chapter 6, we note that the estimate (7.57) is consistent with the earlier result (6.29) if the radius R_0 of the container is replaced by the oscillator length.

7.3 Collective modes in traps

Calculating the properties of modes in a homogeneous gas is relatively straightforward because there are only two length scales in the problem, the coherence length and the wavelength of the excitation. For a gas in a trap there is an additional length, the spatial extent of the cloud, and moreover the coherence length varies in space. However, we have seen in Chapter 6 that static properties of clouds may be calculated rather precisely if the number of atoms is sufficiently large, $Na/\bar{a} \gg 1$, since under these conditions the kinetic energy associated with the confinement of atoms within the cloud, which gives rise to the quantum pressure, may be neglected. It is therefore of interest to explore the properties of modes when the quantum pressure term in the equation of motion is neglected. For such an approximation to be reliable, a mode must not be concentrated in the boundary layer of thickness $\sim \delta$, and must vary in space only on length scales large compared with the local coherence length. In this approach one can describe collective modes, but not excitations which are free-particle-like.

The basic equation for linear modes was derived earlier in Eq. (7.28). When the quantum pressure term is neglected, the quantity $\tilde{\mu}$ reduces to $nU_0 + V$, and therefore

$$\delta\tilde{\mu} = U_0 \delta n. \tag{7.58}$$

Inserting this result into Eq. (7.28), we find that the density disturbance satisfies the equation

$$m\frac{\partial^2 \delta n}{\partial t^2} = U_0 \boldsymbol{\nabla} \cdot (n \boldsymbol{\nabla} \delta n). \tag{7.59}$$

If we consider oscillations with time dependence $\delta n \propto e^{-i\omega t}$, the differential equation (7.59) simplifies to

$$-\omega^2 \delta n = \frac{U_0}{m}(\boldsymbol{\nabla} n \cdot \boldsymbol{\nabla} \delta n + n \nabla^2 \delta n). \tag{7.60}$$

The equilibrium density is given by

$$n = \frac{\mu - V(\mathbf{r})}{U_0}, \tag{7.61}$$

and therefore the equation (7.60) reduces to

$$\omega^2 \delta n = \frac{1}{m}\{\boldsymbol{\nabla} V \cdot \boldsymbol{\nabla} \delta n - [\mu - V(\mathbf{r})]\nabla^2 \delta n\}. \tag{7.62}$$

In the following two subsections we discuss solutions to (7.62) and the associated mode frequencies [3].

7.3.1 Traps with spherical symmetry

First we consider an isotropic harmonic trap ($\lambda = 1$). The potential is

$$V(\mathbf{r}) = \frac{1}{2}m\omega_0^2 r^2, \tag{7.63}$$

and in the Thomas–Fermi approximation the chemical potential and the radius of the cloud are related by the equation $\mu = m\omega_0^2 R^2/2$. It is natural to work in spherical polar coordinates $r, \theta,$ and φ. Equation (7.62) then becomes

$$\omega^2 \delta n = \omega_0^2 r \frac{\partial}{\partial r} \delta n - \frac{\omega_0^2}{2}(R^2 - r^2)\nabla^2 \delta n. \tag{7.64}$$

Because of the spherical symmetry, the general solution for the density deviation is a sum of terms of the form

$$\delta n = D(r) Y_{lm}(\theta, \varphi), \tag{7.65}$$

where Y_{lm} is a spherical harmonic. In a quantum-mechanical description, l is the quantum number for the magnitude of the total angular momentum and m that for its projection on the polar axis.

One simple solution of Eq. (7.64) is

$$\delta n = C r^l Y_{lm}(\theta, \varphi), \tag{7.66}$$

where C is an arbitrary constant. With increasing l these modes become more localized near the surface of the cloud, and they correspond to surface waves. They will be studied in more detail in Sec. 7.4. Since the function (7.66) satisfies the Laplace equation, the last term in Eq. (7.64) vanishes, and one finds $\omega^2 = l\omega_0^2$. The $l = 0$ mode is trivial, since it represents a change in the density which is constant everywhere. The resulting change in the chemical potential is likewise the same at all points in the cloud, and therefore there is no restoring force, and the frequency of the mode is zero. The three $l = 1$ modes correspond to translation of the cloud with no change in the internal structure. Consider the $l = 1, m = 0$ mode. The density variation is proportional to $rY_{10} \propto z$. In equilibrium, the density profile is $n(r) \propto (1 - r^2/R^2)$, and therefore if the centre of the cloud is moved in the z direction a distance ζ, the change in the density is given by $\delta n = -\zeta \partial n/\partial z \propto z$. The physics of the $l = 1$ modes is that, for a harmonic external potential, the centre-of-mass and relative motions are separable for interactions that depend only on the relative coordinates of the particles. The motion of the centre of mass, $\mathbf{r}_{\text{cm}}$, is that of a free particle of mass Nm moving in an external potential $Nm\omega_0^2 r_{\text{cm}}^2/2$, and this has the same frequency as that for the motion of a single particle. These modes are sometimes referred to as Kohn modes. They represent a general feature of the motion, which is unaffected by interactions as well as temperature. Modes with higher values of l have larger numbers of nodes, and higher frequencies.

To investigate more general modes it is convenient to separate out the radial dependence due to the 'centrifugal barrier', the $l(l+1)/r^2$ term in the Laplacian, by defining a new radial function $G(r) = D(r)/r^l$, and to introduce the dimensionless variable

$$\epsilon = \frac{\omega^2}{\omega_0^2}. \tag{7.67}$$

The differential equation for the radial function $G(r)$ is

$$\epsilon G(r) = lG(r) + rG'(r) - \frac{1}{2}(R^2 - r^2)\left[G''(r) + \frac{2(l+1)G'(r)}{r}\right], \tag{7.68}$$

where a prime denotes a derivative. To solve this eigenvalue problem we first introduce the new variable $u = r^2/R^2$. The differential equation satisfied by $G(u)$ is seen to be

$$u(1-u)G''(u) + \left(\frac{2l+3}{2} - \frac{2l+5}{2}u\right)G'(u) + \frac{(\epsilon - l)}{2}G(u) = 0, \tag{7.69}$$

which is in the standard form for the hypergeometric function $F(\alpha, \beta, \gamma, u)$,

$$u(1-u)F''(u) + [\gamma - (\alpha + \beta + 1)u]F'(u) - \alpha\beta F(u) = 0. \tag{7.70}$$

For the function to be well behaved, either α or β must be a negative integer, $-n$. The hypergeometric function is symmetrical under interchange of α and β, and for definiteness we put $\alpha = -n$. Comparing Eqs. (7.69) and (7.70), one sees that $\beta = l + n + 3/2$, $\gamma = l + 3/2$, and that the eigenvalue is given by $\epsilon - l = 2n(l + n + 3/2)$ or

$$\omega^2 = \omega_0^2(l + 3n + 2nl + 2n^2). \tag{7.71}$$

The index n specifies the number of radial nodes.

The normal modes of the cloud are therefore given by

$$\delta n(\mathbf{r}, t) = Cr^l F(-n, l+n+3/2, l+3/2, r^2/R^2)Y_{lm}(\theta, \varphi)e^{-i\omega t}, \tag{7.72}$$

C being an arbitrary constant. Solutions for low values of n and l may be evaluated conveniently by using the standard series expansion in powers of $u = r^2/R^2$ [4]:

$$F(\alpha, \beta, \gamma, u) = 1 + \frac{\alpha\beta}{\gamma}\frac{u}{1!} + \frac{\alpha(\alpha+1)\beta(\beta+1)}{\gamma(\gamma+1)}\frac{u^2}{2!} + \cdots. \tag{7.73}$$

As an alternative, the solutions may be written in terms of the Jacobi polynomials $J_n^{(0,l+1/2)}(2r^2/R^2 - 1)$.

For $n = 0$, the modes have no radial nodes and they are the surface modes given by Eq. (7.66). The velocity field associated with a mode may be obtained from Eqs. (7.27) and (7.58). The mode with $l = 0$ and $n = 1$ is

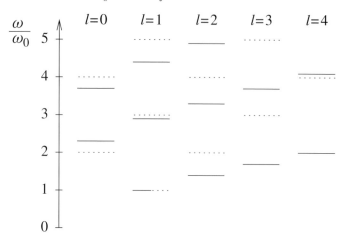

Fig. 7.3. Excitation frequencies of a condensate in an isotropic harmonic trap according to Eq. (7.71) (full lines). The dotted lines indicate the result in the absence of interactions. The degeneracy occurring at $\omega = \omega_0$ for $l = 1$ is due to the fact that these modes correspond to pure centre-of-mass motion.

spherically symmetric and the radial velocity has the same sign everywhere. It is therefore referred to as the *breathing mode*.

The low-lying excitation frequencies in the spectrum (7.71) are shown in Fig. 7.3. For an ideal gas in a harmonic trap, mode frequencies correspond to those of a free particle if the mean free path is large compared with the size of the cloud, and these are shown for comparison. The results exhibit clearly how mode frequencies of the condensate differ from those of an ideal gas.

7.3.2 Anisotropic traps

Next we consider anisotropic traps. Most experimental traps are harmonic and anisotropic, but with an axis of symmetry which we shall take to be the z axis. We write the potential in the form

$$V(x,y,z) = \frac{1}{2}m\omega_0^2\rho^2 + \frac{1}{2}m\omega_3^2 z^2 = \frac{1}{2}m\omega_0^2(\rho^2 + \lambda^2 z^2), \quad (7.74)$$

where $\rho^2 = x^2 + y^2$. The anisotropy parameter $\lambda = \omega_3/\omega_0$ is unity for a spherically symmetric trap, and $\sqrt{8}$ for the TOP trap discussed in Sec. 4.1.2. For traps of the Ioffe–Pritchard type, λ can be adjusted continuously by varying the currents in the coils.

For such a trap the equilibrium density is given in the Thomas–Fermi approximation by

$$n = \frac{\mu}{U_0}\left(1 - \frac{\rho^2}{R^2} - \frac{\lambda^2 z^2}{R^2}\right), \quad (7.75)$$

where the radius R of the cloud in the xy plane is given by $\mu - V(R,0,0) = 0$ or

$$R^2 = \frac{2\mu}{m\omega_0^2}. \quad (7.76)$$

Note that the central density $n(\mathbf{r}=0)$ equals μ/U_0. Equation (7.62) for the mode function is thus

$$\omega^2 \delta n = \omega_0^2(\rho\frac{\partial}{\partial \rho} + \lambda^2 z\frac{\partial}{\partial z})\delta n - \frac{\omega_0^2}{2}(R^2 - \rho^2 - \lambda^2 z^2)\nabla^2 \delta n. \quad (7.77)$$

Because of the axial symmetry there are solutions proportional to $e^{im\varphi}$, where m is an integer. One simple class of solutions is of the form

$$\delta n \propto \rho^l \exp(\pm il\varphi) = (x \pm iy)^l \propto r^l Y_{l,\pm l}(\theta, \varphi), \quad (7.78)$$

which is the same as for surface modes of an isotropic trap. Since $\nabla^2 \delta n = 0$, the frequencies are given by

$$\omega^2 = l\omega_0^2. \quad (7.79)$$

Likewise one may show (Problem 7.2) that there are solutions of the form

$$\delta n \propto z(x \pm iy)^{l-1} \propto r^l Y_{l,\pm(l-1)}(\theta, \varphi), \quad (7.80)$$

whose frequencies are given by

$$\omega^2 = (l-1)\omega_0^2 + \omega_3^2 = (l-1+\lambda^2)\omega_0^2. \quad (7.81)$$

Low-lying modes

In the following we investigate some of the low-lying modes, since these are the ones that have been observed experimentally [5]. As we shall see, the modes have velocity fields of the simple form $\mathbf{v} = (ax, by, cz)$, where a, b and c are constants. One example is $\delta n \propto \rho^2 \exp(\pm i2\varphi) = (x \pm iy)^2$, which corresponds to Eq. (7.78) for $l = 2$. The mode frequency is given by

$$\omega^2 = 2\omega_0^2. \quad (7.82)$$

A second is $\rho z \exp(\pm i\varphi) = z(x \pm iy) \propto r^2 Y_{2,\pm 1}(\theta, \varphi)$ with a frequency given by $\omega^2 = (1+\lambda^2)\omega_0^2$. For traps with spherical symmetry these two types of modes are degenerate, since they both have angular symmetry corresponding to $l=2$, but with different values of the index m equal to ± 2 and ± 1, respectively. There is yet another mode, with $l=2, m=0$, which is degenerate with the others if the trap is spherically symmetric, but which mixes with the lowest $l=0, m=0$ mode, the breathing mode, if the spherical symmetry is broken. To see this explicitly we search for a solution independent of φ of the form

$$\delta n = a + b\rho^2 + cz^2, \tag{7.83}$$

where a, b and c are constants to be determined. Upon insertion of (7.83) into (7.77) we obtain three linear algebraic equations for a, b and c, which may be written as a matrix equation. The condition for the existence of non-trivial solutions is that the determinant of the matrix vanish. This yields $\omega^2 = 0$ (for $\delta n =$ constant) and

$$(\omega^2 - 4\omega_0^2)(\omega^2 - 3\lambda^2\omega_0^2) - 2\lambda^2\omega_0^4 = 0, \tag{7.84}$$

which has roots

$$\omega^2 = \omega_0^2\left(2 + \frac{3}{2}\lambda^2 \pm \frac{1}{2}\sqrt{16 - 16\lambda^2 + 9\lambda^4}\right). \tag{7.85}$$

For $\lambda^2 = 8$ the smaller root yields

$$\omega = \omega_0\left(14 - 2\sqrt{29}\right)^{1/2} \approx 1.797\omega_0. \tag{7.86}$$

Both this $m=0$ mode and the $l=2, m=2$ mode with frequency $\sqrt{2}\omega_0$ given by (7.82) have been observed experimentally [5]. The two mode frequencies (7.85) are shown in Fig. 7.4 as functions of the anisotropy parameter λ.

The density variations in all modes considered above exhibit a quadratic dependence on the cartesian coordinates. The associated velocity fields are linear in x, y, and z according to the acceleration equation (7.27). The motion of the cloud therefore corresponds to homologous stretching of the cloud by a scale factor that depends on direction.

The scissors mode

In the Thomas–Fermi approximation there are modes having a simple analytical form also for a general harmonic trap

$$V(\mathbf{r}) = \frac{1}{2}m(\omega_1^2 x^2 + \omega_2^2 y^2 + \omega_3^2 z^2), \tag{7.87}$$

where the frequencies ω_1, ω_2 and ω_3 are all different. For traps with rotational symmetry about the z axis we have seen that there exist modes with a density variation proportional to xz or yz, the associated frequency being given by $\omega^2 = \omega_0^2(1+\lambda^2)$. These modes are linear combinations of $r^2 Y_{2,1}$ and

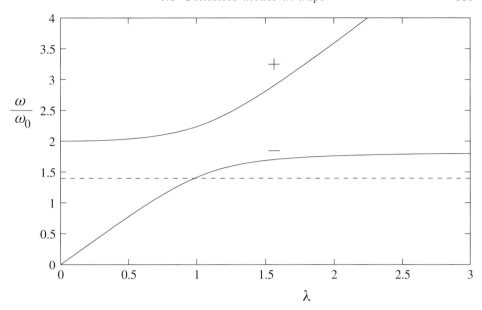

Fig. 7.4. The mode frequencies (7.85) as functions of λ.

$r^2 Y_{2,-1}$, which are degenerate eigenstates. In addition there is a mode with frequency given by $\omega^2 = 2\omega_0^2$ and a density variation proportional to xy, which is a linear combination of the modes with $l = 2, m = \pm 2$. Modes with density variations proportional to xy, yz or zx are purely two-dimensional, and also exist for a general harmonic trap with a potential given by (7.87). Let us consider a density change given by

$$\delta n = Cxy, \qquad (7.88)$$

where C is a coefficient which we assume varies as $e^{-i\omega t}$. In order to derive the velocity field associated with this mode we start with the hydrodynamic equations in their original form. In the Thomas–Fermi approximation, $\delta\tilde{\mu} = U_0 \delta n$ (see Eq. (7.58)) and therefore from Eq. (7.27) the velocity is given by

$$-im\omega \mathbf{v} = -U_0 \boldsymbol{\nabla}(Cxy) = -U_0 C(y, x, 0). \qquad (7.89)$$

Since $\boldsymbol{\nabla}\cdot\mathbf{v} = 0$, the continuity equation (7.26) reduces to

$$-i\omega \delta n = -(\boldsymbol{\nabla} n) \cdot \mathbf{v}. \qquad (7.90)$$

We now use the condition for hydrostatic equilibrium of the unperturbed cloud, $\boldsymbol{\nabla} n = -\boldsymbol{\nabla} V/U_0$ and insert $\mathbf{v}$ from (7.89) on the right hand side of (7.90) which becomes $C(\omega_1^2 + \omega_2^2)xy/i\omega$. This shows that the equations of motion are satisfied provided

$$\omega^2 = \omega_1^2 + \omega_2^2. \tag{7.91}$$

Corresponding results are obtained for density variations proportional to yz and zx, with frequencies given by cyclic permutation of the trap frequencies in (7.91).

The mode (7.88) is sometimes called a *scissors mode*. The reason for this may be seen by considering the density change when the equilibrium cloud in the trap is rotated. The equilibrium density profile is proportional to $1 - x^2/R_1^2 - y^2/R_2^2 - z^2/R_3^2$, where the lengths R_i are given by (6.33). A rotation of the cloud by an angle χ about the z axis corresponds to the transformation $x \to x\cos\chi - y\sin\chi$, $y \to x\sin\chi + y\cos\chi$. The corresponding change in the density for small χ is proportional to xy. The density change in the mode is thus the same as would be produced by a rigid rotation of the cloud. However, the velocity field varies as $(y, x, 0)$, and it is therefore very different from that for rigid rotation about the z axis, which is proportional to $(-y, x, 0)$. The velocity of the condensate must be irrotational, and therefore the latter form of the velocity is forbidden. The scissors mode has recently been observed experimentally in a Bose–Einstein condensate [6]. Its name is taken from nuclear physics: in deformed nuclei, the density distributions of neutrons and protons can execute out-of-phase oscillations of this type which resemble the opening and closing of a pair of scissors. We note that the scissors modes are purely two-dimensional, and therefore their form and their frequencies do not depend on how the trapping potential varies in the third direction, provided it has the general form $m(\omega_1^2 x^2 + \omega_2^2 y^2)/2 + V(z)$.

7.3.3 Collective coordinates and the variational method

In general it is not possible to solve the equations of motion for a trapped Bose gas analytically and one must resort to other approaches, either numerical methods, or approximate analytical ones. In this section we consider low-lying excitations and describe how to calculate properties of modes within two related approximate schemes. We shall illustrate these methods by applying them to the breathing mode of a cloud in a spherically-symmetric trap, with a potential

$$V(r) = \frac{1}{2} m \omega_0^2 r^2. \tag{7.92}$$

In the notation of Sec. 7.3.1, this mode has $l = 0$ and radial index $n = 1$.

Collective coordinates

When interactions play little role, most modes of a many-body system resemble those of single particles. However, in the modes examined above in the Thomas–Fermi approximation the interaction between particles plays an important role, and the motion is highly collective. The idea behind the method of collective coordinates is to identify variables related to many particles that may be used to describe the collective behaviour. A simple example is the centre-of-mass coordinate, which may be used to describe the modes of frequency ω_i in a harmonic trap. In Chapter 6 we showed how the width parameter R of a cloud may be used as a variational parameter in determining an approximate expression for the energy. We now extend this approach to calculate the properties of the breathing mode.

Let us assume that during the motion of the cloud, the density profile maintains its shape, but that its spatial extent depends on time. Rather than adopting the Gaussian trial wave function used previously, we shall take a more general one

$$\psi(r) = \frac{AN^{1/2}}{R^{3/2}} f(r/R) e^{i\phi(r)}, \qquad (7.93)$$

where f is an arbitrary real function, and A, a number, is a normalization constant. The total energy of the cloud obtained by evaluating Eq. (6.9) may be written as

$$E = E_{\text{flow}} + U(R). \qquad (7.94)$$

Here the first term is the kinetic energy associated with particle currents, and is given by

$$E_{\text{flow}} = \frac{\hbar^2}{2m} \int d\mathbf{r} |\psi(\mathbf{r})|^2 (\boldsymbol{\nabla}\phi)^2. \qquad (7.95)$$

The second term is an effective potential energy, and it is equal to the energy of the cloud when the phase does not vary in space. It is made up of a number of terms:

$$U(R) = E_{\text{zp}} + E_{\text{osc}} + E_{\text{int}}, \qquad (7.96)$$

where

$$E_{\text{zp}} = \frac{\hbar^2}{2m} \int d\mathbf{r} \left(\frac{d|\psi|}{dr}\right)^2 = c_{\text{zp}} R^{-2} \qquad (7.97)$$

is the contribution from the zero-point kinetic energy,

$$E_{\text{osc}} = \frac{1}{2} m \omega_0^2 \int d\mathbf{r} r^2 |\psi|^2 = c_{\text{osc}} R^2 \qquad (7.98)$$

is that from the harmonic-oscillator potential, and

$$E_{\text{int}} = \frac{1}{2}U_0 \int d\mathbf{r}|\psi|^4 = c_{\text{int}} R^{-3} \qquad (7.99)$$

is that due to interactions.[3] The coefficients c, which are constants that depend on the choice of f, are defined by these equations. The equilibrium radius of the cloud R_0 is determined by minimizing the total energy. The kinetic energy contribution (7.95) is positive definite, and is zero if ϕ is constant, and therefore the equilibrium condition is that the effective potential be a minimum,

$$\left.\frac{dU}{dR}\right|_{R=R_0} = 0, \qquad (7.100)$$

or, since the contributions to the energy behave as powers of R as shown in Eqs. (7.97)–(7.99),

$$\left.R\frac{dU}{dR}\right|_{R=R_0} = -2E_{\text{zp}} + 2E_{\text{osc}} - 3E_{\text{int}} = 0. \qquad (7.101)$$

When R differs from its equilibrium value there is a force tending to change R. To derive an equation describing the dynamics of the cloud, we need to find the kinetic energy associated with a time dependence of R. Changing R from its initial value to a new value $\tilde{R}$ amounts to a uniform dilation of the cloud, since the new density distribution may be obtained from the old one by changing the radial coordinate of each atom by a factor $\tilde{R}/R$. The velocity of a particle is therefore equal to

$$v(r) = r\frac{\dot{R}}{R}, \qquad (7.102)$$

where $\dot{R}$ denotes the time derivative of R. The kinetic energy of the bulk motion of the gas is thus given by

$$E_{\text{flow}} = \frac{m}{2}\frac{\dot{R}^2}{R^2} \int d\mathbf{r} n(r) r^2$$
$$= \frac{1}{2} m_{\text{eff}} \dot{R}^2, \qquad (7.103)$$

where

$$m_{\text{eff}} = Nm\frac{\overline{r^2}}{R^2}. \qquad (7.104)$$

Here $\overline{r^2} = \int d\mathbf{r} n(r) r^2 / \int d\mathbf{r} n(r)$ is the mean-square radius of the cloud. Note

[3] Note that the total kinetic energy equals $E_{\text{zp}} + E_{\text{flow}}$.

7.3 Collective modes in traps

that m_{eff} is independent of R. For a harmonic trap the integral here is identical with that which occurs in the expression (7.98) for the contribution to the energy due to the oscillator potential, and therefore we may write

$$E_{\text{flow}} = \frac{\dot{R}^2}{\omega_0^2 R^2} E_{\text{osc}} \tag{7.105}$$

or

$$m_{\text{eff}} = \frac{2}{\omega_0^2 R^2} E_{\text{osc}}. \tag{7.106}$$

The total energy of the cloud may thus be written as the sum of the energy of the static cloud Eq. (7.96) and the kinetic energy term Eq. (7.103),

$$E = \frac{1}{2} m_{\text{eff}} \dot{R}^2 + U(R), \tag{7.107}$$

which is the same expression as for a particle of mass m_{eff} moving in a one-dimensional potential $U(R)$. From the condition for energy conservation, $dE/dt = 0$, it follows that the equation of motion is

$$m_{\text{eff}} \ddot{R} = -\frac{\partial U(R)}{\partial R}. \tag{7.108}$$

This equation is not limited to situations close to equilibrium, and in Sec. 7.5 below we shall use it to determine the final velocity of a freely expanding cloud. However, as a first application we investigate the frequency of small oscillations about the equilibrium state. Expanding the effective potential to second order in $R - R_0$, one finds

$$U(R) = U(R_0) + \frac{1}{2} K_{\text{eff}} (R - R_0)^2, \tag{7.109}$$

where

$$K_{\text{eff}} = U''(R_0) \tag{7.110}$$

is the effective force constant. Therefore the frequency of oscillations is given by

$$\omega^2 = \frac{K_{\text{eff}}}{m_{\text{eff}}}. \tag{7.111}$$

This result is in fact independent of the trapping potential, but we shall now specialize our discussion to harmonic traps. From the expressions (7.97)–(7.99) one sees that

$$\begin{aligned} R^2 U'' &= 6 E_{\text{zp}} + 2 E_{\text{osc}} + 12 E_{\text{int}} \\ &= 8 E_{\text{osc}} + 3 E_{\text{int}}, \end{aligned} \tag{7.112}$$

where the latter form follows from the first by using the virial condition Eq. (7.101) to eliminate the zero-point energy. Thus the frequency is given by

$$\omega^2 = \omega_0^2 \left[4 + \frac{3}{2}\frac{E_{\text{int}}(R_0)}{E_{\text{osc}}(R_0)}\right]. \tag{7.113}$$

In Sec. 14.2 we shall apply this result to fermions.

Let us examine a number of limits of this expression. First, when interactions may be neglected one finds $\omega = \pm 2\omega_0$, in agreement with the exact result, corresponding quantum-mechanically to two oscillator quanta. This may be seen from the fact that the Gross–Pitaevskii equation in the absence of interactions reduces to the Schrödinger equation, and the energy eigenvalues are measured with respect to the chemical potential, which is $3\hbar\omega_0/2$. The lowest excited state with spherical symmetry corresponds to two oscillator quanta, since states having a single quantum have odd parity and therefore cannot have spherical symmetry.

Another limit is that of strong interactions, $Na/a_{\text{osc}} \gg 1$. The zero-point energy can be neglected to a first approximation, and the virial condition (7.101) then gives $E_{\text{int}}(R_0) = 2E_{\text{osc}}(R_0)/3$ and therefore

$$\omega^2 = 5\omega_0^2. \tag{7.114}$$

This too agrees with the exact result in this limit, Eq. (7.71) for $n = 1, l = 0$.

It is remarkable that, irrespective of the form of the function f, the mode frequency calculated by the approximate method above agrees with the exact result in the limits of strong interactions and of weak interactions. This circumstance is a special feature of the harmonic-oscillator potential, for which the effective mass is simply related to the potential energy due to the trap.

The method may be applied to anisotropic traps by considering perturbations of the cloud corresponding to transformations of the form $x, y, z \to \alpha x, \beta y, \gamma z$, where the scale factors may be different. This gives three coupled equations for α, β and γ. For an axially-symmetric trap this leads to Eqs. (7.82) and (7.85) for the mode frequencies. Likewise the properties of the scissors modes may be derived by considering displacements of the form $x \to x + ay, y \to y + bx, z \to z$ and cyclic permutations of this.

Variational approach

The calculation of mode frequencies based on the idea of collective coordinates may be put on a more formal footing by starting from the variational principle Eq. (7.4). The basic idea is to take a trial function which depends on a number of time-dependent parameters and to derive equations of motion for these parameters by applying the variational principle [7]. As an example, we again consider the breathing mode of a cloud in a spherical trap. The amplitude of the wave function determines the density distribution, while its phase determines the velocity field. For the amplitude we take a function of the form we considered above in Eq. (7.93). In the calculation for the breathing mode we assumed that the velocity was in the radial direction and proportional to r. Translated into the behaviour of the wave function, this implies that the phase of the wave function varies as r^2, since the radial velocity of the condensate is given by $(\hbar/m)\partial\phi/\partial r$. We therefore write the phase of the wave function as $\beta m r^2/2\hbar$, where β is a second parameter in the wave function. The factor $m/\hbar$ is included to make subsequent equations simpler. The complete trial wave function is thus

$$\psi(r,t) = \frac{AN^{1/2}}{R^{3/2}} f(r/R) e^{i\beta m r^2/2\hbar}. \tag{7.115}$$

We now carry out the integration over r in (7.5) and obtain the Lagrangian as a function of the two independent variables β and R and the time derivative $\dot{\beta}$,

$$L = -[U(R) + \frac{m_{\text{eff}} R^2}{2}(\beta^2 + \dot{\beta})]. \tag{7.116}$$

From the Lagrange equation for β,

$$\frac{d}{dt}\frac{\partial L}{\partial \dot{\beta}} = \frac{\partial L}{\partial \beta} \tag{7.117}$$

we find

$$\beta = \frac{\dot{R}}{R}. \tag{7.118}$$

This is the analogue of the continuity equation for this problem, since it ensures consistency between the velocity field, which is proportional to β and the density profile, which is determined by R. The Lagrange equation for R is

$$\frac{d}{dt}\frac{\partial L}{\partial \dot{R}} = \frac{\partial L}{\partial R}, \tag{7.119}$$

which reduces to $\partial L/\partial R = 0$, since the Lagrangian does not depend on $\dot{R}$. This is

$$m_{\text{eff}} R(\dot{\beta} + \beta^2) = -\frac{\partial U(R)}{\partial R}. \tag{7.120}$$

When Eq. (7.118) for β is inserted into (7.120) we arrive at the equation of motion (7.108) derived earlier. The results of this approach are equivalent to those obtained earlier using more heuristic ideas. However, the variational method has the advantage of enabling one to systematically improve the solution by using trial functions with a greater number of parameters.

Finally, let us compare these results for the breathing mode with those obtained in Sec. 7.3.1 by solving the hydrodynamic equations with the quantum pressure term neglected. The hydrodynamic equations are valid in the limit $E_{\rm zp} \ll E_{\rm int}$, and therefore in the results for the collective coordinate and variational approaches we should take that limit. The lowest $l = 0$ mode is trivial: it corresponds to a uniform change in the density everywhere and has zero frequency because such a density change produces no restoring forces. This corresponds to the index n being zero. The first excited state with $l = 0$ corresponds to $n = 1$, indicating that it has a single radial node in the density perturbation. According to Eq. (7.71) the frequency of the mode is given by $\omega^2 = 5\omega_0^2$, in agreement with the result of the collective coordinate approach, Eq. (7.114). The nature of the mode may be determined either from the general expression in terms of hypergeometric functions or by construction, as we shall now demonstrate.

The s-wave solutions to (7.64) satisfy the equation

$$\omega^2 \delta n = \omega_0^2 r \frac{d}{dr} \delta n - \omega_0^2 \frac{(R^2 - r^2)}{2r} \frac{d^2}{dr^2}(r\delta n). \tag{7.121}$$

Following the method used earlier for anisotropic traps, let us investigate whether there exists a solution of the form

$$\delta n = a + br^2, \tag{7.122}$$

where a and b are constants to be determined. This function is the analogue of Eq. (7.83) for a mode with spherical symmetry ($b = c$). Inserting this expression into (7.121), we find from the terms proportional to r^2 that

$$\omega^2 = 5\omega_0^2. \tag{7.123}$$

Thus the frequency of the mode agrees with the value calculated by other methods. Equating the terms independent of r yields the condition

$$b = -\frac{5}{3}\frac{a}{R^2}. \tag{7.124}$$

The density change δn is thus given by

$$\delta n = a\left(1 - \frac{5r^2}{3R^2}\right). \tag{7.125}$$

This is identical with the density change of the equilibrium cloud produced by a change in R. To show this, we use the fact that in the Thomas–Fermi approximation, when the zero-point kinetic energy is neglected, the equilibrium density is given by

$$n = \frac{C}{R^3}\left(1 - \frac{r^2}{R^2}\right), \tag{7.126}$$

where C is a constant. From (7.126) it follows that a small change δR in the cloud radius R gives rise to a density change

$$\delta n = -\frac{C}{R^4}\left(3 - 5\frac{r^2}{R^2}\right)\delta R, \tag{7.127}$$

which has the same form as Eq. (7.125). The corresponding velocity field may be found from the continuity equation (7.8), which shows that $\mathbf{v} \propto \boldsymbol{\nabla}\delta n$, which is proportional to $\mathbf{r}$. Thus the velocity field is homologous, as was assumed in our discussion of modes in terms of collective coordinates.

7.4 Surface modes

In the previous section we showed that in a spherically-symmetric trap there are modes which are well localized near the surface of the cloud. To shed light on these modes, we approximate the potential in the surface region by a linear function of the coordinates, as we did in our study of surface structure in Sec. 6.3, and write

$$V(\mathbf{r}) = Fx, \tag{7.128}$$

where the coordinate x measures distances in the direction of $\boldsymbol{\nabla} V$. This approximation is good provided the wavelength of the mode is small compared with the linear dimensions of the cloud. Following Ref. [8] we now investigate surface modes for a condensate in the linear ramp potential (7.128). Because of the translational invariance in the y and z directions the solution may be chosen to have the form of a plane wave for these coordinates. We denote the wave number of the mode by q, and take the direction of propagation to be the z axis. Provided the mode is not concentrated in the surface region of thickness δ given by Eq. (6.44), we may use the Thomas–Fermi approximation, in which the equilibrium condensate density n is given by $n(x) = -Fx/U_0$ for $x < 0$, while it vanishes for $x > 0$. Equation (7.62) for the density oscillation in the mode has a solution

$$\delta n = Ce^{qx+iqz}, \tag{7.129}$$

C being an arbitrary constant. This describes a wave propagating on the surface and decaying exponentially in the interior. For the Thomas–Fermi approximation to be applicable the decay length $1/q$ must be much greater than δ. Since (7.129) satisfies Laplace's equation, $\nabla^2 \delta n = 0$, we obtain by inserting (7.129) into (7.62) the dispersion relation

$$\omega^2 = \frac{F}{m} q. \tag{7.130}$$

This has the same form as that for a gravity wave propagating on the surface of an incompressible ideal fluid in the presence of a gravitational field $g = F/m$.

The solution (7.129) is however not the only one which decays exponentially in the interior. To investigate the solutions to (7.62) more generally we insert a function of the form

$$\delta n = f(qx)e^{qx+iqz}, \tag{7.131}$$

and obtain the following second-order differential equation for $f(y)$,

$$y\frac{d^2 f}{dy^2} + (2y+1)\frac{df}{dy} + (1-\epsilon)f = 0, \tag{7.132}$$

where $\epsilon = m\omega^2/Fq$. By introducing the new variable $z = -2y$ one sees that Eq. (7.132) becomes the differential equation for the Laguerre polynomials $L_n(z)$, provided $\epsilon - 1 = 2n$. We thus obtain the general dispersion relation for the surface modes

$$\omega^2 = \frac{F}{m} q(1+2n), \quad n = 0, 1, 2, \ldots . \tag{7.133}$$

The associated density oscillations are given by

$$\delta n(x, z, t) = C L_n(-2qx) e^{qx+iqz-i\omega t}, \tag{7.134}$$

where C is a constant.

Let us now compare the frequencies of these modes with those given in Eq. (7.71) for the modes of a cloud in an isotropic, harmonic trap. For l much greater than n, the dispersion relation becomes $\omega^2 = \omega_0^2 l(1+2n)$. Since the force due to the trap at the surface of the cloud is $F = \omega_0^2 R$ per unit mass and the wave number of the mode at the surface of the cloud is given by $q = l/R$, it follows that $\omega_0^2 l = Fq/m$ and the dispersion relation $\omega^2 = \omega_0^2 l(1+2n)$ is seen to be in agreement with the result (7.133) for the plane surface. For large values of l it is thus a good approximation to replace the harmonic-oscillator potential by the linear ramp. The surface modes are concentrated within a distance of order $(2n+1)R/l$ from the surface, and therefore provided this is smaller than R, it is permissible to approximate the harmonic-oscillator potential by the linear ramp. It should be noted that the frequencies of the $n = 0$ modes for the linear ramp potential agree with the frequencies of the nodeless radial modes (corresponding to $n = 0$) for a harmonic trap at *all* values of l. For modes with radial nodes ($n \neq 0$), the two results agree only for $l \gg n$.

The results above were obtained in the Thomas–Fermi approximation, which is valid only if the depth to which the mode penetrates is much greater than the scale of the surface structure δ, Eq. (6.44). At shorter wavelengths, there are corrections to the dispersion relation (7.130) which may be related to an effective surface tension due to the kinetic energy of matter in the surface region [8]. Finally, we note that the properties of surface excitations have recently been investigated experimentally by exciting them with a moving laser beam [9].

7.5 Free expansion of the condensate

The methods described above are not limited to situations close to equilibrium. One experimentally relevant problem is the evolution of a cloud of condensate when the trap is switched off suddenly. The configuration of the cloud after expansion is used as a probe of the cloud when it is impossible to resolve its initial structure directly. For simplicity, we consider a cloud contained by an isotropic harmonic trap, $V(r) = m\omega_0^2 r^2/2$, which is turned off at time $t = 0$. We employ a trial function of the form (7.115) with $f(r/R) = \exp(-r^2/2R^2)$. As may be seen from Eq. (6.19), the zero-point energy is given by

$$E_{\rm zp} = \frac{3N\hbar^2}{4mR^2} \tag{7.135}$$

and the interaction energy by

$$E_{\rm int} = \frac{N^2 U_0}{2(2\pi)^{3/2} R^3}. \tag{7.136}$$

For the Gaussian trial function the effective mass (Eq. (7.106)) is $m_{\rm eff} = 3Nm/2$. The energy conservation condition therefore yields

$$\frac{3m\dot{R}^2}{4} + \frac{3\hbar^2}{4mR^2} + \frac{1}{2(2\pi)^{3/2}} \frac{NU_0}{R^3} = \frac{3\hbar^2}{4mR(0)^2} + \frac{1}{2(2\pi)^{3/2}} \frac{NU_0}{R(0)^3}, \tag{7.137}$$

where $R(0)$ is the radius at time $t = 0$.

In the absence of interactions ($U_0 = 0$) we may integrate (7.137) with the result

$$R^2(t) = R^2(0) + v_0^2 t^2, \tag{7.138}$$

where v_0, which is equal to the root-mean-square particle velocity, is given by

$$v_0 = \frac{\hbar}{mR(0)}. \tag{7.139}$$

Thus the expansion velocity is the velocity uncertainty predicted by Heisenberg's uncertainty principle for a particle confined within a distance $\sim R(0)$. The initial radius, $R(0)$, is equal to the oscillator length $a_{\mathrm{osc}} = (\hbar/m\omega_0)^{1/2}$, and therefore the result (7.138) may be written as

$$R^2(t) = R^2(0)(1 + \omega_0^2 t^2). \tag{7.140}$$

This result is the exact solution, as one may verify by calculating the evolution of a Gaussian wave packet.

When interactions are present, the development of R as a function of time may be found by numerical integration. However, the asymptotic behaviour for $t \to \infty$ may be obtained using the energy conservation condition (7.137), which yields a final velocity given by

$$v_\infty^2 = \frac{\hbar^2}{m^2 R(0)^2} + \frac{U_0 N}{3(2\pi^3)^{1/2} mR(0)^3}. \tag{7.141}$$

When Na/a_{osc} is large, the initial size of the cloud may be determined by minimizing the sum of the oscillator energy and the interaction energy. According to (6.27) the result for an isotropic trap is

$$R(0) = \left(\frac{2}{\pi}\right)^{1/10} \left(\frac{Na}{a_{\mathrm{osc}}}\right)^{1/5} a_{\mathrm{osc}}. \tag{7.142}$$

At large times, the cloud therefore expands according to the equation

$$\frac{R^2(t)}{R^2(0)} \simeq \frac{U_0 N}{3(2\pi^3)^{1/2} mR(0)^5} t^2 = \frac{2}{3}\omega_0^2 t^2, \tag{7.143}$$

since the final velocity is dominated by the second term in (7.141).

7.6 Solitons

In the dynamical problems addressed so far, we have obtained analytical results for small-amplitude motions, but have had to rely on approximate methods when non-linear effects are important. However the time-dependent Gross–Pitaevskii equation has exact analytical solutions in the non-linear regime. These have the form of solitary waves, or solitons, that is, localized disturbances which propagate without change of form.[4] The subject of solitons has a long history, starting with the observations on shallow water waves made by the British engineer and naval architect John Scott Russell in the decade from 1834 to 1844. Soliton solutions exist for a number of non-linear equations, among them the Korteweg–de Vries equation, which describes the properties of shallow water waves, and the non-linear Schrödinger equation, of which the Gross–Pitaevskii equation (7.1) is a special case.

The physical effects that give rise to the existence of solitons are non-linearity and dispersion. Both of these are present in the Gross–Pitaevskii equation, as one can see by examining the Bogoliubov dispersion relation given by (7.30), $\omega^2 = (nU_0/m)q^2 + \hbar^2 q^4/4m^2$, which exhibits the dependence of the velocity of an excitation on the local density and on the wave number. Solitons preserve their form because the effects of non-linearity compensate for those of dispersion. Before describing detailed calculations, we make some general order-of-magnitude arguments. For definiteness, let us consider a spatially-uniform condensed Bose gas with repulsive interactions. If a localized disturbance of the density has an amplitude Δn and extends over a distance L, it may be seen from the dispersion relation that the velocity of sound within the disturbance is different from the sound velocity in the bulk medium by an amount $\sim s(\Delta n)/n$ due to non-linear effects. One can also see that, since $q \sim 1/L$, dispersion increases the velocity by an amount $\sim s\xi^2/L^2$, where ξ is the coherence length, given by (6.62). For the effects of non-linearity to compensate for those of dispersion, these two contributions must cancel. Therefore the amplitude of the disturbance is related to its length by

$$\frac{\Delta n}{n} \sim -\frac{\xi^2}{L^2}. \qquad (7.144)$$

The velocity u of the disturbance differs from the sound speed by an amount of order the velocity shifts due to dispersion and non-linearity, that is

$$|u - s| \sim s\frac{\xi^2}{L^2}. \qquad (7.145)$$

Note that solitons for this system correspond to density *depressions*, whereas for waves in shallow water, they correspond to elevations in the water level. This difference can be traced to the fact that the dispersion has the opposite sign for waves in shallow water, since $\omega^2 \simeq ghq^2[1 - (qh)^2/3]$, where g is the

---

[4] In the literature, the word 'soliton' is sometimes used to describe solitary waves with special properties, such as preserving their shapes when they collide with each other. We shall follow the usage common in the field of Bose–Einstein condensation of regarding the word 'soliton' as being synonymous with solitary wave.

acceleration due to gravity and h the equilibrium depth of the water. The non-linearity has the same sign in the two cases, since the velocity of surface waves increases with the depth of the water, just as the speed of sound in a condensed Bose gas increases with density.

For repulsive interactions between particles, the simplest example of a soliton is obtained by extending to the whole of space the stationary solution to the Gross–Pitaevskii equation at a wall found in Sec. 6.4 (see Eq. (6.65)),

$$\psi(x) = \psi_0 \tanh\left(\frac{x}{\sqrt{2}\xi}\right) \qquad (7.146)$$

with

$$\xi = \frac{\hbar}{(2mn_0 U_0)^{1/2}} \qquad (7.147)$$

being the coherence length far from the wall. This solution is static, and therefore corresponds to a soliton with velocity zero. It is also referred to as a kink, since the phase of the wave function jumps discontinuously by π as x passes through zero.

The hydrodynamic equations (7.8) and (7.20) possess one-dimensional soliton solutions that depend on the spatial coordinate x and the time t only through the combination $x - ut$. We look for solutions for which the density n approaches a non-zero value n_0 when $x \to \pm\infty$. Since $\partial n/\partial t = -u\partial n/\partial x$, the continuity equation may be rewritten as

$$\frac{\partial}{\partial x}(un - vn) = 0, \qquad (7.148)$$

which upon integration and use of the boundary condition that $v = 0$ at infinity becomes

$$v = u\left(1 - \frac{n_0}{n}\right). \qquad (7.149)$$

Since $\partial v/\partial t = -u\partial v/\partial x$ we may rewrite (7.20) as

$$-mu\frac{\partial v}{\partial x} = -\frac{\partial}{\partial x}\left(nU_0 + \frac{1}{2}mv^2 - \frac{\hbar^2}{2m\sqrt{n}}\frac{\partial^2}{\partial x^2}\sqrt{n}\right). \qquad (7.150)$$

The result of integrating (7.150) is (Problem 7.5)

$$n(x,t) = n_{\min} + (n_0 - n_{\min})\tanh^2[(x - ut)/\sqrt{2}\xi_u], \qquad (7.151)$$

where the width, which depends on velocity, is

$$\xi_u = \frac{\xi}{(1 - (u/s)^2)^{1/2}}. \qquad (7.152)$$

The velocity u is related to the density ratio $n_{\min}/n_0$ by

$$\frac{u^2}{s^2} = \frac{n_{\min}}{n_0}, \quad \text{or} \quad u^2 = \frac{n_{\min} U_0}{m}, \tag{7.153}$$

where $s = (n_0 U_0/m)^{1/2}$ is the velocity of sound in the uniform gas. The velocity of the soliton is therefore equal to the bulk sound velocity evaluated at the density $n_{\min}$. When $u = 0$ the minimum density in the soliton vanishes, and the density profile (7.151) reduces to that associated with (7.146). These analytical results, which were derived by Tsuzuki more than thirty years ago [10], confirm the qualitative estimates (7.144) and (7.145) arrived at earlier.

Another quantity of importance for these solitons is the change in phase across them. This may be found by integrating the expression for the superfluid velocity, since $v(x) = (\hbar/m)\partial\phi/\partial x$. Thus from Eqs. (7.149) and (7.151) one finds

$$\phi(x \to \infty) - \phi(x \to -\infty) = -\frac{mu}{\hbar}\int_{-\infty}^{\infty} dx \frac{n_0 - n_{\min}}{n_0 \cosh^2(x/\sqrt{2}\xi_u) - (n_0 - n_{\min})}$$

$$= -2\cos^{-1}\left(\sqrt{\frac{n_{\min}}{n_0}}\right). \tag{7.154}$$

For a soliton moving in the positive x direction the phase change is negative. Physically this is because the wave is a density depression, and consequently the fluid velocity associated with it is in the negative x direction. In Sec. 13.3 we shall discuss how solitons may be generated experimentally by manipulating the phase of the condensate.

In one spatial dimension there are also soliton solutions for attractive interactions ($U_0 < 0$), the simplest of which is

$$\psi(x,t) = \psi(0)e^{-i\mu t/\hbar}\frac{1}{\cosh[(2m|\mu|/\hbar^2)^{1/2}x]}, \tag{7.155}$$

where the chemical potential μ is given by

$$\mu = \frac{1}{2}U_0|\psi(0)|^2. \tag{7.156}$$

These are self-bound states which are localized in space in the x direction, since ψ vanishes for large $|x|$.

Solitons are also observed in non-linear optics, and the intensity of the light plays a role similar to that of the condensate density in atomic clouds. By analogy, the word *dark* is used to describe solitons that correspond to density depressions. This category of solitons is further divided into *black* ones, for which the minimum density is zero, and *grey* ones, for which it is greater than zero. Solitons with a density maximum are referred to as *bright*.

Dark solitons in external potentials behave essentially like particles if the potential varies sufficiently slowly in space [11]. To determine the soliton velocity, we argue that, if the potential varies sufficiently slowly in space, the soliton moves at the same rate as it would in a uniform medium. The energy of the soliton may be calculated by assuming that the velocity field and the deviation of the density from its equilibrium value are given by the expressions for a background medium with uniform density. It is convenient to consider the quantity $E - \mu N$ rather than the energy itself to allow for the deficit of particles in the soliton. The contribution to this quantity from the soliton consists of four parts. One is the change in the energy due to the external potential, since when a soliton is added, particles are removed from the vicinity of the soliton. The density depression gives rise to a local increase of the fluid velocity, and this leads to the second contribution which is due to the extra kinetic energy that arises because of this. Another effect is that the density reduction leads to a change in the interparticle interaction energy, and this gives the third contribution. The fourth contribution is due to the spatial variation of the magnitude of the condensate wave function. When all these terms are added, one finds that the energy of the soliton, per unit area perpendicular to the direction of propagation, is given by

$$\mathcal{E} = \frac{4\hbar}{3\sqrt{mU_0}}[\mu - V(x_\mathrm{s}) - n_{\min}(x_\mathrm{s})U_0]^{3/2} = \frac{4\hbar}{3\sqrt{mU_0}}[\mu - V(x_\mathrm{s}) - mu^2(x_\mathrm{s})]^{3/2}, \quad (7.157)$$

where x_s is the position of the centre of the soliton, which depends on time. In writing the second form, we have used the fact that the velocity $u(x_\mathrm{s})$ of the soliton is equal to the sound velocity at the minimum density in the soliton, $u^2(x_\mathrm{s}) = n_{\min}(x_\mathrm{s})U_0/m$. It is interesting to note that the energy of a soliton decreases as its velocity increases, whereas for an ordinary particle it increases.

From Eq. (7.157) we can immediately derive a number of conclusions. Since, in the Thomas–Fermi theory, the chemical potential and the external potential are related to the local density $n_0(x_\mathrm{s})$ in the absence of the soliton by Eq. (6.31), $\mu = V(x_\mathrm{s}) + n_0(x_\mathrm{s})U_0$, it follows that the energy of the soliton is proportional to $[n_0(x_\mathrm{s}) - n_{\min}(x_\mathrm{s})]^{3/2}$. The requirement of energy conservation therefore implies that solitons move so that the depression of the density, with respect to its value in the absence of the soliton, is a constant. From the second form of Eq. (7.157) one sees that for energy to be conserved, the quantity $\mu-V(x_s)-mu^2(x_s)$ must be constant, or $V(x_s)+mu^2(x_s)$ must be constant. The latter expression is the energy of a particle of mass $2m$. Consequently, we arrive at the remarkable conclusion that the motion of a soliton in a Bose–Einstein condensate in an external potential is the same as that of a particle of mass $2m$ in the same potential. Thus, for a potential having a minimum, the period of the motion of a soliton is $\sqrt{2}$ times that of a particle of mass m in the potential, the energy of the particle being equal to the value of the external potential at the turning points of the motion.

Because of the inhomogeneity of the potential, the form of the velocity that we have adopted is not an exact solution of the equation of motion, and as the soliton accelerates due to the external potential, it will emit sound waves in much the same way as a charged particle emits electromagnetic radiation when accelerated. However, this effect is small if the potential varies sufficiently slowly. Dissipation tends to make the soliton less dark, that is to reduce the depth of the density depression. In a condensate of finite extent, such as one in a trap, emission of phonons is suppressed by the discrete nature of the spectrum of elementary excitations.

In bulk matter, purely one-dimensional solitons are unstable to perturbations in the other dimensions. This may be shown by studying small departures of the condensate wave function from its form for a soliton, just as we earlier investigated oscillations of a condensate about the ground-state solution. The corresponding equations are the Bogoliubov equations (7.39) and (7.40) with $\psi(\mathbf{r},t)$ put equal to the solution for a soliton. For dark solitons in a condensate with repulsive interactions, this instability was demonstrated in Ref. [12]. When the instability grows, solitons break up into pairs of vortices. Bright solitons in a medium with attractive interactions are unstable with respect to a periodic spatial variation in the transverse direction [13]. This corresponds to a tendency to break up into small clumps, since the attractive interaction favours more compact structures, as discussed in Sec. 6.2.

Problems

PROBLEM 7.1 Derive the hydrodynamic equations (7.8) and (7.19) directly from the action principle (7.4) with the Lagrangian (7.5), by varying the action with respect to the magnitude and the phase of the condensate wave function $\psi(\mathbf{r},t)=f(\mathbf{r},t)e^{i\phi(\mathbf{r},t)}$.

PROBLEM 7.2 Show that for a harmonic trap with an axis of symmetry there exist collective modes of the form (7.80), and verify the result (7.81) for their frequencies.

PROBLEM 7.3 Consider a trap with axial symmetry, corresponding to the potential (7.74) with $\lambda \neq 1$. Use the variational method with a trial function of the form

$$\psi = Ce^{-\rho^2/2R^2}e^{-z^2/2Z^2}e^{i\alpha\rho^2 m/2\hbar}e^{i\beta z^2 m/2\hbar},$$

R, Z, α and β being variational parameters, to obtain equations of motion for R and Z and identify the potential $U(R, Z)$. Determine the frequencies of small oscillations around the equilibrium state and compare the results with the frequencies (7.85) obtained by solving the hydrodynamic equations in the Thomas–Fermi approximation.

PROBLEM 7.4 Use the trial function given in Problem 7.3 to study the free expansion of a cloud of condensate upon release from an axially-symmetric trap of the form (7.74) with $\lambda \neq 1$, by numerically solving the coupled differential equations for R and Z. Check that your numerical results satisfy the energy conservation condition.

PROBLEM 7.5 Show that the equation (7.150) may be integrated to give

$$\frac{\hbar^2}{2m}\left(\frac{\partial\sqrt{n}}{\partial x}\right)^2 = (nU_0 - mu^2)\frac{(n-n_0)^2}{2n},$$

and use this result to derive Eqs. (7.151)–(7.153).

References

[1] N. N. Bogoliubov, *J. Phys. (USSR)* **11**, 23 (1947), reprinted in D. Pines, *The Many-Body Problem*, (W. A. Benjamin, New York, 1961), p. 292.
[2] R. P. Feynman, *Phys. Rev.* **91**, 1301 (1953); **94**, 262 (1954).
[3] S. Stringari, *Phys. Rev. Lett.* **77**, 2360 (1996).
[4] I. S. Gradshteyn and I. M. Ryzhik, *Table of Integrals, Series and Products*, Fifth edition, (Academic Press, 1994), 9.100.
[5] D. S. Jin, J. R. Ensher, M. R. Matthews, C. E. Wieman, and E. A. Cornell, *Phys. Rev. Lett.* **77**, 420 (1996).
[6] O. Maragò, S. A. Hopkins, J. Arlt, E. Hodby, G. Hechenblaikner, and C. J. Foot, *Phys. Rev. Lett.* **84**, 2056 (2000).
[7] Y. Castin and R. Dum, *Phys. Rev. Lett.* **77**, 5315 (1996).
[8] U. Al Khawaja, C. J. Pethick, and H. Smith, *Phys. Rev. A* **60**, 1507 (1999).
[9] R. Onofrio, D. S. Durfee, C. Raman, M. Köhl, C. E. Kuklewicz, and W. Ketterle, *Phys. Rev. Lett.* **84**, 810 (2000).
[10] T. Tsuzuki, *J. Low Temp. Phys.* **4**, 441 (1971).

[11] T. Busch and J. R. Anglin, *Phys. Rev. Lett.* **84**, 2298 (2000).

[12] E. A. Kuznetsov and S. K. Turitsyn, *Zh. Eksp. Teor. Fiz.* **94**, 119 (1988) [*Sov. Phys.-JETP* **67,** 1583 (1988)].

[13] V. E. Zakharov and A. M. Rubenchik, *Zh. Eksp. Teor. Fiz.* **65**, 997 (1973) [*Sov. Phys.-JETP* **38,** 494 (1974)].

8
Microscopic theory of the Bose gas

In Chapter 7 we studied elementary excitations of the condensate using the Gross–Pitaevskii equation, in which the wave function of the condensate is treated as a classical field. In this chapter we develop the microscopic theory of the Bose gas, taking into account the quantum nature of excitations. First we discuss the excitation spectrum of a homogeneous gas at zero temperature (Sec. 8.1) within the Bogoliubov approximation and determine the depletion of the condensate and the change in ground-state energy. Following that we derive in Sec. 8.2 the Bogoliubov equations for inhomogeneous gases, and also consider the weak-coupling limit in which particle interactions may be treated as a perturbation. Excitations at non-zero temperatures are the subject of Sec. 8.3, where we describe the Hartree–Fock and Popov approximations, which are mean-field theories. Interactions between atoms change the frequencies of spectral lines. Such shifts are important in atomic clocks, and an important example in the context of Bose–Einstein condensation is the shift of the 1S–2S transition in hydrogen, which is used to measure the atomic density. Collective effects not taken into account in the Hartree–Fock theory and the other mean-field theories mentioned above are important for understanding these shifts, as will be described in Sec. 8.4.

The starting point for our calculations is the Hamiltonian (6.3). In terms of creation and annihilation operators for bosons, $\hat{\psi}^\dagger(\mathbf{r})$ and $\hat{\psi}(\mathbf{r})$ respectively, it has the form

$$H = \int d\mathbf{r} \left[-\hat{\psi}^\dagger(\mathbf{r}) \frac{\hbar^2}{2m} \nabla^2 \hat{\psi}(\mathbf{r}) + V(\mathbf{r}) \hat{\psi}^\dagger(\mathbf{r}) \hat{\psi}(\mathbf{r}) + \frac{U_0}{2} \hat{\psi}^\dagger(\mathbf{r}) \hat{\psi}^\dagger(\mathbf{r}) \hat{\psi}(\mathbf{r}) \hat{\psi}(\mathbf{r}) \right]. \tag{8.1}$$

In the Gross–Pitaevskii equation one works not with creation and annihilation operators but with the wave function for the condensed state, which is a classical field. The Gross–Pitaevskii approach is thus analogous to the classical theory of electrodynamics, in which a state is characterized by classical electric and magnetic fields, rather than by creation and annihilation operators for photons.

To take into account quantum fluctuations about the state in which all atoms are condensed in a single quantum state it is natural to write[1]

$$\hat{\psi}(\mathbf{r}) = \psi(\mathbf{r}) + \delta\hat{\psi}(\mathbf{r}). \tag{8.2}$$

If the fluctuation term $\delta\hat{\psi}(\mathbf{r})$ is neglected, the Hamiltonian is equivalent to the energy expression which leads to the Gross–Pitaevskii equation.

8.1 Excitations in a uniform gas

As a first illustration we consider a uniform gas of interacting bosons contained in a box of volume V. The Hamiltonian (8.1) then becomes

$$H = \sum_{\mathbf{p}} \epsilon_p^0 a_{\mathbf{p}}^\dagger a_{\mathbf{p}} + \frac{U_0}{2V} \sum_{\mathbf{p},\mathbf{p}',\mathbf{q}} a_{\mathbf{p}+\mathbf{q}}^\dagger a_{\mathbf{p}'-\mathbf{q}}^\dagger a_{\mathbf{p}'} a_{\mathbf{p}} , \tag{8.3}$$

where $\epsilon_p^0 = p^2/2m$. Here the operators $a_{\mathbf{p}}$ and $a_{\mathbf{p}}^\dagger$ that destroy and create bosons in the state with momentum $\mathbf{p}$ satisfy the usual Bose commutation relations

$$[a_{\mathbf{p}}, a_{\mathbf{p}'}^\dagger] = \delta_{\mathbf{p},\mathbf{p}'}, \quad [a_{\mathbf{p}}, a_{\mathbf{p}'}] = 0, \quad \text{and} \quad [a_{\mathbf{p}}^\dagger, a_{\mathbf{p}'}^\dagger] = 0. \tag{8.4}$$

We assume that in the interacting system the lowest-lying single-particle state is macroscopically occupied, that is N_0/N tends to a non-zero value in the thermodynamic limit when N and V tend to infinity in such a way that the density N/V remains constant. In the unperturbed system we have

$$a_0^\dagger|N_0\rangle = \sqrt{N_0+1}|N_0+1\rangle \text{ and } a_0|N_0\rangle = \sqrt{N_0}|N_0-1\rangle, \tag{8.5}$$

and in the Hamiltonian we therefore replace a_0 and $a_0^\dagger$ by $\sqrt{N_0}$, as was first done by Bogoliubov [1]. This is equivalent to using Eq. (8.2) with the wave function for the condensed state given by $\psi = \sqrt{N_0}\phi_0$, where $\phi_0 = V^{-1/2}$ is the wave function for the zero-momentum state.

Within the Bogoliubov approach one assumes that $\delta\hat{\psi}(\mathbf{r})$ is small and retains in the interaction all terms which have (at least) two powers of $\psi(\mathbf{r})$ or $\psi^*(\mathbf{r})$. This is equivalent to including terms which are no more than

---

[1] In this chapter we use the notation $\hat{\psi}$ to distinguish the annihilation operator from the wave function ψ. When ambiguities do not exist (as with the annihilation operator $a_{\mathbf{p}}$) we omit the 'hat'.

quadratic in $\delta\hat{\psi}(\mathbf{r})$ and $\delta\hat{\psi}^\dagger(\mathbf{r})$, that is in $a_\mathbf{p}$ and $a_\mathbf{p}^\dagger$ for $\mathbf{p} \neq 0$. One finds

$$H = \frac{N_0^2 U_0}{2V} + \sum_{\mathbf{p}(\mathbf{p}\neq 0)} (\epsilon_p^0 + 2n_0 U_0) a_\mathbf{p}^\dagger a_\mathbf{p} + \frac{n_0 U_0}{2} \sum_{\mathbf{p}(\mathbf{p}\neq 0)} (a_\mathbf{p}^\dagger a_{-\mathbf{p}}^\dagger + a_\mathbf{p} a_{-\mathbf{p}}), \quad (8.6)$$

where $n_0 = N_0/V$ is the density of particles in the zero-momentum state.[2] The first term is the energy of N_0 particles in the zero-momentum state, and the second is that of independent excitations with energy $\epsilon_p^0 + 2n_0 U_0$, which is the energy of an excitation moving in the Hartree–Fock mean field produced by interactions with other atoms. To see this it is convenient to consider an interaction $U(r)$ with non-zero range instead of the contact one, and introduce its Fourier transform $U(p)$ by

$$U(p) = \int d\mathbf{r}\, U(r) \exp(-i\mathbf{p}\cdot\mathbf{r}/\hbar). \quad (8.7)$$

When the operators a_0 and $a_0^\dagger$ in the Hamiltonian are replaced by c numbers, the term in the interaction proportional to N_0 is

$$\sum_{\mathbf{p}(\mathbf{p}\neq 0)} n_0[U(0) + U(p)] a_\mathbf{p}^\dagger a_\mathbf{p} + \frac{1}{2}\sum_{\mathbf{p}(\mathbf{p}\neq 0)} n_0 U(p)(a_\mathbf{p}^\dagger a_{-\mathbf{p}}^\dagger + a_\mathbf{p} a_{-\mathbf{p}}). \quad (8.8)$$

The $a_\mathbf{p}^\dagger a_\mathbf{p}$ term has two contributions. The first, $n_0 U(0)$, is the Hartree energy, which comes from the direct interaction of a particle in the state $\mathbf{p}$ with the N_0 atoms in the zero-momentum state. The second is the exchange, or Fock, term, in which an atom in the state $\mathbf{p}$ is scattered into the zero-momentum state, while a second atom is simultaneously scattered from the condensate to the state $\mathbf{p}$. These identifications will be further elucidated in Sec. 8.3.1 below where we consider the Hartree–Fock approximation in greater detail. For a contact interaction the Fourier transform of the interaction $U(p)$ is independent of p, and therefore the Hartree and Fock terms are both equal to $n_0 U_0$. The final terms in Eqs. (8.6) and (8.8) correspond to the scattering of two atoms in the condensate to states with momenta $\pm\mathbf{p}$ and the inverse process in which two atoms with momenta $\pm\mathbf{p}$ are scattered into the condensate.

The task now is to find the eigenvalues of the Hamiltonian (8.6). The original Hamiltonian conserved the number of particles, and therefore we wish to find the eigenvalues of the new Hamiltonian for a fixed average

---

[2] In this chapter and the following ones it is important to distinguish between the condensate density and the total density, and we shall denote the condensate density by n_0 and the total density by n.

particle number. The operator for the total particle number is given by

$$\hat{N} = \sum_{\mathbf{p}} a_{\mathbf{p}}^{\dagger} a_{\mathbf{p}}, \tag{8.9}$$

which on treating the zero-momentum-state operators as c numbers becomes

$$\hat{N} = N_0 + \sum_{\mathbf{p}(\mathbf{p}\neq 0)} a_{\mathbf{p}}^{\dagger} a_{\mathbf{p}}. \tag{8.10}$$

Expressed in terms of the total number of particles, the Hamiltonian (8.6) may be written

$$H = \frac{N^2 U_0}{2V} + \sum_{\mathbf{p}(\mathbf{p}\neq 0)} \left[(\epsilon_p^0 + n_0 U_0) a_{\mathbf{p}}^{\dagger} a_{\mathbf{p}} + \frac{n_0 U_0}{2} (a_{\mathbf{p}}^{\dagger} a_{-\mathbf{p}}^{\dagger} + a_{\mathbf{p}} a_{-\mathbf{p}}) \right], \tag{8.11}$$

where in the first term we have replaced $\hat{N}$ by its expectation value. This is permissible since the fluctuation in the particle number is small. Since we consider states differing little from the state with all particles in the condensed state it makes no difference whether the condensate density or the total density appears in the terms in the sum. The reduction of the coefficient of $a_{\mathbf{p}}^{\dagger} a_{\mathbf{p}}$ from $\epsilon_p^0 + 2n_0 U_0$ to $\epsilon_p^0 + n_0 U_0$, is due to the condition that the total number of particles be fixed. In the classical treatment of excitations in Chapter 7 this corresponds to the subtraction of the chemical potential, since for the uniform Bose gas at zero temperature, the chemical potential is $n_0 U_0$, Eq. (6.12).

The energy $\epsilon_p^0 + n_0 U_0$ does not depend on the direction of $\mathbf{p}$, and therefore we may write the Hamiltonian (8.11) in the symmetrical form

$$H = \frac{N^2 U_0}{2V} + {\sum_{\mathbf{p}(\mathbf{p}\neq 0)}}' [(\epsilon_p^0 + n_0 U_0)(a_{\mathbf{p}}^{\dagger} a_{\mathbf{p}} + a_{-\mathbf{p}}^{\dagger} a_{-\mathbf{p}}) + n_0 U_0 (a_{\mathbf{p}}^{\dagger} a_{-\mathbf{p}}^{\dagger} + a_{\mathbf{p}} a_{-\mathbf{p}})], \tag{8.12}$$

where the prime on the sum indicates that it is to be taken only over one half of momentum space, since the terms corresponding to $\mathbf{p}$ and $-\mathbf{p}$ must be counted only once.

8.1.1 The Bogoliubov transformation

The structure of the Hamiltonian is now simple, since it consists of a sum of independent terms of the form

$$\epsilon_0 (a^{\dagger} a + b^{\dagger} b) + \epsilon_1 (a^{\dagger} b^{\dagger} + ba). \tag{8.13}$$

Here ϵ_0 and ϵ_1 are c numbers. The operators $a^\dagger$ and a create and annihilate bosons in the state with momentum $\mathbf{p}$, and $b^\dagger$ and b are the corresponding operators for the state with momentum $-\mathbf{p}$.

The eigenvalues and eigenstates of this Hamiltonian may be obtained by performing a canonical transformation, as Bogoliubov did in the context of liquid helium [1]. This method has proved to be very fruitful, and it is used extensively in the theory of superconductivity and of magnetism as well as in other fields. We shall use it again in Chapter 14 when we consider pairing of fermions. The basic idea is to introduce a new set of operators α and β such that the Hamiltonian has only terms proportional to $\alpha^\dagger \alpha$ and $\beta^\dagger \beta$.

Creation and annihilation operators for bosons obey the commutation relations

$$[a, a^\dagger] = [b, b^\dagger] = 1, \quad \text{and} \quad [a, b^\dagger] = [b, a^\dagger] = 0. \tag{8.14}$$

We introduce new operators α and β by the transformation

$$\alpha = ua + vb^\dagger, \quad \beta = ub + va^\dagger, \tag{8.15}$$

where u and v are coefficients to be determined. We require that also these operators satisfy Bose commutation rules,

$$[\alpha, \alpha^\dagger] = [\beta, \beta^\dagger] = 1, \quad [\alpha, \beta^\dagger] = [\beta, \alpha^\dagger] = 0. \tag{8.16}$$

Since the phases of u and v are arbitrary, we may take u and v to be real. By inserting (8.15) into (8.16) and using (8.14) one sees that u and v must satisfy the condition

$$u^2 - v^2 = 1. \tag{8.17}$$

The inverse transformation corresponding to (8.15) is

$$a = u\alpha - v\beta^\dagger, \quad b = u\beta - v\alpha^\dagger. \tag{8.18}$$

We now substitute (8.18) in (8.13) and obtain the result

$$H = 2v^2 \epsilon_0 - 2uv\epsilon_1 + [\epsilon_0(u^2 + v^2) - 2uv\epsilon_1](\alpha^\dagger \alpha + \beta^\dagger \beta)$$
$$+ [\epsilon_1(u^2 + v^2) - 2uv\epsilon_0](\alpha\beta + \beta^\dagger \alpha^\dagger). \tag{8.19}$$

The term proportional to $\alpha\beta + \beta^\dagger \alpha^\dagger$ can be made to vanish by choosing u and v so that its coefficient is zero:

$$\epsilon_1(u^2 + v^2) - 2uv\epsilon_0 = 0. \tag{8.20}$$

The sign of u is arbitrary, and if we adopt the convention that it is positive, the normalization condition (8.17) is satisfied by the following parametrization of u and v,

$$u = \cosh t, \quad v = \sinh t, \tag{8.21}$$

which in turn implies that the condition (8.20) may be written as

$$\epsilon_1(\cosh^2 t + \sinh^2 t) - 2\epsilon_0 \sinh t \cosh t = 0, \tag{8.22}$$

or

$$\tanh 2t = \frac{\epsilon_1}{\epsilon_0}. \tag{8.23}$$

From this result one finds

$$u^2 = \frac{1}{2}\left(\frac{\epsilon_0}{\epsilon} + 1\right) \quad \text{and} \quad v^2 = \frac{1}{2}\left(\frac{\epsilon_0}{\epsilon} - 1\right), \tag{8.24}$$

where

$$\epsilon = \sqrt{\epsilon_0^2 - \epsilon_1^2}. \tag{8.25}$$

It is necessary to choose the positive branch of the square root, since otherwise u and v would be imaginary, contrary to our initial assumption. Solving for $u^2 + v^2$ and $2uv$ in terms of the ratio ϵ_1/ϵ_0 and inserting the expressions into (8.19) leads to the result

$$H = \epsilon(\alpha^\dagger \alpha + \beta^\dagger \beta) + \epsilon - \epsilon_0. \tag{8.26}$$

The ground-state energy is $\epsilon - \epsilon_0$, which is negative, and the excited states correspond to the addition of two independent kinds of bosons with energy ϵ, created by the operators $\alpha^\dagger$ and $\beta^\dagger$. For ϵ to be real, the magnitude of ϵ_0 must exceed that of ϵ_1. If $|\epsilon_1| > |\epsilon_0|$, the excitation energy is imaginary, corresponding to an instability of the system.

8.1.2 Elementary excitations

We may now use the results of the previous subsection to bring the Hamiltonian (8.12) into diagonal form. We make the transformation

$$a_{\mathbf{p}} = u_p \alpha_{\mathbf{p}} - v_p \alpha^\dagger_{-\mathbf{p}}, \quad a_{-\mathbf{p}} = u_p \alpha_{-\mathbf{p}} - v_p \alpha^\dagger_{\mathbf{p}}, \tag{8.27}$$

where $a_{\mathbf{p}}$ corresponds to a in the simple model, $a_{-\mathbf{p}}$ to b, $\alpha_{\mathbf{p}}$ to α, and $\alpha_{-\mathbf{p}}$ to β. The result is

$$H = \frac{N^2 U_0}{2V} + \sum_{\mathbf{p}(\mathbf{p}\neq 0)} \epsilon_p \alpha^\dagger_{\mathbf{p}} \alpha_{\mathbf{p}} - \frac{1}{2}\sum_{\mathbf{p}(\mathbf{p}\neq 0)} (\epsilon_p^0 + n_0 U_0 - \epsilon_p) \tag{8.28}$$

with
$$\epsilon_p = \sqrt{(\epsilon_p^0 + n_0 U_0)^2 - (n_0 U_0)^2} = \sqrt{(\epsilon_p^0)^2 + 2\epsilon_p^0 n_0 U_0}. \tag{8.29}$$

The energy spectrum (8.29) agrees precisely with the result (7.48) derived in the previous chapter. For small p the energy is $\epsilon_p = sp$, where
$$s^2 = \frac{n_0 U_0}{m}. \tag{8.30}$$

The operators that create and destroy elementary excitations are given by
$$\alpha_{\mathbf{p}}^\dagger = u_p a_{\mathbf{p}}^\dagger + v_p a_{-\mathbf{p}}. \tag{8.31}$$

The coefficients satisfy the normalization condition
$$u_p^2 - v_p^2 = 1 \tag{8.32}$$

corresponding to Eq. (8.17) and are given explicitly by
$$u_p^2 = \frac{1}{2}\left(\frac{\xi_p}{\epsilon_p} + 1\right) \quad \text{and} \quad v_p^2 = \frac{1}{2}\left(\frac{\xi_p}{\epsilon_p} - 1\right), \tag{8.33}$$

where $\xi_p = \epsilon_p^0 + n_0 U_0$ is the difference between the Hartree–Fock energy of a particle and the chemical potential, Eq. (6.12).

Thus the system behaves as a collection of non-interacting bosons with energies given by the Bogoliubov spectrum previously derived from classical considerations in Chapter 7. In the ground state of the system there are no excitations, and thus $\alpha_{\mathbf{p}}|0\rangle = 0$.

Depletion of the condensate

The particle number is given by Eq. (8.10) which, rewritten in terms of $\alpha_{\mathbf{p}}^\dagger$ and $\alpha_{\mathbf{p}}$, has the form
$$\begin{aligned}\hat{N} =\ & N_0 + \sum_{\mathbf{p}(p\neq 0)} v_p^2 + \sum_{\mathbf{p}(p\neq 0)} (u_p^2 + v_p^2)\alpha_{\mathbf{p}}^\dagger \alpha_{\mathbf{p}} \\ & - \sum_{\mathbf{p}(p\neq 0)} u_p v_p (\alpha_{\mathbf{p}}^\dagger \alpha_{-\mathbf{p}}^\dagger + \alpha_{-\mathbf{p}} \alpha_{\mathbf{p}}).\end{aligned} \tag{8.34}$$

In deriving this expression we used the Bose commutation relations to reorder operators so that the expectation value of the operator terms gives zero in the ground state. The physical interpretation of this expression is that the first term is the number of atoms in the condensate. The second term represents the depletion of the condensate by interactions when no real excitations are present. In the ground state of the interacting gas, not all particles are in the zero-momentum state because the two-body interaction mixes into the ground-state components with atoms in other states. Consequently, the probability of an atom being in the zero-momentum state is reduced. The last terms correspond to the depletion of the condensate due to the presence of real excitations. For states which contain a definite number of elementary excitations, the expectation value of $a_{\mathbf{p}}^\dagger a_{-\mathbf{p}}^\dagger$ and its Hermitian conjugate vanish, and therefore the number operator may equivalently be written as

$$\hat{N} = N_0 + \sum_{\mathbf{p}(\mathbf{p}\neq 0)} v_p^2 + \sum_{\mathbf{p}(\mathbf{p}\neq 0)} (u_p^2 + v_p^2) a_{\mathbf{p}}^\dagger a_{\mathbf{p}}. \tag{8.35}$$

This shows that when an excitation with non-zero momentum $\mathbf{p}$ is added to the gas, keeping N_0 fixed, the number of particles changes by an amount

$$\nu_p = u_p^2 + v_p^2 = \frac{\xi_p}{\epsilon_p}, \tag{8.36}$$

where, as before, $\xi_p = \epsilon_p^0 + n_0 U_0$. Thus, when an excitation is added keeping the total number of particles fixed, N_0 must be reduced by the corresponding amount. At large momenta the particle number associated with an excitation tends to unity, since then excitations are just free particles, while for small momenta the effective particle number diverges as ms/p.

The depletion of the ground state at zero temperature may be calculated by evaluating the second term in Eq. (8.34) explicitly and one finds for the number of particles per unit volume in excited states[3]

$$n_{\text{ex}} = \frac{1}{V} \sum_{\mathbf{p}(\mathbf{p}\neq 0)} v_p^2 = \int \frac{d\mathbf{p}}{(2\pi\hbar)^3} v_p^2 = \frac{1}{3\pi^2} \left(\frac{ms}{\hbar}\right)^3, \tag{8.37}$$

which is of order one particle per volume ξ^3, where ξ is the coherence length, Eq. (6.62). Physically this result may be understood by noting that v_p^2 is of order unity for momenta $p \sim \hbar/\xi$, and then falls off rapidly at larger momenta. The number density of particles in excited states is thus of order the number of states per unit volume with wave number less than $1/\xi$, that is, $1/\xi^3$ in three dimensions. The depletion may also be expressed in terms of the scattering length by utilizing the result (8.30) with $U_0 = 4\pi\hbar^2 a/m$, and one finds

$$\frac{n_{\text{ex}}}{n} = \frac{8}{3\sqrt{\pi}} (na^3)^{1/2}. \tag{8.38}$$

In deriving this result we have assumed that the depletion of the condensate

---

[3] When transforming sums to integrals we shall use the standard prescription $\sum_{\mathbf{p}} \ldots = V \int d\mathbf{p}/(2\pi\hbar)^3 \ldots$.

is small, and (8.38) is therefore only valid when the particle spacing is large compared with the scattering length, or $n_{\text{ex}} \ll n$. In most experiments that have been carried out, the ground-state depletion is of the order of one per cent. Recent experiments on ^{85}Rb near a Feshbach resonance achieved very large values of the scattering length corresponding to a depletion of 10%, thus opening up the possibility of measuring effects beyond the validity of the mean-field approximation [2].

Ground-state energy

The calculation of higher-order contributions to the energy requires that one go beyond the simple approximation in which the effective interaction is replaced by $U_0 = 4\pi\hbar^2 a/m$. The difficulty with the latter approach is seen by considering the expression for the ground-state energy E_0 that one obtains from Eq. (8.28),

$$E_0 = \frac{N^2 U_0}{2V} - \frac{1}{2}\sum_{\mathbf{p}} \left(\epsilon_p^0 + n_0 U_0 - \epsilon_p\right). \quad \text{(wrong!)} \quad (8.39)$$

Formally the sum is of order U_0^2, as one can see by expanding the summand for large p. However, the sum diverges linearly at large p: the leading terms in the summand are of order $1/p^2$, and the sum over momentum space when converted to an integral gives a factor $p^2 dp$. This difficulty is due to the fact that we have used the effective interaction U_0, which is valid only for small momenta, to calculate high-momentum processes. In perturbation theory language, the effective interaction takes into account transitions to intermediate states in which the two interacting particles have arbitrarily high momenta. If the sum in Eq. (8.39) is taken over all states, contributions from these high-energy intermediate states are included twice. To make a consistent calculation of the ground-state energy one must use an effective interaction $U(p_c)$ in which all intermediate states with momenta in excess of some cut-off value p_c are taken into account, and then evaluate the energy omitting in the sum in the analogue of Eq. (8.39) all intermediate states with momenta in excess of this cut off. The ground-state energy is therefore

$$E_0 = \frac{N^2 U(p_c)}{2V} - \frac{1}{2}\sum_{\mathbf{p}(p<p_c)} \left(\epsilon_p^0 + n_0 U_0 - \epsilon_p\right). \quad (8.40)$$

The effective interaction $\tilde{U} = U(p_c)$ for zero energy E and for small values of p_c may be obtained from (5.40) by replacing T by U_0, which is the effective interaction for $p_c = 0$ and $E = 0$. The imaginary part of the effective interaction, which is due to the $i\delta$ term in the energy denominator in (5.40), is proportional to $E^{1/2}$, and therefore it vanishes at zero energy. The effective interaction for small p_c and zero energy is thus given by

$$U(p_c) = U_0 + \frac{U_0^2}{V} \sum_{\mathbf{p}(p<p_c)} \frac{1}{2\epsilon_p^0}. \tag{8.41}$$

With this expression for the effective interaction $U(p_c)$ one finds

$$E_0 = \frac{N^2 U_0}{2V} - \frac{1}{2} \sum_{\mathbf{p}(p<p_c)} \left[\epsilon_p^0 + n_0 U_0 - \epsilon_p - \frac{(nU_0)^2}{2\epsilon_p^0} \right]. \tag{8.42}$$

If one chooses the cut-off momentum to be large compared with ms but small compared with $\hbar/a$ the result does not depend on p_c, and, using the fact that $n_0 \simeq n$, one finds

$$\begin{aligned} \frac{E_0}{V} &= \frac{n^2 U_0}{2} + \frac{8}{15\pi^2} \left(\frac{ms}{\hbar} \right)^3 ms^2 \\ &= \frac{n^2 U_0}{2} \left[1 + \frac{128}{15\pi^{1/2}} (na^3)^{1/2} \right]. \end{aligned} \tag{8.43}$$

The first form of the correction term indicates that the order of magnitude of the energy change is the number of states having wave numbers less than the inverse coherence length, times the typical energy of an excitation with this wave number, as one would expect from the form of the integral. This result was first obtained by Lee and Yang [3].

States with definite particle number

The original microscopic Hamiltonian (8.3) conserves the total number of particles. The assumption that the annihilation operator for a particle has a non-zero expectation value, as indicated in Eq. (8.2), implies that the states we are working with are not eigenstates of the particle number operator. In an isolated cloud of gas, the number of particles is fixed, and therefore the expectation value of the particle annihilation operator vanishes. Assuming that the annihilation operator for a particle has a non-zero expectation value is analogous to assuming that the operator for the electromagnetic field due to photons may be treated classically. In both cases one works with coherent states, which are superpositions of states with different numbers of particles or photons. It is possible to calculate the properties of a Bose gas containing a definite particle number by introducing the operators [4]

$$\tilde{a}_{\mathbf{p}} = a_0^\dagger (\hat{N}_0 + 1)^{-1/2} a_{\mathbf{p}}, \quad \tilde{a}_{\mathbf{p}}^\dagger = a_{\mathbf{p}}^\dagger (\hat{N}_0 + 1)^{-1/2} a_0, \quad (\mathbf{p} \neq 0), \tag{8.44}$$

where $\hat{N}_0 = a_0^\dagger a_0$ is the operator for the number of particles in the zero-momentum state. By evaluating the commutators explicitly, one can show that these operators obey Bose commutation relations when they act on any state which has a non-vanishing number of particles in the zero-momentum state. In the presence of a Bose–Einstein condensate, components of the many-particle state having no particles in the zero-momentum state play essentially no role, so we shall not consider this restriction further. In addition, the operator $\tilde{a}_\mathbf{p}^\dagger \tilde{a}_\mathbf{p}$ is identically equal to $a_\mathbf{p}^\dagger a_\mathbf{p}$ for $\mathbf{p} \neq 0$. Retaining only terms no more than quadratic in the operators $\tilde{a}_\mathbf{p}$ and $\tilde{a}_\mathbf{p}^\dagger$, one may write the Hamiltonian for states that deviate little from the fully-condensed state containing a definite number of particles N as

$$H = \frac{N(N-1)U_0}{2V} + \sum_{\mathbf{p}(\mathbf{p}\neq 0)}{}' \left\{ (\epsilon_p^0 + \frac{\hat{N}_0}{V}U_0)(\tilde{a}_\mathbf{p}^\dagger \tilde{a}_\mathbf{p} + \tilde{a}_{-\mathbf{p}}^\dagger \tilde{a}_{-\mathbf{p}}) \right.$$

$$\left. + \frac{U_0}{V}[(\hat{N}_0 + 2)^{1/2}(\hat{N}_0 + 1)^{1/2} \tilde{a}_\mathbf{p}^\dagger \tilde{a}_{-\mathbf{p}}^\dagger + \tilde{a}_\mathbf{p} \tilde{a}_{-\mathbf{p}}(\hat{N}_0 + 2)^{1/2}(\hat{N}_0 + 1)^{1/2}] \right\}.$$
(8.45)

When one replaces $\hat{N}_0$ by its expectation value N_0 and neglects terms of relative order $1/N_0$ and $1/N$, this equation becomes identical with Eq. (8.12) apart from the replacement of $a_\mathbf{p}$ and $a_\mathbf{p}^\dagger$ by $\tilde{a}_\mathbf{p}$ and $\tilde{a}_\mathbf{p}^\dagger$. In terms of new operators defined by

$$\tilde{\alpha}_\mathbf{p}^\dagger = u_p \tilde{a}_\mathbf{p}^\dagger + v_p \tilde{a}_{-\mathbf{p}} = u_p a_\mathbf{p}^\dagger (\hat{N}_0 + 1)^{-1/2} a_0 + v_p a_0^\dagger (\hat{N}_0 + 1)^{-1/2} a_{-\mathbf{p}}, \quad (8.46)$$

which is analogous to Eq. (8.31), the Hamiltonian reduces to Eq. (8.28), but with the operators $\tilde{\alpha}_\mathbf{p}$ instead of $\alpha_\mathbf{p}$. This shows that the addition of an elementary excitation of momentum $\mathbf{p}$ is the superposition of the addition of a particle of momentum $\mathbf{p}$ together with the removal of a particle from the condensate, and the removal of a particle with momentum $-\mathbf{p}$ accompanied by the addition of a particle to the condensate. The fact that the total number of particles remains unchanged is brought out explicitly. The physical character of long-wavelength excitations may be seen by using the fact that $u_p \simeq v_p$ in this limit. Therefore, for large N_0, $\tilde{\alpha}_\mathbf{p}^\dagger$ is proportional to $a_\mathbf{p}^\dagger a_0 + a_0^\dagger a_{-\mathbf{p}}$, which is the condensate contribution to the operator $\sum_{\mathbf{p}'} a_{\mathbf{p}+\mathbf{p}'}^\dagger a_{\mathbf{p}'}$ that creates a density fluctuation. This confirms the phonon nature of long-wavelength excitations.

8.2 Excitations in a trapped gas

In Chapter 7 we calculated properties of excitations using a classical approach. The analogous quantum-mechanical theory can be developed along similar lines. It parallels the treatment for the uniform system given in

Sec. 8.1 and we describe it here. Instead of starting from a functional for the energy as we did in the classical case, we consider the Hamiltonian operator (8.1) and the expression (8.2) which corresponds to separating out the condensed state. Also, since we wish to conserve particle number on average, it is convenient to work with the operator $K = H - \mu \hat{N}$, rather than the Hamiltonian itself. The term with no fluctuation operators is the Gross–Pitaevskii functional (6.9). The terms with a single fluctuation operator vanish if ψ satisfies the time-independent Gross–Pitaevskii equation (6.11), since the latter follows from the condition that the variations of the energy should vanish to first order in variations in ψ. To second order in the fluctuations, the Hamiltonian may be written

$$K = H - \mu \hat{N} = E_0 - \mu N_0 + \int d\mathbf{r} \left(-\delta\hat{\psi}^\dagger(\mathbf{r}) \frac{\hbar^2}{2m} \nabla^2 \delta\hat{\psi}(\mathbf{r}) \right.$$
$$+ [V(\mathbf{r}) + 2U_0|\psi(\mathbf{r})|^2 - \mu]\delta\hat{\psi}^\dagger(\mathbf{r})\delta\hat{\psi}(\mathbf{r})$$
$$\left. + \frac{U_0}{2} \{ \psi(\mathbf{r})^2 [\delta\hat{\psi}^\dagger(\mathbf{r})]^2 + \psi^*(\mathbf{r})^2 [\delta\hat{\psi}(\mathbf{r})]^2 \} \right), \quad (8.47)$$

which should be compared with the similar expression (8.11) for the energy of a gas in a constant potential. To find the energies of elementary excitations we adopt an approach similar to that used in Sec. 7.2, where we calculated the properties of excitations from the time-dependent Gross–Pitaevskii equation. The equations of motion for the operators $\delta\hat{\psi}$ and $\delta\hat{\psi}^\dagger$ in the Heisenberg picture are

$$i\hbar \frac{\partial \delta\hat{\psi}}{\partial t} = [\delta\hat{\psi}, K] \quad \text{and} \quad i\hbar \frac{\partial \delta\hat{\psi}^\dagger}{\partial t} = [\delta\hat{\psi}^\dagger, K], \quad (8.48)$$

which upon substitution of K from (8.47) become

$$i\hbar \frac{\partial \delta\hat{\psi}}{\partial t} = \left[-\frac{\hbar^2}{2m}\nabla^2 + V(\mathbf{r}) + 2n_0(\mathbf{r})U_0 - \mu \right] \delta\hat{\psi} + U_0 \psi(\mathbf{r})^2 \delta\hat{\psi}^\dagger \quad (8.49)$$

and

$$-i\hbar \frac{\partial \delta\hat{\psi}^\dagger}{\partial t} = \left[-\frac{\hbar^2}{2m}\nabla^2 + V(\mathbf{r}) + 2n_0(\mathbf{r})U_0 - \mu \right] \delta\hat{\psi}^\dagger + U_0 \psi^*(\mathbf{r})^2 \delta\hat{\psi}. \quad (8.50)$$

In order to solve these coupled equations we carry out a transformation analogous to (7.41)

$$\delta\hat{\psi}(\mathbf{r}, t) = \sum_i [u_i(\mathbf{r})\alpha_i e^{-i\epsilon_i t/\hbar} - v_i^*(\mathbf{r})\alpha_i^\dagger e^{i\epsilon_i t/\hbar}], \quad (8.51)$$

where the operators $\alpha_i^\dagger$ and α_i create and destroy bosons in the excited state i. In the ground states we may take $\psi(\mathbf{r})$ to be real, and therefore we may write $\psi(\mathbf{r})^2 = \psi^*(\mathbf{r})^2 = n_0(\mathbf{r})$. By substitution one sees that u_i and v_i must satisfy the Bogoliubov equations

$$\left[-\frac{\hbar^2}{2m}\nabla^2 + V(\mathbf{r}) + 2n_0(\mathbf{r})U_0 - \mu - \epsilon_i\right] u_i(\mathbf{r}) - n_0(\mathbf{r})U_0 v_i(\mathbf{r}) = 0 \quad (8.52)$$

and

$$\left[-\frac{\hbar^2}{2m}\nabla^2 + V(\mathbf{r}) + 2n_0(\mathbf{r})U_0 - \mu + \epsilon_i\right] v_i(\mathbf{r}) - n_0(\mathbf{r})U_0 u_i(\mathbf{r}) = 0, \quad (8.53)$$

just as in the classical treatment. The only difference compared with Eqs. (7.42) and (7.43) is that we have replaced the classical frequency by $\epsilon_i/\hbar$. By a generalization of the usual discussion for the Schrödinger equation, one may show that the eigenstates with different energies are orthogonal in the sense that

$$\int d\mathbf{r}[u_i(\mathbf{r})u_j^*(\mathbf{r}) - v_i^*(\mathbf{r})v_j(\mathbf{r})] = 0. \quad (8.54)$$

The sign difference between the u and v terms reflects the fact that the energy ϵ_i occurs with different signs in the two Bogoliubov equations. The requirement that $\alpha_i^\dagger$ and α_i satisfy Bose commutation relations gives the condition

$$\int d\mathbf{r}[|u_i(\mathbf{r})|^2 - |v_i(\mathbf{r})|^2] = 1. \quad (8.55)$$

This choice of normalization agrees with that of $u(\mathbf{r})$ and $v(\mathbf{r})$ in the classical theory for the uniform system, as given by Eqs. (7.44) and (7.51). Once the eigenvalues ϵ_i and the associated solutions u_i and v_i have been determined we may express the operator K in terms of them, and the result is $K = \sum_i \epsilon_i \alpha_i^\dagger \alpha_i + \text{constant}$.

The system has also a trivial zero-energy mode. When the overall phase of the condensate wave function is changed, the energy is unaltered, and therefore there is no restoring force. Thus this mode has zero frequency. It corresponds to a change of the condensate wave function $\delta\psi = i\psi\delta\phi$, where $\delta\phi$ is the change in phase.

8.2.1 Weak coupling

The theory of collective modes developed above is quite general, but analytical results can be obtained only in limiting cases. The hydrodynamic theory described in Chapter 7 is applicable provided interactions are strong enough that the Thomas–Fermi approximation for the static structure is valid. This requires that $Na/\bar{a}$ be large compared with unity. In addition the method is restricted to modes that are mainly in the interior of the cloud, since the Thomas–Fermi method fails in the boundary layer of thickness δ at the surface of the cloud (Sec. 6.3). In the opposite limit, when $Na/\bar{a} \ll 1$, properties of modes may be investigated by perturbation theory, and we now describe this approach. This regime has not yet been explored experimentally, but theoretical study of it has brought to light a number of interesting results.

For simplicity, let us consider atoms in a harmonic trap with no axis of symmetry and assume the effective interaction between the atoms to be of the contact form, $U(\mathbf{r} - \mathbf{r}') \propto \delta(\mathbf{r} - \mathbf{r}')$. In the ground state in the absence of interactions, all atoms are in the lowest level of the oscillator. Elementary excitations of the system correspond to promoting one or more atoms to excited states of the oscillator. If the oscillator frequencies are not commensurate, excitations associated with motions along the three coordinate directions are independent if the interaction is weak. The simplest excitation one can make is to take one atom from the ground state of the oscillator and put it into, say, the state with n oscillator quanta for motion in the z direction. If we suppress the x and y degrees of freedom, the ground state may be written as $|0^N\rangle$ and the excited state by $|0^{N-1}n^1\rangle$. The expectation value of the total energy in the ground state is given to first order in the interaction by

$$E_0 = \frac{N}{2}\hbar(\omega_1 + \omega_2 + \omega_3) + \frac{N(N-1)}{2}\langle 00|U|00\rangle, \quad (8.56)$$

where $\langle ij|U|kl\rangle$ is the matrix element of the two-body interaction between oscillator states, see Eq. (8.70). The energy in the excited state may be evaluated directly, and it is

$$\begin{aligned} E_n &= \frac{N}{2}\hbar(\omega_1 + \omega_2 + \omega_3) + n\hbar\omega_3 + \frac{(N-1)(N-2)}{2}\langle 00|U|00\rangle \\ &\quad + 2(N-1)\langle 0n|U|0n\rangle. \end{aligned} \quad (8.57)$$

The factor of 2 in the last term on the right hand side of (8.57) appears because, for a contact interaction, the Hartree and Fock terms contribute equal amounts, as explained in more detail in Sec. 8.3.1 below (see Eq. (8.72)). The excitation energy is therefore

$$\epsilon_n = E_n - E_0 = n\hbar\omega_3 + (N-1)(2\langle 0n|U|0n\rangle - \langle 00|U|00\rangle). \quad (8.58)$$

A simple calculation gives

$$\langle 01|U|01\rangle = \frac{1}{2}\langle 00|U|00\rangle, \quad \text{and} \quad \langle 02|U|02\rangle = \frac{3}{8}\langle 00|U|00\rangle. \quad (8.59)$$

Thus the frequency of the $n = 1$ excitation is the same as in the absence of interactions. This is because this mode corresponds to an excitation of the centre-of-mass motion alone, as we have seen earlier. For the $n = 2$ mode one finds for $N \gg 1$

$$\epsilon_2 = 2\hbar\omega_3 - \frac{N}{4}\langle 00|U|00\rangle. \tag{8.60}$$

This result may also be obtained by using the collective coordinate approach which we illustrated in Sec. 7.3.3 for the breathing mode. In the weak-coupling limit the mode corresponds to a uniform dilation along the z direction.

In the treatment above we have not taken into account the degeneracy of low-lying excitations in the absence of interactions. For example, the states $|0^{N-1}2^1\rangle$ and $|0^{N-2}1^2\rangle$ are degenerate. Interactions break this degeneracy, but they shift the excitation energy by terms which are independent of N. When the number of particles is large, these contributions are negligible compared with the leading term, which is proportional to N. When many excitations are present the problem becomes more complex. In Chapter 9 we shall develop this approach to consider the properties of a weakly-interacting Bose gas under rotation.

8.3 Non-zero temperature

At non-zero temperatures, elementary excitations will be present. At temperatures well below the transition temperature their number is small and interactions between them may be neglected. In equilibrium, the distribution function for excitations is thus the usual Bose–Einstein one evaluated with the energies calculated earlier for the Bogoliubov approximation. An important observation is that addition of one of these excitations does not change the total particle number and, consequently, there is no chemical potential term in the Bose distribution,

$$f_i = \frac{1}{\exp(\epsilon_i/kT) - 1}, \tag{8.61}$$

where i labels the state. In this respect these excitations resemble phonons and rotons in liquid ^{4}He. From the distribution function for excitations we may calculate the thermodynamic properties of the gas. For example, for a uniform Bose gas the thermal contribution to the energy is

$$E(T) - E(T=0) = V \int \frac{d\mathbf{p}}{(2\pi\hbar)^3} \epsilon_p f_p, \tag{8.62}$$

and the thermal depletion of the condensate density is given by

$$n_{\text{ex}}(T) - n_{\text{ex}}(T=0) = \int \frac{d\mathbf{p}}{(2\pi\hbar)^3} \frac{\xi_p}{\epsilon_p} f_p, \tag{8.63}$$

since according to (8.35) and (8.36) the addition of an excitation keeping the total number of particles fixed reduces the number of particles in the condensate by an amount ξ_p/ϵ_p. When the temperature is less than the characteristic temperature T_* defined by

$$kT_* = ms^2 \simeq nU_0, \tag{8.64}$$

the excitations in a uniform gas are phonon-like, with an energy given by $\epsilon = sp$. In this limit the thermal contribution to the energy, Eq. (8.62), is proportional to T^4, as opposed to the $T^{5/2}$ behaviour found for the non-interacting gas, Eq. (2.55) with $\alpha = 3/2$. Likewise the entropy and specific heat obtained from (8.62) are both proportional to T^3, while the corresponding result for the non-interacting gas is $T^{3/2}$. Due to the fact that the particle number ξ_p/ϵ_p becomes inversely proportional to p for $p \ll ms$, the thermal depletion of the condensate density obtained from (8.63) is proportional to T^2 which should be contrasted with the $T^{3/2}$ behaviour of n_{ex} for a non-interacting gas exhibited in Eq. (2.30).

8.3.1 The Hartree–Fock approximation

In the calculations described above we assumed the system to be close to its ground state, and the excitations we found were independent of each other. At higher temperatures, interactions between excitations become important. These are described by the terms in the Hamiltonian with lower powers of the condensate wave function. Deriving expressions for the properties of a Bose gas at arbitrary temperatures is a difficult task, but fortunately there is a useful limit in which a simple physical picture emerges. In a homogeneous gas not far from its ground state, the energy of an excitation with high momentum is given by Eq. (7.50). This approximate form is obtained from the general theory by neglecting v: excitations correspond to addition of a particle of momentum $\pm\mathbf{p}$ and removal of one from the condensate. The wave function of the state is a product of single-particle states, symmetrized with respect to interchange of the particle coordinates to take into account the Bose statistics. The neglect of v is justifiable if the $a^\dagger a^\dagger$ and aa terms in the Bogoliubov Hamiltonian are negligible for the excitations of interest. This requires that $\epsilon_p^0 \gg n_0 U_0$. Expressed in terms of the temperature, this corresponds to $T \gg T_*$, Eq. (8.64).

To explore the physics further, consider a Bose gas in a general spatially-dependent potential. Let us assume that the wave function has the form of a product of single-particle states symmetrized with respect to interchange of particles:

$$\Psi(\mathbf{r}_1, \mathbf{r}_2, \ldots, \mathbf{r}_N) = c_N \sum_{\text{sym}} \phi_1(\mathbf{r}_1)\phi_2(\mathbf{r}_2) \ldots \phi_N(\mathbf{r}_N). \tag{8.65}$$

Here the wave functions of the occupied single-particle states are denoted by ϕ_i. If, for example, there are N_α particles in the state α, that state will occur N_α times in the sequence of single-particle states in the product. The sum denotes symmetrization with respect to interchange of particle coordinates, and

$$c_N = \left(\frac{\prod_i N_i!}{N!}\right)^{1/2} \tag{8.66}$$

is a normalization factor. This wave function is the natural generalization of the wave function (6.1) when all particles are in the same state. To bring out the physics it is again helpful to consider a non-zero-range interaction $U(\mathbf{r} - \mathbf{r}')$, where $\mathbf{r}$ and $\mathbf{r}'$ are the coordinates of the two atoms, rather than a contact one (cf. Eq. (8.7)). In the expression for the interaction energy for the wave function (8.65) there are two sorts of terms. The first are ones which would occur if the wave function had not been symmetrized. These are the so-called direct, or Hartree, terms and they contain contributions of the form

$$U_{ij}^{\text{Hartree}} = \int d\mathbf{r} d\mathbf{r}' U(\mathbf{r} - \mathbf{r}')|\phi_i(\mathbf{r})|^2|\phi_j(\mathbf{r}')|^2, \tag{8.67}$$

which is the energy of a pair of particles in the state $\phi_i(\mathbf{r})\phi_j(\mathbf{r}')$. The second class of terms, referred to as exchange or Fock terms, arise because of the symmetrization of the wave function, and have the form

$$U_{ij}^{\text{Fock}} = \int d\mathbf{r} d\mathbf{r}' U(\mathbf{r} - \mathbf{r}')\phi_i^*(\mathbf{r})\phi_j^*(\mathbf{r}')\phi_i(\mathbf{r}')\phi_j(\mathbf{r}). \tag{8.68}$$

For fermions, the Fock term has the opposite sign because of the antisymmetry of the wave function. Calculation of the coefficients of these terms may be carried out using the wave function (8.65) directly, but it is much more convenient to use the formalism of second quantization which is designed expressly for calculating matrix elements of operators between wave functions of the form (8.65). The general expression for the interaction energy operator in this notation is [5]

$$U = \frac{1}{2} \sum_{ijkl} \langle ij|U|kl\rangle a_i^\dagger a_j^\dagger a_l a_k, \tag{8.69}$$

where

$$\langle ij|U|kl\rangle = \int d\mathbf{r}d\mathbf{r}' U(\mathbf{r}-\mathbf{r}')\phi_i^*(\mathbf{r})\phi_j^*(\mathbf{r}')\phi_l(\mathbf{r}')\phi_k(\mathbf{r}) \tag{8.70}$$

is the matrix element of the two-body interaction. To evaluate the potential energy we thus need to calculate the expectation value of $a_i^\dagger a_j^\dagger a_l a_k$. Since the single-particle states we work with are assumed to be orthogonal, the expectation value of this operator vanishes unless the two orbitals in which particles are destroyed are identical with those in which they are created. There are only two ways to ensure this: either $i=k$ and $j=l$, or $i=l$ and $j=k$. The matrix element for the first possibility is $N_i(N_j-\delta_{ij})$, δ_{ij} being the Kronecker delta function, since for bosons a_i ($a_i^\dagger$) acting on a state with N_i particles in the single-particle state i yields $\sqrt{N_i}$ ($\sqrt{N_i+1}$) times the state with one less (more) particle in that single-particle state. For fermions the corresponding factors are $\sqrt{N_i}$ and $\sqrt{1-N_i}$. The matrix element for the second possibility is N_iN_j if we exclude the situation when all four states are the same, which has already been included in the first case. For bosons, the expectation value of the interaction energy is therefore

$$\begin{aligned}U &= \frac{1}{2}\sum_{ij}\langle ij|U|ij\rangle N_i(N_j-\delta_{ij}) + \frac{1}{2}\sum_{ij(i\neq j)}\langle ij|U|ji\rangle N_iN_j \\ &= \frac{1}{2}\sum_i \langle ii|U|ii\rangle N_i(N_i-1) + \sum_{i<j}(\langle ij|U|ij\rangle + \langle ij|U|ji\rangle)N_iN_j.\end{aligned} \tag{8.71}$$

The contributions containing $\langle ij|U|ij\rangle$ are direct terms, and those containing $\langle ij|U|ji\rangle$ are exchange terms.

For a contact interaction, the matrix elements $\langle ij|U|ij\rangle$ and $\langle ij|U|ji\rangle$ are identical, as one can see from Eq. (8.70), and therefore the interaction energy is

$$U = \frac{1}{2}\sum_i \langle ii|U|ii\rangle N_i(N_i-1) + 2\sum_{i<j}\langle ij|U|ij\rangle N_iN_j. \tag{8.72}$$

For bosons, the effect of exchange is to double the term proportional to N_iN_j, whereas for fermions in the same internal state, the requirement of antisymmetry of the wave function leads to a cancellation and the total potential energy vanishes for a contact interaction.

Let us now turn to the homogeneous Bose gas. From translational invariance it follows that the single-particle wave functions must be plane waves. The matrix elements of the interaction are U_0/V, provided the initial and final states have the same total momentum, and the energy of a state of the form (8.65) is

$$E = \sum_{\mathbf{p}} \epsilon_p^0 N_{\mathbf{p}} + \frac{U_0}{2V} \sum_{\mathbf{p},\mathbf{p}'} N_{\mathbf{p}}(N_{\mathbf{p}'} - \delta_{\mathbf{p},\mathbf{p}'}) + \frac{U_0}{2V} \sum_{\mathbf{p},\mathbf{p}'(\mathbf{p}\neq\mathbf{p}')} N_{\mathbf{p}} N_{\mathbf{p}'} \quad (8.73)$$

$$= \sum_{\mathbf{p}} \epsilon_p^0 N_{\mathbf{p}} + \frac{U_0}{2V} N(N-1) + \frac{U_0}{2V} \sum_{\mathbf{p},\mathbf{p}'(\mathbf{p}\neq\mathbf{p}')} N_{\mathbf{p}} N_{\mathbf{p}'} \quad (8.74)$$

$$= \sum_{\mathbf{p}} \epsilon_p^0 N_{\mathbf{p}} + \frac{U_0}{V}\left(N^2 - \frac{1}{2}\sum_{\mathbf{p}} N_{\mathbf{p}}^2 - \frac{N}{2}\right). \quad (8.75)$$

If the zero-momentum state is the only macroscopically-occupied one and we take only terms of order N^2, the interaction energy is $(N^2 - N_0^2/2)U_0/V$. Thus for a system with a given number of particles, the interaction energy above T_c is twice that at zero temperature, due to the existence of the exchange term, which is not present for a pure condensate.

The energy $\epsilon_{\mathbf{p}}$ of an excitation of momentum $\mathbf{p}$ is obtained from (8.75) by changing $N_{\mathbf{p}}$ to $N_{\mathbf{p}} + 1$, thereby adding a particle to the system. It is

$$\epsilon_{\mathbf{p}} = \epsilon_p^0 + \frac{N}{V}U_0 + \frac{N - N_{\mathbf{p}}}{V}U_0. \quad (8.76)$$

The first interaction term is the Hartree contribution, which represents the direct interaction of the added particle with the N particles in the original system. The second interaction term, the Fock one, is due to exchange, and it is proportional to the number of particles in states different from that of the added particle, since there is no exchange contribution between atoms in the same single-particle state.[4]

Let us now calculate the energy of an excitation when one state, which we take to be the zero-momentum state, is macroscopically occupied, while all others are not. To within terms of order $1/N$ we may therefore write

$$\epsilon_{\mathbf{p}=0} = (2n - n_0)U_0 = (n_0 + 2n_{\text{ex}})U_0, \quad (8.77)$$

where n is the total density of particles, n_0 is the density of particles in the

---

[4] The interaction that enters the usual Hartree and Hartree–Fock approximations is the *bare* interaction between particles. Here we are using effective interactions, not bare ones, and the approximation is therefore more general than the conventional Hartree–Fock one. It is closer to the Brueckner–Hartree–Fock method in nuclear physics, but we shall nevertheless refer to it as the Hartree–Fock approximation.

condensate, and $n_{ex} = n - n_0$ is the number of non-condensed particles. For other states one has

$$\epsilon_\mathbf{p} = \epsilon_p^0 + 2nU_0. \tag{8.78}$$

We shall now apply these results to calculate equilibrium properties.

Thermal equilibrium

To investigate the thermodynamics in the Hartree–Fock approximation we follow the same path as for the non-interacting gas, but with the Hartree–Fock expression for the energy rather than the free-particle one. The way in which one labels states in the Hartree–Fock approximation is the same as for free particles, and therefore the entropy S is given by the usual result for bosons

$$S = k \sum_\mathbf{p} [(1 + f_\mathbf{p}) \ln(1 + f_\mathbf{p}) - f_\mathbf{p} \ln f_\mathbf{p}], \tag{8.79}$$

where $f_\mathbf{p}$ is the average occupation number for states with momenta close to $\mathbf{p}$. The equilibrium distribution is obtained by maximizing the entropy subject to the condition that the total energy and the total number of particles be fixed. The excitations we work with here correspond to adding a single atom to the gas. Thus in maximizing the entropy, one must introduce the chemical potential term to maintain the particle number at a constant value. This is to be contrasted with the Bogoliubov approximation where we implemented the constraint on the particle number explicitly. The equilibrium distribution is thus

$$f_\mathbf{p} = \frac{1}{\exp[(\epsilon_\mathbf{p} - \mu)/kT] - 1}, \tag{8.80}$$

where the excitation energies are given by the Hartree–Fock expressions (8.77) and (8.78).

For the zero-momentum state to be macroscopically occupied, the energy to add a particle to that state must be equal to the chemical potential, to within terms of order $1/N$. Thus according to Eq. (8.77),

$$\mu = (n_0 + 2n_{ex})U_0, \tag{8.81}$$

and the average occupancy of the other states is given by

$$f_\mathbf{p} = \frac{1}{\exp[(\epsilon_p^0 + n_0 U_0)/kT] - 1}. \tag{8.82}$$

For consistency the number of non-condensed particles must be given by

$$N - N_0 = \sum_{\mathbf{p}(\mathbf{p}\neq 0)} f_{\mathbf{p}}$$

$$= \sum_{\mathbf{p}(\mathbf{p}\neq 0)} \frac{1}{\exp[(\epsilon_p^0 + n_0 U_0)/kT] - 1}. \quad (8.83)$$

This provides a self-consistency condition for the number of particles in the condensate, since $n_0 = N_0/V$ occurs in the distribution function.

Above the transition temperature energies of all states are shifted by the same amount, and consequently the thermodynamic properties of the gas are simply related to those of the non-interacting gas. In particular the energy and the free energy are the same as that of a perfect Bose gas apart from an additional term $N^2 U_0/V$. Since this does not depend on temperature, interactions have no effect on the transition temperature.

We now formulate the theory in terms of creation and annihilation operators. If the zero-momentum state is the only one which is macroscopically occupied, the terms in the Hamiltonian (8.3) that contribute to the expectation value (8.73) of the energy for a state whose wave function is a symmetrized product of plane-wave states may be written as

$$H = \frac{N_0^2 U_0}{2V} + \sum_{\mathbf{p}(\mathbf{p}\neq 0)} (\epsilon_p^0 + 2n_0 U_0) a_{\mathbf{p}}^\dagger a_{\mathbf{p}} + \frac{U_0}{V} \sum_{\mathbf{pp'}(\mathbf{p},\mathbf{p'}\neq 0)} a_{\mathbf{p}}^\dagger a_{\mathbf{p}} a_{\mathbf{p'}}^\dagger a_{\mathbf{p'}}, \quad (8.84)$$

where we have neglected terms of order $1/N$.

We now imagine that the system is in a state close to one with average occupation number $f_{\mathbf{p}}$, and ask what the Hamiltonian is for small changes in the number of excitations. We therefore write

$$a_{\mathbf{p}}^\dagger a_{\mathbf{p}} = f_{\mathbf{p}} + (a_{\mathbf{p}}^\dagger a_{\mathbf{p}} - f_{\mathbf{p}}), \quad (8.85)$$

and expand the Hamiltonian to first order in the fluctuation term. As we did in making the Bogoliubov approximation, we neglect fluctuations in the occupation of the zero-momentum state. The result is

$$H = \frac{N_0^2 U_0}{2V} - \frac{U_0}{V} \sum_{\mathbf{pp'}(\mathbf{p},\mathbf{p'}\neq 0)} f_{\mathbf{p}} f_{\mathbf{p'}} + \sum_{\mathbf{p}(\mathbf{p}\neq 0)} (\epsilon_p^0 + 2n U_0) a_{\mathbf{p}}^\dagger a_{\mathbf{p}}. \quad (8.86)$$

This shows that the energy to add a particle to a state with non-zero momentum is $\epsilon_p^0 + 2nU_0$, which is the Hartree–Fock expression for the excitation energy derived in Eq. (8.78).

It is also of interest to calculate the chemical potential, which is the energy to add a particle, the entropy being held constant:

$$\mu = \left.\frac{\partial E}{\partial N}\right|_S. \tag{8.87}$$

The entropy associated with the zero-momentum state is zero, and therefore a simple way to evaluate the derivative is to calculate the energy change when a particle is added to the condensate, keeping the distribution of excitations fixed:

$$\mu = \left.\frac{\partial E}{\partial N_0}\right|_{f_{\mathbf{p}(\mathbf{p}\neq 0)}} = (n_0 + 2n_{\text{ex}})U_0. \tag{8.88}$$

This result agrees with Eq. (8.81).

Since the energy ϵ_p of an excitation tends to $2nU_0$ for $p \to 0$, the excitation energy measured with respect to the chemical potential has a gap $n_0 U_0$ in this approximation. The long-wavelength excitations in the Hartree–Fock approximation are particles moving in the mean field of the other particles, whereas physically one would expect them to be sound waves, with no gap in the spectrum, as we found in the Bogoliubov theory. We now show how to obtain a phonon-like spectrum at non-zero temperatures.

8.3.2 The Popov approximation

To go beyond the Hartree–Fock approximation one must allow for the mixing of particle-like and hole-like excitations due to the interaction, which is reflected in the coupling of the equations for u and v. This effect is important for momenta for which ϵ_p^0 is comparable with or less than $n_0 U_0$. A simple way to do this is to add to the Hartree–Fock Hamiltonian (8.86) the terms in the Bogoliubov Hamiltonian (8.6) that create and destroy pairs of particles. The Hamiltonian is therefore[5]

$$H = \frac{N_0^2 U_0}{2V} - \frac{U_0}{V} \sum_{\mathbf{p}\mathbf{p}'(\mathbf{p},\mathbf{p}'\neq 0)} f_{\mathbf{p}} f_{\mathbf{p}'} + \sum_{\mathbf{p}(\mathbf{p}\neq 0)} (\epsilon_p^0 + 2nU_0) a_{\mathbf{p}}^\dagger a_{\mathbf{p}}$$
$$+ n_0 U_0 {\sum_{\mathbf{p}(\mathbf{p}\neq 0)}}' (a_{\mathbf{p}}^\dagger a_{-\mathbf{p}}^\dagger + a_{\mathbf{p}} a_{-\mathbf{p}}). \tag{8.89}$$

This approximation is usually referred to as the Popov approximation [6]. Rather than working with the Hamiltonian itself, we consider the quantity

---

[5] In this approximation the effect of the thermal excitations on the 'pairing' terms proportional to $a_{\mathbf{p}}^\dagger a_{-\mathbf{p}}^\dagger + a_{\mathbf{p}} a_{-\mathbf{p}}$ has been neglected. More generally one could replace in the last term of (8.89) n_0 by $n_0 + V^{-1} \sum_{\mathbf{p}\neq 0} A_{\mathbf{p}}$ where $A_{\mathbf{p}}$ is the average value of $a_{\mathbf{p}} a_{-\mathbf{p}}$, which is non-zero when excitations are present. However, this approximation suffers from the disadvantage that the energy of a long-wavelength elementary excitation does not tend to zero.

$H - \mu \hat{N}$ to take care of the requirement that particle number be conserved. One finds

$$H - \mu \hat{N} = -\frac{N_0^2 U_0}{2V} - \frac{2N_0 N_{\text{ex}} U_0}{V} - \frac{N_{\text{ex}}^2 U_0}{V}$$
$$+ \sum_{\mathbf{p}(\mathbf{p}\neq 0)}{}' \left[(\epsilon_p^0 + n_0 U_0)(a_{\mathbf{p}}^\dagger a_{\mathbf{p}} + a_{-\mathbf{p}}^\dagger a_{-\mathbf{p}}) + n_0 U_0 (a_{\mathbf{p}}^\dagger a_{-\mathbf{p}}^\dagger + a_{\mathbf{p}} a_{-\mathbf{p}}) \right],$$
(8.90)

where

$$N_{\text{ex}} = \sum_{\mathbf{p}(\mathbf{p}\neq 0)} <a_{\mathbf{p}}^\dagger a_{\mathbf{p}}> \qquad (8.91)$$

is the expectation value of the number of excited particles. Remarkably, the form of the Hamiltonian is identical with the Bogoliubov one for zero temperature, Eq. (8.12), except that the c-number term is different. Also, the occupancy of the zero-momentum state must be determined self-consistently. The spectrum is thus given by the usual Bogoliubov expression, with the density of the condensate being the one at the temperature of interest, not the zero-temperature value. Again the long-wavelength excitations are phonons, with a speed s given by

$$s(T)^2 = \frac{n_0(T) U_0}{m}, \qquad (8.92)$$

and the coherence length that determines the transition to free-particle behaviour is given by

$$\xi(T) = \left[\frac{\hbar^2}{2m n_0(T) U_0} \right]^{1/2} = \frac{\hbar}{\sqrt{2} m s(T)}. \qquad (8.93)$$

This is to be contrasted with the Hartree–Fock spectrum, where the excitation energy at long wavelengths differs from the chemical potential by an amount $n_0 U_0$.

8.3.3 Excitations in non-uniform gases

The Hartree–Fock and Popov approximations may be applied to excitations in trapped gases. In the Hartree–Fock approximation the wave function ϕ_i for an excited state satisfies the equation

$$\left[-\frac{\hbar^2}{2m} \nabla^2 + V(\mathbf{r}) + 2n(\mathbf{r}) U_0 \right] \phi_i(\mathbf{r}) = \epsilon_i \phi_i(\mathbf{r}), \qquad (8.94)$$

and the corresponding equation for the wave function ϕ_0 for the condensed state is

$$\left\{-\frac{\hbar^2}{2m}\nabla^2 + V(\mathbf{r}) + [n_0(\mathbf{r}) + 2n_{\text{ex}}(\mathbf{r})]U_0\right\}\phi_0(\mathbf{r}) = \mu\phi_0(\mathbf{r}), \quad (8.95)$$

the absence of a factor of 2 in the condensate density term reflecting the fact that there is no exchange term for two atoms in the same state. Here $n_0(\mathbf{r}) = N_0|\phi_0(\mathbf{r})|^2$ is the density of atoms in the condensed state and $n_{\text{ex}}(\mathbf{r}) = \sum_{i\neq 0} N_i|\phi_i(\mathbf{r})|^2$ is the density of non-condensed particles. The wave functions and occupation numbers are determined self-consistently by imposing the conditions $N_i = \{\exp[(\epsilon_i - \mu)/kT] - 1\}^{-1}$ for $i \neq 0$ and $N = N_0 + \sum_{i\neq 0} N_i$.

The equations for the Popov approximation are

$$\left[-\frac{\hbar^2}{2m}\nabla^2 + V(\mathbf{r}) + 2n(\mathbf{r})U_0 - \mu - \epsilon_i\right]u_i(\mathbf{r}) - n_0(\mathbf{r})U_0 v_i(\mathbf{r}) = 0 \quad (8.96)$$

and

$$\left[-\frac{\hbar^2}{2m}\nabla^2 + V(\mathbf{r}) + 2n(\mathbf{r})U_0 - \mu + \epsilon_i\right]v_i(\mathbf{r}) - n_0(\mathbf{r})U_0 u_i(\mathbf{r}) = 0, \quad (8.97)$$

for the excitations. The wave function for the condensed state satisfies the generalized Gross–Pitaevskii equation

$$\left\{-\frac{\hbar^2}{2m}\nabla^2 + V(\mathbf{r}) + [n_0(\mathbf{r}) + 2n(\mathbf{r})]U_0 - \mu\right\}\phi_0(\mathbf{r}) = 0, \quad (8.98)$$

in which there is an extra contribution to the potential due to interaction of the condensate with the non-condensed particles. The density of non-condensed particles is given by $n_{\text{ex}}(\mathbf{r}) = \sum_{i\neq 0} N_i(|u_i(\mathbf{r})|^2 + |v_i(\mathbf{r})|^2)$.

In the Hartree–Fock and Popov approximations the only effect of interactions between particles is to provide static mean fields that couple either to the density of particles or create or destroy pairs of particles. A difficulty with this approach may be seen by considering a cloud of gas in a harmonic trap. This has collective modes associated with the motion of the centre of mass of the cloud, and they have frequencies which are sums of multiples of the oscillator frequencies. In the Hartree–Fock and Popov approximations the corresponding modes have different frequencies since the static potential acting on the excitations is affected by particle interactions. To cure these difficulties it is necessary to allow for coupling between the motion of the condensate and that of the thermal excitations. We shall describe this effect for uniform systems in Sec. 10.4, where we consider first and second sound. A related effect is that at non-zero temperature the effective two-body interaction is affected by the other excitations. This is connected with the problem mentioned in footnote 5 of how to treat 'pairing' terms consistently. These effects can be important for low-lying excitations, but are generally of little consequence for higher-energy excitations. For a discussion of them we refer to Refs. [7] and [8].

8.3.4 The semi-classical approximation

In Sec. 2.3.1 we showed how the properties of a trapped cloud of non-interacting particles may be described semi-classically. This approximation holds provided the typical de Broglie wavelengths of particles are small compared with the length scales over which the trapping potential and the particle density vary significantly. Locally the gas may then be treated as uniform. Properties of non-condensed particles may be calculated using a semi-classical distribution function $f_\mathbf{p}(\mathbf{r})$ and particle energies given by

$$\epsilon_\mathbf{p}(\mathbf{r}) = \frac{p^2}{2m} + V(\mathbf{r}). \tag{8.99}$$

When particles interact, the properties of the excitations may still be described semi-classically subject to the requirement that spatial variations occur over distances large compared with the wavelengths of typical excitations. The properties of the condensed state must generally be calculated from the Gross–Pitaevskii equation generalized to allow for the interaction of the condensate with the thermal excitations. With a view to later applications we now describe the semi-classical versions of the Hartree–Fock, Bogoliubov, and Popov approximations.

Within Hartree–Fock theory, the semi-classical energies are given by

$$\epsilon_\mathbf{p}(\mathbf{r}) = p^2/2m + 2n(\mathbf{r})U_0 + V(\mathbf{r}), \tag{8.100}$$

where we have generalized the result (8.78) by adding to it the potential energy $V(\mathbf{r})$. In determining thermodynamic properties, the energy of an excitation enters in the combination $\epsilon_i - \mu$. A simple expression for this may be found if the Thomas–Fermi approximation is applicable for the condensate, which is the case if length scales for variations of the condensate density and the potential are large compared with the coherence length. In the Thomas–Fermi approximation the chemical potential is given by adding the contribution $V(\mathbf{r})$ to the result (8.88) and is

$$\mu = V(\mathbf{r}) + [n_0(\mathbf{r}) + 2n_{\mathrm{ex}}(\mathbf{r})]U_0. \tag{8.101}$$

8.3 Non-zero temperature

The semi-classical limit of the Bogoliubov approximation is obtained by replacing in the coupled differential equations (8.52) and (8.53) the kinetic energy operator by $p^2/2m$. The energies $\epsilon_\mathbf{p}(\mathbf{r})$ are found by setting the determinant equal to zero, with the result

$$\epsilon_\mathbf{p}(\mathbf{r}) = \{[p^2/2m + 2n_0(\mathbf{r})U_0 + V(\mathbf{r}) - \mu]^2 - [n_0(\mathbf{r})U_0]^2\}^{1/2}. \quad (8.102)$$

Here the chemical potential μ is that which enters the zero-temperature Gross–Pitaevskii equation. In the Thomas–Fermi approximation (see Sec. 6.2.2) the chemical potential is obtained by neglecting the kinetic energy in the Gross–Pitaevskii equation, which yields $\mu = V(\mathbf{r}) + n_0(\mathbf{r})U_0$. The semi-classical Bogoliubov excitation energies are therefore

$$\epsilon_\mathbf{p}(\mathbf{r}) = \left[(p^2/2m)^2 + (p^2/m)n_0(\mathbf{r})U_0\right]^{1/2}. \quad (8.103)$$

Finally, the semi-classical limit of the Popov approximation is obtained by generalizing the equations (8.52) and (8.53) to higher temperatures. Inspection of the Hamiltonian (8.89) shows that the term $2n_0(\mathbf{r})U_0$, which occurs in both (8.52) and (8.53), should be replaced by $2n(\mathbf{r})U_0$. The semi-classical energies within the Popov approximation therefore become

$$\epsilon_\mathbf{p}(\mathbf{r}) = \left([p^2/2m + 2n(\mathbf{r})U_0 + V(\mathbf{r}) - \mu]^2 - [n_0(\mathbf{r})U_0]^2\right)^{1/2}, \quad (8.104)$$

where the condensate density n_0 is to be determined self-consistently. If the Thomas–Fermi expression (8.101) for the chemical potential is valid, the excitation spectrum is identical with the result of the Bogoliubov theory, Eq. (8.103), except that the condensate density n_0 now depends on temperature. Note that in the Hartree–Fock approximation the excitations correspond to addition of a particle, while in the Bogoliubov and Popov ones the excitation energies given above are evaluated for no change in the total particle number. If the term $[n_0(\mathbf{r})U_0]^2$ in Eqs. (8.102) and (8.104) is neglected, the excitation energy becomes equal to the Hartree–Fock result for $\epsilon_\mathbf{p}(\mathbf{r}) - \mu$.

The density of non-condensed atoms is given in the Bogoliubov approximation by

$$n_{\text{ex}}(\mathbf{r}) = \int \frac{d\mathbf{p}}{(2\pi\hbar)^3} \frac{p^2/2m + 2n_0(\mathbf{r})U_0 + V(\mathbf{r}) - \mu}{\epsilon_\mathbf{p}(\mathbf{r})} \frac{1}{e^{\epsilon_\mathbf{p}(\mathbf{r})/kT} - 1}, \quad (8.105)$$

in the Hartree–Fock approximation by

$$n_{\text{ex}}(\mathbf{r}) = \int \frac{d\mathbf{p}}{(2\pi\hbar)^3} \frac{1}{e^{(\epsilon_\mathbf{p}(\mathbf{r}) - \mu)/kT} - 1}, \quad (8.106)$$

and in the Popov approximation by

$$n_{\rm ex}({\bf r}) = \int \frac{d{\bf p}}{(2\pi\hbar)^3} \frac{p^2/2m + 2n({\bf r})U_0 + V({\bf r}) - \mu}{\epsilon_{\bf p}({\bf r})} \frac{1}{e^{\epsilon_{\bf p}({\bf r})/kT} - 1}. \quad (8.107)$$

The factors multiplying the distribution functions in the Bogoliubov and Popov approximations are the numbers of non-condensed particles associated with an excitation. In the Hartree–Fock and Popov approximations the density of non-condensed atoms and the number of particles in the condensate must be determined self-consistently. Applications of the results of this section to atoms in traps will be described in Chapters 10 and 11.

8.4 Collisional shifts of spectral lines

Because interactions between atoms depend on the internal atomic states, the frequency of a transition between two different states of an atom in a dilute gas differs from the frequency for the free atom. These shifts are referred to as *collisional shifts* or, because the accuracy of atomic clocks is limited by them, *clock shifts*. The basic physical effect is virtual scattering processes like those that give rise to the effective interaction U_0 used earlier in our discussion of interactions between two bosons in the same internal state, and the magnitude of the effect is proportional to atomic scattering lengths. This is to be contrasted with the rate of real scattering processes, which is proportional to U_0^2.

As described in Sec. 4.7, the collisional shift of the 1S–2S transition excited by absorption of two photons is used to measure the gas density in experiments on spin-polarized hydrogen. We now describe the physics of collisional shifts, using this transition as an example. The theory of them was originally developed in the context of hydrogen masers [9]. To understand the effect it is necessary to go beyond the Hartree–Fock picture described above and consider collective effects, but to set the stage we shall describe Hartree–Fock theory.

Hartree–Fock theory

For simplicity, we assume the gas to be uniform. In the spirit of the Hartree–Fock approximation, we take the initial state of the gas before excitation of one of the hydrogen atoms to the 2S state to be a symmetrized product of single-particle states, as given by Eq. (8.65), where we shall take the particle states to be plane waves, $\phi_{\bf p}({\bf r}) = V^{-1/2} \exp(i{\bf p} \cdot {\bf r}/\hbar)$. If we denote the momenta of the particles by $\mathbf{p}_a \ldots \mathbf{p}_l$, the wave function is

$$\Psi(\mathbf{r}_1, \mathbf{r}_2, \ldots, \mathbf{r}_N) = c_N \sum_{\text{sym}} \phi_{\mathbf{p}_a}(\mathbf{r}_1) \phi_{\mathbf{p}_b}(\mathbf{r}_2) \ldots \phi_{\mathbf{p}_l}(\mathbf{r}_N), \tag{8.108}$$

and the energy of the initial state in the Hartree–Fock approximation is given by Eq. (8.74).

Next we consider a state with one atom in the 2S state and the remainder in the 1S state, and we shall assume that the momenta of the atoms are the same as in the state (8.108). To be specific we assume that the momentum of the 2S atom is $\mathbf{p}_a$. In the Hartree–Fock approximation the wave function for the gas is the wave function for the 2S atom multiplied by that for $N-1$ 1S atoms, Eq. (8.108), but with N replaced by $N-1$:

$$\Psi(\mathbf{r}_1; \mathbf{r}_2, \ldots, \mathbf{r}_N) = \phi^{2S}_{\mathbf{p}_a}(\mathbf{r}_1) c_{N-1} \sum_{\text{sym}} \phi_{\mathbf{p}_b}(\mathbf{r}_2) \ldots \phi_{\mathbf{p}_l}(\mathbf{r}_N). \tag{8.109}$$

Here, the coordinate $\mathbf{r}_1$ before the semicolon is that of the 2S atom, and $\mathbf{r}_2, \ldots, \mathbf{r}_N$ are the coordinates of the 1S atoms. The momenta of the single-particle states are denoted by $\mathbf{p}_a, \mathbf{p}_b, \ldots, \mathbf{p}_l$, just as for the state with N 1S atoms.

The Hamiltonian for a mixture of 1S and 2S atoms is given by the sum of the energies of the isolated atoms at rest, the kinetic energy of the atoms, the interaction energy of the 1S atoms with each other, and the interaction between 1S and 2S atoms. Since we shall consider weak excitation of the gas, the density of 2S atoms is much lower than that of 1S atoms, and the interaction between two 2S atoms will not enter. The relative difference between the mass m of a 1S atom and that of a 2S one is of order one part in 10^8, and this may be neglected in evaluating the kinetic energy. We shall take the interaction between the 2S atom and the 1S ones to be of the contact form similar to that between like atoms,

$$H_{12} = U_{12} \int d\mathbf{r} \hat{\psi}^\dagger(\mathbf{r}) \hat{\psi}^\dagger_{2S}(\mathbf{r}) \hat{\psi}_{2S}(\mathbf{r}) \hat{\psi}(\mathbf{r}), \tag{8.110}$$

where $U_{12} = 4\pi\hbar^2 a_{12}/m$, a_{12} being the scattering length for interactions between a 1S atom and a 2S one, and the operators $\hat{\psi}^\dagger_{2S}(\mathbf{r})$ and $\hat{\psi}_{2S}(\mathbf{r})$ create and destroy 2S atoms. At the end of the section we shall comment on the validity of this approximation.

The energy E' of the state (8.109) may be calculated by the methods used earlier for a single component. Measured with respect to the energy of N stationary, isolated atoms, one in the 2S state and the others in the 1S state, it is

$$E' = E_{\text{kin}} + \frac{U_0}{2V}(N-1)(N-2) + \frac{U_0}{2V}\sum_{\mathbf{p}\neq\mathbf{p}'}N_{\mathbf{p}}N_{\mathbf{p}'}$$
$$- \frac{U_0}{V}(N-N_{\mathbf{p}_a}) + (N-1)\frac{U_{12}}{V}. \qquad (8.111)$$

Note that, while $\mathbf{p}_a$ is the momentum of the 2S atom, $N_{\mathbf{p}_a}$ is the number of 1S atoms with the same momentum $\mathbf{p}_a$ as that of the 2S atom. The first and third terms are the kinetic energy and the exchange energy of the original state with N 1S atoms, and the second term is the Hartree energy of $N-1$ 1S atoms. The last term is the Hartree energy due to the interaction of the single 2S atom with $N-1$ 1S atoms, and the next-to-last term is the reduction of the exchange energy due to replacing one 1S atom by a 2S one. The energy difference between the states (8.109) and (8.108) is therefore the difference between the rest-mass energies of a 2S atom and a 1S one, which is the energy of the transition in an isolated atom, plus the differences between Eqs. (8.111) and (8.74), which is given by

$$\Delta E = nU_{12} - U_0(2n - \frac{N_{\mathbf{p}_a}}{V}), \qquad (8.112)$$

where we have neglected terms of relative order $1/N$. The quantity ΔE represents the energy shift of the line due to interactions between atoms. The first term here is the interaction contribution to the Hartree energy of the added 2S atom, and the second term is the reduction of the Hartree and Fock energies due to removal of a 1S atom.

It is instructive to consider limiting cases of this expression. For a pure condensate with all atoms initially in the zero-momentum state, only atoms in that state can be excited and therefore $N_{\mathbf{p}_a} = N_0 = N$. The energy difference is then given by

$$\Delta E_{\text{cond}} = n(U_{12} - U_0). \qquad (8.113)$$

This is the difference of the Hartree energies. Exchange terms are absent for 1S–1S interactions, since all 1S atoms are in the same momentum state. There are also no exchange contributions for 1S–2S interactions because 1S atoms and 2S ones are in different internal states, and are therefore distinguishable. In the opposite limit, when there is no condensate, we find an energy difference

$$\Delta E_{\text{nc}}^{\text{HF}} = n(U_{12} - 2U_0), \qquad (8.114)$$

the extra factor of 2 for the U_0 term being due to exchange interactions between the 1S atoms.

Collective effects

At first sight one might expect the line shift to be given by the energy difference calculated above. This is not true for the thermal gas, since the product wave function is then a bad approximation for the final state after optical excitation of an atom in the gas. To understand why this is so, we describe the excitation process in greater detail. In the experiments, two counterpropagating laser beams with frequencies equal to one-half that of the 1S–2S transition are applied. Excitation may occur by absorption of two photons propagating either in the same direction, or in opposite directions. We focus attention on the latter process, which is particularly important because the total momentum of the two photons is zero, and consequently the line has no first-order Doppler shift. It is therefore very sharp, and relatively small frequency shifts can be measured.

The effective two-photon interaction is local in space: it destroys a 1S atom and creates a 2S one at essentially the same point. Also, for absorption of two photons propagating in opposite directions, the effective interaction is independent of position, because the phase factors $e^{\pm i\mathbf{q}\cdot\mathbf{r}}$ from the two photons cancel. The effective coupling between the atom and the photons is spatially independent, and it may therefore be represented by an operator which in second-quantized notation may be written as

$$\mathcal{O} = \int d\mathbf{r}\, \hat{\psi}_{2S}^{\dagger}(\mathbf{r}) \hat{\psi}(\mathbf{r}) \tag{8.115}$$

apart from a multiplicative constant. When this operator acts on a state with N 1S atoms, it generates N terms, one for each way of assigning the point $\mathbf{r}$ to one of the particle coordinates. If the initial state is of the form of a symmetrized product of single-particle states, each of these terms individually will be of the form of a single-particle wave function for the 2S atom, multiplied by the wave function for $N-1$ atoms in the 1S state. For definiteness, let us denote the state with N ground-state atoms by

$$|\Psi\rangle = |\{N_{\mathbf{p}_i}\}\rangle. \tag{8.116}$$

The operator $\mathcal{O}$ acting on this state gives a linear combination of states,

$$\mathcal{O}|\Psi\rangle = \sum \sqrt{N_{\mathbf{p}'}}|\mathbf{p}'; N_{\mathbf{p}_a}, \ldots, N_{\mathbf{p}'} - 1, \ldots\rangle, \tag{8.117}$$

where the sum is over all momenta which correspond to occupied single-particle states initially. This demonstrates that two-photon absorption in general produces a state which is not a simple product state of the form assumed in Hartree–Fock theory, but rather a linear combination of such states. In the language of quantum optics, it is an entangled state.

To find the energy eigenvalues for a 2S atom in the presence of $N-1$ 1S atoms we must diagonalize the Hamiltonian in a basis of states of the form $|\mathbf{p}'; N_{\mathbf{p}_a}, \ldots, N_{\mathbf{p}'} - 1, \ldots\rangle$. The problem is trivial if all atoms in the initial state are in the same single-particle state, since then the sum in Eq. (8.117) reduces to the single configuration with the $N-1$ 1S atoms and the 2S atom in the zero-momentum state. The energy of this state is given by the Hartree result, and the difference between the energies of the initial and final states is given by Eq. (8.113).

The other situation that is simple to analyse is when there is no condensate. Remarkably, the state (8.117) is then an energy eigenstate, but its energy is not equal to the Hartree–Fock energy. We may write the Hamiltonian for the 1S and 2S atoms as

$$H = \sum_{\mathbf{p}} \epsilon_p^0 (a_{\mathbf{p}}^\dagger a_{\mathbf{p}} + \bar{a}_{\mathbf{p}}^\dagger \bar{a}_{\mathbf{p}}) + \frac{U_0}{2V} \sum_{\mathbf{p},\mathbf{p}',\mathbf{q}} a_{\mathbf{p}+\mathbf{q}}^\dagger a_{\mathbf{p}'-\mathbf{q}}^\dagger a_{\mathbf{p}'} a_{\mathbf{p}}$$

$$+ \frac{U_{12}}{V} \sum_{\mathbf{p},\mathbf{p}',\mathbf{q}} \bar{a}_{\mathbf{p}+\mathbf{q}}^\dagger a_{\mathbf{p}'-\mathbf{q}}^\dagger a_{\mathbf{p}'} \bar{a}_{\mathbf{p}}, \qquad (8.118)$$

where the operators $\bar{a}_{\mathbf{p}}^\dagger$ and $\bar{a}_{\mathbf{p}}$, which are the Fourier transforms of $\hat{\psi}_{2S}^\dagger(\mathbf{r})$ and $\hat{\psi}_{2S}(\mathbf{r})$, create and destroy 2S atoms in momentum states. It is left as an exercise (Problem 8.4) to show that the state (8.117) is an eigenstate of the energy in the limit of a large number of particles, provided no state is macroscopically occupied, and here we shall simply calculate the expectation value of the energy.

In the expectation value of the energy of the state (8.117) two sorts of terms arise, diagonal ones in which the 2S atom remains in the same state, and cross terms in which the states of the 2S atom are different. The first class of terms gives the Hartree–Fock result above, Eq. (8.114). The others are interference terms, which arise because the initial state containing N identical bosons is symmetric under interchange of the coordinates of the atoms. The two-photon perturbation creates a state of one 2S atom and $N-1$ 1S atoms which is symmetric under interchange of all N coordinates, irrespective of whether a coordinate refers to a 1S atom or a 2S one. There is no general symmetry requirement legislating that the wave function for bosons in different internal states be symmetric under interchange of coordinates. The interference terms have precisely the same structure as the Fock terms for identical particles. The off-diagonal matrix elements of the interaction Hamiltonian are

$$\langle \mathbf{p}'; N_{\mathbf{p}_a}, \ldots, N_{\mathbf{p}'}-1, \ldots | H | \mathbf{p}; N_{\mathbf{p}_a}, \ldots, N_{\mathbf{p}}-1, \ldots \rangle = \frac{U_{12}}{V} \sqrt{N_{\mathbf{p}} N_{\mathbf{p}'}}, \qquad (8.119)$$

8.4 Collisional shifts of spectral lines

and the interference terms contribute to the expectation value of the interaction energy between 1S and 2S atoms an amount

$$\frac{\langle\Psi|\mathcal{O}^\dagger H\mathcal{O}|\Psi\rangle}{\langle\Psi|\mathcal{O}^\dagger\mathcal{O}|\Psi\rangle}\bigg|_{\text{interference}} = \frac{U_{12}}{NV}\sum_{\mathbf{p}\neq\mathbf{p}'} N_\mathbf{p} N_{\mathbf{p}'}. \qquad (8.120)$$

If no state is macroscopically occupied, the result (8.120) reduces to nU_{12} for large N. The total energy difference between the two states of the gas when there is no condensate is obtained by adding this to the Hartree–Fock result, and is

$$\Delta E_{\text{nc}} = 2n(U_{12} - U_0). \qquad (8.121)$$

Comparison of this result with that for a condensate, Eq. (8.113), shows that the frequency shift of the line in a gas with no condensate is twice that for a fully Bose–Einstein condensed gas of the same density. As we shall explain in Sec. 13.2, this reflects the fact that for a Bose gas with no condensate, the probability of finding two atoms at the same point is *twice* that for a pure condensate of the same density. The energies (8.113) and (8.121) are equal to the changes of the expectation values of the energy of the system when a 1S atom is replaced by a 2S one, keeping the wave function otherwise unchanged. The difference between the results for a condensate and a thermal gas was first brought out clearly in the work of Oktel and Levitov [10]. In the limits considered here, their calculation using the random phase approximation is equivalent to the one described above.

The origin of the change in the energy shift compared with the Hartree–Fock result is a collective effect that arises because the state created by the excitation of a 1S atom consists of a superposition of many single-particle configurations. A similar effect in the electron gas is responsible for the long-wavelength density fluctuation spectrum being dominated by plasmons, not single particle–hole pairs. The calculation above is analogous to the one for the schematic model for collective motion in nuclei [11].

For intermediate situations, when both condensed and non-condensed particles are present, Oktel and Levitov showed that there are two excitation frequencies. These correspond to two coupled modes of the condensate and the non-condensed particles that result from the interaction between the 1S and 2S atoms hybridizing the two modes discussed above for a pure condensate and a gas with no condensate.

A basic assumption made in the calculations above is that atomic interactions are well approximated by pseudopotentials. This is expected to be good as long as the distance between atoms is much greater than both the magnitude of the scattering length and the distance out to which the bare atom–atom interaction is important. However, an atom can be excited when it is close to another atom, in which case the line shifts may be substantial. As a consequence, the absorption spectrum is expected to have a sharp peak, due to transitions occurring when atoms are far apart, and an incoherent background due to excitations when atoms are close together. Even though at any given frequency the incoherent background is small, its contribution to the frequency-weighted sum rule analogous to Eq. (3.41) is comparable with that from the sharp peak. A more detailed consideration of the problem shows that the results derived above using the pseudopotential are a good approximation for the shift of the sharp part of the line [12].

The above results for the frequency shift are not specific to two-photon transitions, and they agree with those obtained from the standard theory of collisional shifts [9], which uses the quantum kinetic equation. An important application is to one-photon transitions, such as hyperfine lines in alkali atoms and hydrogen that are used as atomic clocks. The advances in understanding of interactions between cold alkali atoms have led to the conclusion that, because of their smaller collisional shifts, the rubidium isotopes ^{85}Rb and ^{87}Rb offer advantages for use as atomic clocks compared with ^{133}Cs, the atom currently used [13]. This has recently been confirmed experimentally for ^{87}Rb [14].

Problems

PROBLEM 8.1 The long-wavelength elementary excitations of a dilute, uniform Bose gas are phonons. Determine the specific heat at low temperatures and compare the result with that obtained in Chapter 2 for the ideal, uniform Bose gas at low temperatures. Estimate the temperature at which the two results are comparable.

PROBLEM 8.2 Calculate the sound velocity in the centre of a cloud of 10^4 atoms of ^{87}Rb in a harmonic-oscillator trap, $V(r) = m\omega_0^2 r^2/2$, for $\omega_0/2\pi = 150$ Hz. Evaluate the characteristic wave number at which the frequency of an excitation changes from a linear to a quadratic dependence on wave number.

PROBLEM 8.3 Calculate the thermal depletion of the condensate of a uniform Bose gas at temperatures well below T_c. Give limiting expressions for temperatures $T \ll nU_0/k$ and $T \gg nU_0/k$ and interpret the results in terms of the number of thermal excitations and their effective particle numbers.

PROBLEM 8.4 Show that if no single-particle state is macroscopically occupied, the many-particle state Eq. (8.117) is an eigenstate of the Hamiltonian

(8.118), with an energy different from that of the original state by an amount given by Eq. (8.121).

References

[1] N. N. Bogoliubov, *J. Phys.* (*USSR*) **11**, 23 (1947), reprinted in D. Pines, *The Many-Body Problem*, (W. A. Benjamin, New York, 1961), p. 292.

[2] S. L. Cornish, N. R. Claussen, J. L. Roberts, E. A. Cornell, and C. A. Wieman, *Phys. Rev. Lett.* **85**, 1795 (2000).

[3] T. D. Lee and C. N. Yang, *Phys. Rev.* **105**, 1119 (1957). See also T. D. Lee, K. Huang, and C. N. Yang, *Phys. Rev.* **106**, 1135 (1957).

[4] M. D. Girardeau and R. Arnowitt, *Phys. Rev.* **113**, 755 (1959); M. D. Girardeau, *Phys. Rev. A* **58**, 775 (1998).

[5] L. D. Landau and E. M. Lifshitz, *Quantum Mechanics*, Third edition, (Pergamon, Oxford, 1977), §64.

[6] V. N. Popov, *Functional Integrals and Collective Excitations*, (Cambridge Univ. Press, Cambridge, 1987). See also A. Griffin, *Phys. Rev. B* **53**, 9341 (1996).

[7] K. Burnett, in *Bose–Einstein Condensation in Atomic Gases*, Proceedings of the Enrico Fermi International School of Physics, Vol. CXL, ed. M. Inguscio, S. Stringari, and C. E. Wieman, (IOS Press, Amsterdam, 1999), p. 265.

[8] A. Griffin, in *Bose–Einstein Condensation in Atomic Gases*, Proceedings of the Enrico Fermi International School of Physics, Vol. CXL, ed. M. Inguscio, S. Stringari, and C. E. Wieman, (IOS Press, Amsterdam, 1999), p. 591.

[9] B. J. Verhaar, J. M. V. A. Koelman, H. T. C. Stoof, O. J. Luiten, and S. B. Crampton, *Phys. Rev. A* **35**, 3825 (1987); J. M. V. A. Koelman, S. B. Crampton, H. T. C. Stoof, O. J. Luiten, and B. J. Verhaar, *Phys. Rev. A* **38**, 3535 (1988).

[10] M. Ö. Oktel and L. S. Levitov, *Phys. Rev. Lett.* **83**, 6 (1999).

[11] G. E. Brown, *Unified Theory of Nuclear Models*, (North-Holland, Amsterdam, 1964), p. 29.

[12] C. J. Pethick and H. T. C. Stoof, *Phys. Rev. A* **64**, 013618 (2001).

[13] S. J. J. M. F. Kokkelmans, B. J. Verhaar, K. Gibble, and D. J. Heinzen, *Phys. Rev. A* **56**, 4389 (1997).

[14] C. Fertig and K. Gibble, *Phys. Rev. Lett.* **85**, 1622 (2000); Y. Sortais, S. Bize, C. Nicolas, A. Clairon, C. Salomon, and C. Williams, *Phys. Rev. Lett.* **85**, 3117 (2000).

9
Rotating condensates

One of the hallmarks of a superfluid is its response to rotation, or for charged superfluids, to a magnetic field [1]. The special properties of superfluids are a consequence of their motions being constrained by the fact that the velocity of the condensate is proportional to the gradient of the phase of the wave function. Following the discovery of Bose–Einstein condensation in atomic gases, much work has been devoted to the properties of rotating condensates in traps, and these developments have been reviewed in [2]. We begin by demonstrating that the circulation around a closed contour in the condensate is quantized (Sec. 9.1). Following that we consider properties of a single vortex line (Sec. 9.2). In Sec. 9.3 we then study conditions for equilibrium for a condensate in a rotating trap. The next section is devoted to vortex motion and includes a derivation of the Magnus force (Sec. 9.4). Finally we consider clouds of bosons with weak interactions, and develop a picture of the response to rotation in terms of elementary excitations with non-zero angular momentum (Sec. 9.5).

9.1 Potential flow and quantized circulation

The fact that according to Eq. (7.14) the velocity of the condensate is the gradient of a scalar,

$$\mathbf{v} = \frac{\hbar}{m}\boldsymbol{\nabla}\phi, \tag{9.1}$$

has far-reaching consequences for the possible motions of the fluid. From Eq. (9.1) it follows immediately that

$$\boldsymbol{\nabla}\times\mathbf{v} = 0, \tag{9.2}$$

that is, the velocity field is irrotational, unless the phase of the order parameter has a singularity. Possible motions of the condensate are therefore very restricted. Quite generally, from the single-valuedness of the condensate wave function it follows that around a closed contour the change $\Delta\phi$ in the phase of the wave function must be a multiple of 2π, or

$$\Delta\phi = \oint \boldsymbol{\nabla}\phi \cdot d\mathbf{l} = 2\pi\ell, \qquad (9.3)$$

where ℓ is an integer. Thus the *circulation* Γ around a closed contour is given by

$$\Gamma = \oint \mathbf{v} \cdot d\mathbf{l} = \frac{\hbar}{m} 2\pi\ell = \ell\frac{h}{m}, \qquad (9.4)$$

which shows that it is quantized in units of h/m. The magnitude of the quantum of circulation is approximately $(4.0 \times 10^{-7}/A)$ m^2 s^{-1} where A is the mass number.

As a simple example of such a flow, consider purely azimuthal flow in a trap invariant under rotation about the z axis. To satisfy the requirement of single-valuedness, the condensate wave function must vary as $e^{i\ell\varphi}$, where φ is the azimuthal angle. If ρ is the distance from the axis of the trap, it follows from Eq. (9.4) that the velocity is

$$v_\varphi = \ell \frac{h}{2\pi m \rho}. \qquad (9.5)$$

The circulation is thus $\ell h/m$ if the contour encloses the axis, and zero otherwise. If $\ell \neq 0$, the condensate wave function must vanish on the axis of the trap, since otherwise the kinetic energy due to the azimuthal motion would diverge. The structure of the flow pattern is thus that of a vortex line. Quantized vortex lines were first proposed in the context of superfluid liquid ^{4}He by Onsager [3]. Feynman independently proposed quantization of circulation in liquid ^{4}He and investigated its consequences for flow experiments [4].

For an external potential with axial symmetry, and for a state that has a singularity only on the axis, each particle carries angular momentum $\ell\hbar$ about the axis, and therefore the total angular momentum L about the axis is $N\ell\hbar$. If the singularity in the condensate wave function lies off the axis of the trap, the angular momentum will generally differ from $N\ell\hbar$. For a state having a density distribution with axial symmetry, the quantization of circulation is equivalent to quantization of the angular momentum per particle about the axis of symmetry. However, for other states circulation is still quantized, even though angular momentum per particle is not. The generalization of Eq. (9.2) to a state with a vortex lying along the z axis is

$$\nabla \times \mathbf{v} = \hat{z}\frac{\ell h}{m}\delta^2(\boldsymbol{\rho}), \qquad (9.6)$$

where δ^2 is a two-dimensional Dirac delta function in the xy plane, $\boldsymbol{\rho} = (x,y)$, and $\hat{z}$ is a unit vector in the z direction. When there are many vortices, the right hand side of this equation becomes a vector sum of two-dimensional delta functions on planes perpendicular to the direction of the vortex line. The strength of the delta function is a vector directed along the vortex line and with a magnitude equal to the circulation associated with the vortex. We now turn to a more detailed description of single vortices.

9.2 Structure of a single vortex

Consider a trap with axial symmetry, and let us assume that the wave function varies as $e^{i\ell\varphi}$. If we write the condensate wave function in cylindrical polar coordinates as

$$\psi(\mathbf{r}) = f(\rho, z)e^{i\ell\varphi}, \qquad (9.7)$$

where f is real, the energy Eq. (6.9) is

$$E = \int d\mathbf{r}\left\{\frac{\hbar^2}{2m}\left[\left(\frac{\partial f}{\partial \rho}\right)^2 + \left(\frac{\partial f}{\partial z}\right)^2\right] + \frac{\hbar^2}{2m}\ell^2\frac{f^2}{\rho^2} + V(\rho, z)f^2 + \frac{U_0}{2}f^4\right\}. \qquad (9.8)$$

The only difference between this result and the one for a condensate with a phase that does not depend on position is the addition of the $1/\rho^2$ term. This is a consequence of the azimuthal motion of the condensate which gives rise to a kinetic energy density $mf^2v_\varphi^2/2 = \hbar^2\ell^2 f^2/2m\rho^2$. The equation for the amplitude f of the condensate wave function may be obtained from the Gross–Pitaevskii equation (6.11). It is

$$-\frac{\hbar^2}{2m}\left[\frac{1}{\rho}\frac{d}{d\rho}\left(\rho\frac{df}{d\rho}\right) + \frac{d^2f}{dz^2}\right] + \frac{\hbar^2}{2m\rho^2}\ell^2 f + V(\rho, z)f + U_0 f^3 = \mu f. \qquad (9.9)$$

Alternatively, Eq. (9.9) may be derived by inserting Eqs. (9.7) and (9.8) into the variational principle $\delta(E - \mu N) = 0$. It forms the starting point for determining the energy of a vortex in a uniform medium as well as in a trap.

9.2.1 A vortex in a uniform medium

First we consider an infinite medium with a uniform potential, which we take to be zero, $V(\rho, z) = 0$. In the ground state the wave function does not depend on z, so terms with derivatives with respect to z vanish. Because of the importance of vortex lines with a single quantum of circulation we specialize to that case and put $\ell = 1$. At large distances from the axis the radial derivative and the centrifugal barrier term $\propto 1/\rho^2$ become unimportant, and therefore the magnitude of the condensate wave function becomes $f = f_0 \equiv (\mu/U_0)^{1/2}$. Close to the axis the derivative and centrifugal terms dominate, and the solution regular on the axis behaves as ρ, as it does for a free particle with unit angular momentum in two dimensions. Comparison of terms in the Gross–Pitaevskii equation (9.9) shows that the crossover between the two sorts of behaviour occurs at distances from the axis of order the coherence length in matter far from the axis, in agreement with the general arguments given in Chapter 6. It is therefore convenient to scale lengths to the coherence length ξ defined by (see Eq. (6.61))

$$\frac{\hbar^2}{2m\xi^2} = nU_0 = \mu, \tag{9.10}$$

where $n = f_0^2$ is the density far from the vortex, and we introduce the new variable $x = \rho/\xi$. We also scale the amplitude of the condensate wave function to its value f_0 far from the vortex by introducing the variable $\chi = f/f_0$. The energy density $\mathcal{E}$ then has the form

$$\mathcal{E} = n^2 U_0 \left[\left(\frac{d\chi}{dx}\right)^2 + \frac{\chi^2}{x^2} + \frac{1}{2}\chi^4 \right], \tag{9.11}$$

and the Gross–Pitaevskii equation (9.9) becomes

$$-\frac{1}{x}\frac{d}{dx}\left(x\frac{d\chi}{dx}\right) + \frac{\chi}{x^2} + \chi^3 - \chi = 0. \tag{9.12}$$

This equation may be solved numerically, and the solution is shown in Fig. 9.1.

Let us now calculate the energy of the vortex. One quantity of interest is the extra energy associated with the presence of a vortex, compared with the energy of the same number of particles in the uniform state. The energy ϵ per unit length of the vortex is

$$\epsilon = \int_0^b 2\pi\rho d\rho \left[\frac{\hbar^2}{2m}\left(\frac{df}{d\rho}\right)^2 + \frac{\hbar^2}{2m}\frac{f^2}{\rho^2} + \frac{U_0}{2}f^4 \right]. \tag{9.13}$$

The second term in the integrand is the kinetic energy of the azimuthal motion, and it varies as f^2/ρ^2. Consequently, its integral diverges logarithmically at large distances. This is similar to the logarithmic term in the

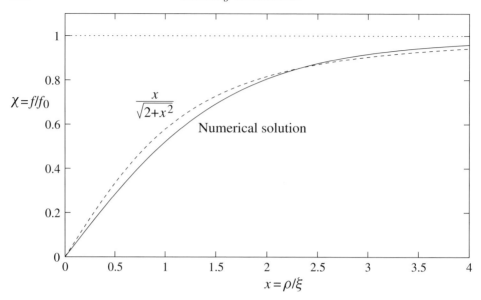

Fig. 9.1. The condensate wave function for a singly-quantized vortex as a function of radius. The numerical solution is given by the full line and the approximate function $x/(2+x^2)^{1/2}$ by the dashed line.

electrostatic energy of a charged rod. In order to obtain well-defined answers, we therefore consider the energy of atoms within a finite distance b of the vortex, and we shall further take b to be large compared with ξ.

To find the energy associated with the vortex, we must subtract from the total energy that of a uniform gas with the same number of particles ν per unit length contained within a cylinder of radius b. The energy per unit volume of a uniform gas is $\tilde{n}^2 U_0/2$, where $\tilde{n} = \nu/\pi b^2$ is the average density. The average density in the reference system is not equal to the density far from the axis in the vortex state, since the vortex state has a 'hole' in the density distribution near the axis. Because the number of particles is the same for the two states, the density of the vortex state at large distances from the axis is greater than that of the uniform system. The number of particles per unit length is given by

$$\nu = \int_0^b 2\pi\rho d\rho f^2 = \pi b^2 f_0^2 - \int_0^b 2\pi\rho d\rho (f_0^2 - f^2). \tag{9.14}$$

Thus the energy per unit length of the uniform system is given by

$$\epsilon_0 = \frac{1}{2}\pi b^2 f_0^4 U_0 - f_0^2 U_0 \int_0^b 2\pi\rho d\rho (f_0^2 - f^2), \tag{9.15}$$

where we have neglected terms proportional to the square of the last term in (9.14). These are of order $f_0^4 U_0 \xi^4 / b^2$ and are therefore unimportant because of the assumption that $b \gg \xi$. Thus ϵ_v, the total energy per unit length associated with the vortex, is the difference between Eqs. (9.13) and (9.15) and it is given by

$$\epsilon_v = \int_0^b 2\pi \rho d\rho \left[\frac{\hbar^2}{2m} \left(\frac{df}{d\rho} \right)^2 + \frac{\hbar^2}{2m} \frac{f^2}{\rho^2} + \frac{U_0}{2}(f_0^2 - f^2)^2 \right] \quad (9.16)$$

$$= \frac{\pi \hbar^2}{m} n \int_0^{b/\xi} x dx \left[\left(\frac{d\chi}{dx} \right)^2 + \frac{\chi^2}{x^2} + \frac{1}{2}(1 - \chi^2)^2 \right]. \quad (9.17)$$

If this expression is evaluated for the numerical solution of the Gross–Pitaevskii equation one finds

$$\epsilon_v = \pi n \frac{\hbar^2}{m} \ln \left(1.464 \frac{b}{\xi} \right). \quad (9.18)$$

This result was first obtained by Ginzburg and Pitaevskii in the context of their phenomenological theory of liquid ^{4}He close to T_λ [5]. The mathematical form of the theory is identical with that of Gross–Pitaevskii theory for the condensate at zero temperature, but the physical significance of the coefficients that appear in it is different.

The expression (9.17) may be used as the basis for a variational solution for the condensate wave function. In the usual way, one inserts a trial form for f and minimizes the energy expression with respect to the parameters in the trial function. For example, if one takes the trial solution [6]

$$\chi = \frac{x}{(\alpha + x^2)^{1/2}}, \quad (9.19)$$

which has the correct properties at both small and large distances, the optimal value of α is 2, and this solution is also shown in Fig. 9.1. We comment that minimizing the energy of the vortex, Eq. (9.17), is equivalent to minimizing the quantity $E - \mu N$, keeping the chemical potential μ fixed. Here E is the total energy. For the variational solution (9.19) with $\alpha = 2$ one finds the result $\epsilon_v = \pi(n\hbar^2/m)\ln(1.497b/\xi)$ which is very close to the exact result (9.18).

With the condensate wave function (9.7), each particle carries one unit of angular momentum, and therefore the total angular momentum per unit length is given by

$$\mathcal{L} = \nu \hbar. \quad (9.20)$$

We caution the reader that the simple expression for the angular momentum for this problem is a consequence of the rotational symmetry. As we shall see later, for a cloud with a vortex not on the axis of the trap, the angular momentum depends on the position of the vortex. Also, for a trap which is not rotationally invariant about the axis of rotation, the angular momentum is not conserved, and therefore does not have a definite value.

Multiply-quantized vortices

One may also consider vortices with more than one quantum of circulation, $|\ell| > 1$. The only change in the Gross–Pitaevskii equation is that the centrifugal potential term $\sim 1/x^2$ must be multiplied by ℓ^2. The velocity field at large distances from the centre of the vortex varies as $\ell\hbar/m\rho$, and therefore the kinetic energy of the azimuthal motion is

$$\epsilon_{\rm v} \approx \ell^2 \pi n \frac{\hbar^2}{m} \ln \frac{b}{\xi} \qquad (9.21)$$

to logarithmic accuracy. To calculate the numerical factor in the logarithm one must determine the structure of the vortex core by solving the Gross–Pitaevskii equation, which is Eq. (9.12) with the second term multiplied by ℓ^2. For small ρ the condensate wave function behaves as $\rho^{|\ell|}$, as does the wave function of a free particle in two dimensions with azimuthal angular momentum $\ell\hbar$. The result (9.21) indicates that the energy of a vortex with more than one unit of circulation is greater than the energy of a collection of singly-quantized vortices with the same total circulation. This suggests that vortices with more than a single unit of circulation will not exist in equilibrium. To make this argument more convincingly one must allow for the effects of interaction between vortices. For example, for two parallel vortex lines with ℓ_1 and ℓ_2 units of circulation separated by a distance d, the energy of interaction per unit length is given to logarithmic accuracy by (see Problem 9.3)

$$\epsilon_{\rm int} = \frac{2\pi\ell_1\ell_2\hbar^2 n}{m} \ln \frac{R}{d}, \qquad (9.22)$$

where R is a measure of the distance of the vortices from the boundary of the container. This expression holds provided $R \gg d$ and $d \gg \xi$. We therefore conclude that the interaction energy is small compared with the energy of two isolated vortex lines provided their separation is small compared with the size of the container. It should be mentioned that for a rotationally invariant system we should compare energies of states with the same angular momentum to determine the most stable state. However, in practice this latter constraint does not alter the conclusion.

We stress that the considerations given above assume that the vortices appear in an otherwise uniform medium, and the conclusions may therefore change when inhomogeneity is allowed for. Multiply-quantized vortices can be energetically favourable, if the extent of the gas perpendicular to the rotation axis is less than or comparable with the coherence length, or if the condensate is multiply-connected due to a repulsive potential applied in the vicinity of the rotation axis.

9.2.2 A vortex in a trapped cloud

We now calculate the energy of a vortex in a Bose–Einstein condensed cloud in a trap, following Ref. [7]. This quantity is important for estimating the lowest angular velocity at which it is energetically favourable for a vortex to enter the cloud. We consider a harmonic trapping potential which is rotationally invariant about the z axis, and we shall imagine that the number of atoms is sufficiently large that the Thomas–Fermi approximation gives a good description of the non-rotating cloud. The radius of the core of a vortex located on the z axis of the trap, which is determined by the coherence length there, is then small compared with the size of the cloud. This may be seen from the fact that the coherence length ξ_0 at the centre of the cloud is given by Eq. (9.10), which may be rewritten as

$$\frac{\hbar^2}{2m\xi_0^2} = \mu, \tag{9.23}$$

where $\mu = n(0)U_0$ is the chemical potential, $n(0)$ being the density at the centre in the absence of rotation. In addition, the chemical potential is related to the radius R of the cloud in the xy plane by Eq. (6.33), which for a harmonic oscillator potential is

$$\mu = m\omega_\perp^2 R^2/2, \tag{9.24}$$

where $\omega_\perp$ is the oscillator frequency for motions in the plane. Combining Eqs. (9.23) and (9.24), we are led to the result

$$\frac{\xi}{R} = \frac{\hbar\omega_\perp}{2\mu}. \tag{9.25}$$

This shows that the coherence length is small compared with the radius of the cloud if the chemical potential is large compared with the oscillator quantum of energy, a condition satisfied when the Thomas–Fermi approximation holds. Outside the vortex core the density varies on a length scale $\sim R$. Thus the energy of the vortex out to a radius ρ_1 intermediate between the core size and the radius of the cloud ($\xi \ll \rho_1 \ll R$) may be calculated using the result (9.18) for the energy of a vortex in a uniform medium. At larger radii, the density profile is essentially unaltered compared with that for the cloud without a vortex, but the condensate moves with a velocity determined by the circulation of the vortex. The extra energy in the region at large distances is thus the kinetic energy of the condensate, which may be calculated from hydrodynamics.

To begin, let us consider the two-dimensional problem, in which we neglect the trapping force in the z direction. The cloud is cylindrical, with radius ρ_2, and the energy per unit length is then given by

$$\epsilon_{\rm v} = \pi n_0 \frac{\hbar^2}{m} \ln\left(1.464 \frac{\rho_1}{\xi_0}\right) + \frac{1}{2} \int_{\rho_1}^{\rho_2} mn(\rho) v^2(\rho) 2\pi \rho d\rho. \tag{9.26}$$

Here n_0 is the particle density for $\rho \to 0$ in the absence of a vortex, while ξ_0 is the coherence length evaluated for that density. Since the magnitude of the velocity v is $v = h/2\pi\rho m$, and the density in a harmonic trap varies as $1 - \rho^2/R^2$ in the Thomas–Fermi approximation, one finds

$$\begin{aligned}\epsilon_{\rm v} &= \pi n_0 \frac{\hbar^2}{m} \left[\ln\left(1.464 \frac{\rho_1}{\xi_0}\right) + \int_{\rho_1}^{\rho_2} \frac{\rho d\rho}{\rho^2}\left(1 - \frac{\rho^2}{\rho_2^2}\right)\right] \\ &\simeq \pi n_0 \frac{\hbar^2}{m} \left[\ln\left(1.464 \frac{\rho_2}{\xi_0}\right) - \frac{1}{2}\right],\end{aligned} \tag{9.27}$$

where the integral has been evaluated for $\rho_1 \ll \rho_2$, with terms of higher order in ρ_1/ρ_2 being neglected. The logarithmic term is the result for a medium of uniform density, while the $-1/2$ reflects the lowering of the kinetic energy due to the reduction of particle density caused by the trapping potential. Thus the energy per unit length is given by an expression similar to (9.18) but with a different numerical constant $1.464/e^{1/2} \approx 0.888$,

$$\epsilon_{\rm v} = \pi n_0 \frac{\hbar^2}{m} \ln\left(0.888 \frac{\rho_2}{\xi_0}\right). \tag{9.28}$$

The angular momentum $\mathcal{L}$ per unit length is $\hbar$ times the total number of particles per unit length. For $\rho_2 \gg \xi$ the latter may be evaluated in the Thomas–Fermi approximation, and one finds

$$\mathcal{L} = n_0 \hbar \int_0^{\rho_2} \left(1 - \frac{\rho^2}{\rho_2^2}\right) 2\pi \rho d\rho = \frac{1}{2} n_0 \pi \rho_2^2 \hbar. \tag{9.29}$$

Let us now consider the three-dimensional problem. If the semi-axis, Z, of the cloud in the z direction is much greater than the coherence length, one may estimate the energy of the cloud by adding the energy of horizontal slices of the cloud, thus neglecting the kinetic energy term due to the vertical gradient of the condensate wave function. The total energy is then given by (9.28), integrated over the vertical extent of the cloud,

$$E = \frac{\pi\hbar^2}{m} \int_{-Z}^{Z} dz n_0(z) \ln\left[0.888\frac{\rho_2(z)}{\xi(z)}\right]. \tag{9.30}$$

For a harmonic trap the density on the z axis is $n_0(z) = n(0)(1 - z^2/Z^2)$, while $\rho_2(z) = R(1 - z^2/Z^2)^{1/2}$ and $\xi(z) = \xi_0[n(0)/n_0(z)]^{1/2}$, where ξ_0 is now the coherence length at the centre of the cloud. The energy is then given by

$$E = \frac{\pi\hbar^2 n(0)}{m} \int_{-Z}^{Z} dz \left(1 - \frac{z^2}{Z^2}\right) \ln\left[0.888\frac{R}{\xi_0}\left(1 - \frac{z^2}{Z^2}\right)\right]. \tag{9.31}$$

Using the fact that $\int_0^1 dy (1-y^2) \ln(1-y^2) = (12\ln 2 - 10)/9$, we obtain the final result

$$E = \frac{4\pi n(0)}{3}\frac{\hbar^2}{m} Z \ln\left(0.671\frac{R}{\xi_0}\right), \tag{9.32}$$

which is in very good agreement with numerical calculations for large clouds [8].

The total angular momentum is

$$\begin{aligned} L &= \hbar \int d\mathbf{r} n(\mathbf{r}) = n(0)\hbar \int_{-Z}^{Z} dz \int_0^{\rho_2(z)} 2\pi\rho d\rho (1 - \frac{\rho^2}{R^2} - \frac{z^2}{Z^2}) \\ &= \frac{8\pi}{15} n(0) R^2 Z \hbar. \end{aligned} \tag{9.33}$$

These results will be used in Sec. 9.3.2 below to discuss the critical angular velocity for a vortex state.

9.2.3 Off-axis vortices

The angular momentum of a state with a vortex line parallel to the axis of the trap, but not coincident with it, generally is not an integral number of quanta per particle. To demonstrate this explicitly, consider a vortex line with a single quantum of circulation in a condensate confined by a cylindrical container whose cross section is a circle of radius R. To begin with, imagine that the density is constant everywhere, except in a small region in the core of the vortex. The angular momentum per unit length about the axis of the cylinder is given by

$$\mathcal{L} = nm \int \rho d\rho d\varphi v_\varphi \rho. \tag{9.34}$$

The angular integral is

$$\int d\varphi v_\varphi \rho = \oint \mathbf{v}\cdot d\mathbf{l}, \tag{9.35}$$

which is the circulation. This is equal to h/m if the vortex line lies within the contour, and is zero otherwise. Thus, if the centre of the vortex line is at a distance b from the axis of the cylinder, the angular momentum per unit length is given by

$$\mathcal{L} = nh\int_b^R \rho d\rho = \nu\hbar\left(1 - \frac{b^2}{R^2}\right), \tag{9.36}$$

where $\nu = \pi R^2 n$ is the number of particles per unit length. When the density depends on ρ, as it does for a dilute gas in a trap, the dependence of $\mathcal{L}$ on b will be different.

Next we calculate the energy of the state. The flow pattern for the velocity may be calculated from the hydrodynamic equations. If the density is constant except in the core of the vortex, it follows from the equation of continuity that $\boldsymbol{\nabla}\cdot\mathbf{v} = 0$. Consequently, the phase of the condensate wave function obeys Laplace's equation. At the cylinder the radial component of the velocity must vanish. The problem is therefore equivalent to that of determining the electrostatic potential due to a charged rod inside a conducting cylinder and parallel to the axis of the cylinder. This may be solved by the method of images, by introducing a second vortex with the opposite circulation at a distance R^2/b from the axis of the cylinder at the same azimuthal angle as the original vortex. The total velocity field is thus obtained by superimposing that due to a vortex with circulation κ at radius b and that due to a vortex of circulation $-\kappa$ at radius R^2/b. The total kinetic energy per unit length is thus given by

$$\epsilon_{\text{kin}} = \frac{1}{2}\int \rho d\rho d\varphi nmv^2. \tag{9.37}$$

To take into account the reduction of the density in the vortex core we cut off the integral at a distance ξ from the centre of the vortex. Evaluation of the integral gives (see Problem 9.4)

$$\epsilon_v = \frac{mn\kappa^2}{4\pi}[\ln(R/\xi) + \ln(1 - b^2/R^2)], \tag{9.38}$$

where only the leading logarithmic dependence on R/ξ has been retained. Using a more detailed model of the core of the vortex changes only the coefficient of R/ξ in the logarithm. We shall use this result to discuss motion of an off-axis vortex in Sec. 9.4.

9.3 Equilibrium of rotating condensates

The equilibrium state of a rotating condensate depends on the symmetry of the trap. In the following we first discuss traps with axial symmetry, which implies that the component of the angular momentum about the symmetry axis is conserved. Subsequently we consider traps with no axis of symmetry.

9.3.1 Traps with an axis of symmetry

The calculations above show that one way to add angular momentum to a condensate of bosons in a trap with an axis of symmetry is to put all atoms into a state with non-zero angular momentum (the vortex state). In Sec. 7.3.1 we found that another way to add angular momentum to a cloud is to create elementary excitations, such as surface waves. In general, more complicated states may be created by combining the two processes by, e.g., adding elementary excitations to a vortex state. An interesting question is what the lowest-energy state is for a given angular momentum. Following nuclear physics terminology, this state is sometimes referred to as the *yrast* state.[1]

We begin by considering a cloud with angular momentum much less than $\hbar$ per atom. Intuitively one would expect the state to be close to that of the non-rotating ground state, and therefore it is natural to anticipate that the lowest-energy state would be the ground state plus a number of elementary excitations. Let us assume for the moment that the interaction is repulsive. If the number of particles is sufficiently large that the Thomas–Fermi approximation is valid, excitation energies may be calculated from the results of Sec. 7.3.1. For a harmonic trap, the lowest-energy elementary excitations with angular momentum $l\hbar$ are surface waves. Their energies $\hbar\omega$ may be obtained from the frequencies for modes obtained in Chapter 7: for isotropic traps by Eq. (7.71) for $n = 0$ and for anisotropic ones by Eq. (7.79). The energies are therefore in both cases equal to $\hbar\omega_0 l^{1/2}$. Thus the energy per unit angular momentum is $\omega_0/l^{1/2}$. The lowest energy cost per unit angular momentum is achieved if modes with high l are excited. However with increasing l, the modes penetrate less and less into the bulk of the cloud, and the simple hydrodynamic picture of modes developed in Chapter 7 fails when the penetration depth of the surface wave, $1/q \sim R/l$ (see Sec. 7.4) becomes comparable with the thickness of the surface, δ, given by Eq. (6.45). At higher values of l, the modes become free-particle like, with an energy approaching $\hbar^2 q^2/2m \propto l^2$, and the energy per unit angular

---

[1] The word *yrast* is the superlative of *yr*, which in Swedish means 'dizzy'.

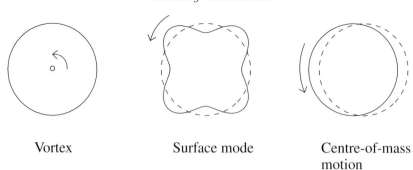

Fig. 9.2. Schematic representation of ways of adding angular momentum to a Bose–Einstein condensed cloud in a trap.

momentum increases. Consequently, the lowest energy cost per unit angular momentum is for surface waves with wave numbers of order $1/\delta$, or $l \sim R/\delta$, and has a value

$$\frac{\epsilon_l}{l\hbar} \sim \omega_0 \left(\frac{\delta}{R}\right)^{1/2} \sim \omega_0 \left(\frac{a_{\mathrm{osc}}}{R}\right)^{2/3}, \qquad (9.39)$$

since $\delta \sim a_{\mathrm{osc}}^{4/3}/R^{1/3}$ with $a_{\mathrm{osc}} = (\hbar/m\omega_0)^{1/2}$.

For $L \approx N\hbar$, states like the vortex state for one quantum of circulation are energetically most favourable. As the angular momentum increases further, states with a pair of vortices, or a vortex array have the lowest energy. Explicit calculations within the Gross–Pitaevskii approach may be found in Refs. [9] and [10] for weak coupling.

When the interparticle interaction is attractive, the picture is completely different. In the lowest-energy state for a given value of the angular momentum all the angular momentum is associated with the centre-of-mass motion of the cloud, and all internal correlations are identical with those in the ground state. This corresponds to excitation of a surface mode with $l = 1$. The dependence of properties on the sign of the interaction is due to the fact that the lowest-energy state with a fixed, non-zero angular momentum undergoes a phase transition as the interaction passes through zero. This will be demonstrated explicitly in Sec. 9.5.

In summary, for repulsive interactions the lowest-energy state of a Bose–Einstein condensate in a trap is generally a superposition of vortices and elementary excitations, especially surface waves. When the interaction is attractive, the lowest energy is achieved by putting all the angular momentum into the centre-of-mass motion. The three ways of adding angular momentum to a cloud are illustrated schematically in Fig. 9.2.

9.3.2 Rotating traps

In the previous discussion we assumed that the trap has an axis of symmetry, and therefore the angular momentum about that axis is conserved. Another situation arises for traps which have no axis of symmetry, since angular momentum is then not conserved. As an example one may consider an anisotropic trap rotating about some axis. The question we now address is what the equilibrium state is under such conditions.

The difficulty with rotating traps is that in the laboratory frame the trapping potential is generally time-dependent. It is therefore convenient to approach the problem of finding the equilibrium state by transforming to the frame rotating with the trapping potential, since in that frame the potential is constant in time, and thus the standard methods for finding equilibrium may be employed. According to the well-known result from mechanics, in the frame rotating with the potential the energy E' of a cloud of atoms is given in terms of the energy E in the non-rotating frame by [11]

$$E' = E - \mathbf{L} \cdot \mathbf{\Omega}, \tag{9.40}$$

where $\mathbf{L}$ is the angular-momentum vector and $\mathbf{\Omega}$ is the angular-velocity vector describing the rotation of the potential. In a trap with no axis of symmetry, angular momentum about the axis of rotation is not conserved, and therefore the angular momentum $\mathbf{L}$ of the system must be identified with the quantum-mechanical expectation value of the angular momentum. The problem is then to find the state with lowest energy in the rotating frame, that is, with the lowest value of E'.

An important conclusion may be drawn from Eq. (9.40). If a state with angular-momentum component L along the rotation axis has energy E_L, it will be energetically favourable compared with the ground state if the angular velocity of the trap exceeds a critical value Ω_c, given by

$$\Omega_c = \frac{E_L - E_0}{L}. \tag{9.41}$$

The value of Ω_c depends on the character of the excited state. For the vortex state in a cloud with a number of particles large enough that the Thomas–Fermi approximation is valid, the energy of the vortex state relative to the ground state is given by (9.32), while L is given by (9.33). Their ratio determines Ω_c according to (9.41), and we therefore obtain a critical angular velocity given by

$$\Omega_c = \frac{5}{2}\frac{\hbar}{mR^2}\ln\left(0.671\frac{R}{\xi_0}\right) = \frac{5}{2}\omega_0\left(\frac{a_{\text{osc}}}{R}\right)^2\ln\left(0.671\frac{R}{\xi_0}\right). \tag{9.42}$$

Apart from the logarithmic term and a numerical factor, this is the angular velocity of a particle at the edge of the cloud with angular momentum $\hbar$. Since the minimum value of the ratio of energy to angular momentum for surface waves is given by Eq. (9.39), the critical angular velocity for such waves is given by

$$\Omega_c \sim \omega_0 \left(\frac{a_{\text{osc}}}{R}\right)^{2/3}. \tag{9.43}$$

Thus in the Thomas–Fermi regime, vortex states can be in equilibrium in a rotating trap with an angular velocity lower than that required for surface waves. The smallest value of the critical angular velocity allowing for all sorts of possible excitations is referred to as the lower critical angular velocity. Below this angular velocity, the non-rotating ground-state has the lowest energy in the rotating frame, while at higher angular velocities, other states are favoured. It is usually denoted by Ω_{c1}, by analogy with the lower critical magnetic field in type II superconductors, at which the Meissner state ceases to be energetically favourable. At higher magnetic fields, flux lines, the charged analogues of vortices in uncharged systems, are created. As the angular velocity increases past Ω_{c1}, the angular momentum of the equilibrium state changes discontinuously from zero to $\hbar$ per particle.

As the rotation rate is increased, the nature of the equilibrium state changes, first to a state with two vortices rotating around each other, then to three vortices in a triangle, and subsequently to arrays of more and more vortices. Calculations of such structures for a trapped Bose gas with weak interaction have been carried out in Ref. [9]. All the vortices have a single quantum of circulation since, as argued in Sec. 9.2.1, vortices with multiple quanta of circulation are unstable with respect to decay into vortices with a single quantum.

For a bulk superfluid in rotation at an angular velocity high compared with the minimum angular velocity Ω_{c1} at which it is energetically favourable to have a vortex, the state with lowest energy in the rotating frame has a uniform array of vortices arranged on a triangular lattice [12]. For a Bose–Einstein condensed gas in a trap, one would expect a similar conclusion to hold provided the density of the fluid does not vary appreciably on a length scale equal to the spacing between vortices.

Let us contrast the velocity field for an array of vortices with that for an ordinary fluid. For a fluid in equilibrium in a frame rotating at angular velocity $\boldsymbol{\Omega}$, the velocity locally is that for uniform rotation, that is $\mathbf{v} = \boldsymbol{\Omega} \times \mathbf{r}$. Therefore $\boldsymbol{\nabla} \times \mathbf{v}$ is uniform and equal to $2\boldsymbol{\Omega}$. The velocity field for a condensate can be made similar to this if it contains an array of vortices aligned in the direction of the angular velocity. If the number n_v of vortices per unit area in the plane perpendicular to $\mathbf{\Omega}$ is uniform and equal to

$$n_v = \frac{2m\Omega}{h}, \tag{9.44}$$

the average circulation per unit area in the condensate is 2Ω. However, because the quantum of circulation is non-zero, it is impossible to create a velocity field $\mathbf{v} = \mathbf{\Omega} \times \mathbf{r}$ everywhere simply from an array of vortices, since the velocity diverges in the immediate vicinity of a vortex.

The importance of quantized vortices as a direct manifestation of the quantum nature of a Bose–Einstein condensate has stimulated a number of experimental investigations of rotating condensates. The basic idea in most such experiments is to use a laser beam to 'stir' the condensate. Due to the interaction of atoms with radiation, as described in Sec. 4.2, a laser beam, whose axis has a fixed direction and whose centre rotates in space, creates a rotating potential for atoms. Since the spatial scale of a vortex core in a trapped cloud is typically less than 1 μm, it is impossible to resolve structures optically. To magnify the spatial scale of vortex structure it is therefore common to switch off the trap and allow clouds to expand ballistically before examining them optically. This method has been applied in a number of experiments on one-component condensates. However, the first observation of vortices in dilute gases was made using a two-component condensate [13].

The idea of using two-component condensates arose because of concerns about the time required to reach equilibrium in one-component condensates, and the specific technique was proposed by Williams and Holland [14]. The experiment was performed with ^{87}Rb atoms, and the two components corresponded to atoms in the two hyperfine states $|F = 1, m_F = -1\rangle$ and $|F = 2, m_F = +1\rangle$. In brief, the basic idea is to use a laser-beam stirrer, as described above, together with a spatially-uniform microwave field, which couples the two hyperfine states by a two-photon transition. When the two perturbations operate on, say, a non-rotating condensate in the state $|F = 1, m_F = -1\rangle$, they can produce a rotating condensate in the state $|F = 2, m_F = +1\rangle$. If the frequency of the microwave field and the rotation frequency are chosen appropriately, atoms from the non-rotating condensate are transferred resonantly to a vortex state of atoms in the second hyperfine state. In the first experiment, the core of the vortex contained a non-rotating condensate of atoms in the state $|F = 1, m_F = -1\rangle$. Because of the repulsive interaction between atoms in unlike hyperfine states, the vortex core was so large that it could be imaged directly, without ballistic expansion. In subsequent experiments, the condensate of non-rotating atoms was removed, thereby yielding a rotating condensate in a one-component system [15].

The difficulties with nucleating vortices in single-component condensates turned out not to be serious. Single vortices and vortex arrays have been created by using laser beams to stir a gas of ^{87}Rb atoms [16], and the angular momentum of a rotating Bose–Einstein condensed state has been measured [17]. Recently, regular triangular arrays containing up to 130 vortices have been observed in a condensate of ^{23}Na atoms [18]. These results confirm the basic picture of the equilibrium states of rotating superfluids described above.

The study of rotating Bose–Einstein condensates in traps has given new insights into the behaviour of vortex lines in inhomogeneous systems, and for further details we refer to Ref. [2]. There are many outstanding problems, among them understanding quantitatively the response of a rotating superfluid to perturbations, such as the presence of thermal excitations. A particular example is the decay of vortex states. In the experiments reported in Ref. [18], the number of vortices had decreased markedly after 10 s, but individual vortices were sometimes present near the axis of rotation after as long as 40 s.

9.4 Vortex motion

One of the remarkable results of the classical hydrodynamics of an ideal fluid is the law of conservation of circulation, or Kelvin's theorem. This states that the circulation around a contour moving with the fluid is a constant in time [19, §8]. Expressed in terms of the motion of vortex lines, the theorem states that vortex lines move with the local fluid velocity.

As a first example, consider an off-axis vortex in an incompressible fluid contained in a cylinder, as discussed in Sec. 9.2.3. As a consequence of the interaction between the vortex and its image vortex, the azimuthal angle of their positions will change, their radial coordinates remaining constant. The angular velocity may be determined by classical mechanics, since this is given by

$$\dot{\varphi} = \Omega = \frac{\partial E}{\partial L}. \qquad (9.45)$$

Since the energy (9.38) and the angular momentum (9.36) both depend parametrically on b, the distance of the vortex line from the axis of the cylinder, the frequency is given by

$$\Omega = \frac{\partial E}{\partial L} = \frac{\partial E/\partial b}{\partial L/\partial b} = \frac{\hbar}{mR^2} \frac{1}{1 - b^2/R^2}$$
$$= \frac{1}{b} \frac{\hbar}{m(R^2/b - b)}. \tag{9.46}$$

Thus the velocity of the vortex at radius b from the axis is $\hbar/m(R^2/b - b)$. The fluid velocity at the position of the original vortex is due to the image vortex, and it is equal to $\hbar/md$, where $d = R^2/b - b$ is the separation of the two vortices. Thus the velocity at which the vortex advances is precisely the flow velocity of the fluid at the position of the vortex, in agreement with Kelvin's theorem.

9.4.1 Force on a vortex line

Insight into Kelvin's theorem may be gained by considering the force acting on a vortex line. We do this by calculating the flux of momentum into the region in the vicinity of the vortex line. Consider a surface surrounding the vortex line, and moving with it. The momentum flux inwards across the surface is given by[2]

$$F_i = -\int dS_j \Pi_{ij}, \tag{9.47}$$

where Π_{ij} is the momentum flux density tensor, and the element of area of the surface, considered as a vector in the direction of the outward normal, is denoted by dS_j.

We imagine that the surface lies well outside the core of the vortex, so terms in the momentum flux density tensor involving the gradient of the amplitude of the condensate wave function may be neglected. We shall also neglect the effect of external potentials. Thus the momentum flux density tensor is [19, §7]

$$\Pi_{ij} = p\delta_{ij} + nmv_iv_j, \tag{9.48}$$

where p is the pressure, and the force is

$$F_i = -\int dS_j (p\delta_{ij} + nmv_iv_j). \tag{9.49}$$

The force thus comes partly from pressure variations over the surface, and partly from transport of momentum by the bulk motion of fluid crossing the surface. Due to the presence of the vortex, the velocity field close to

---
[2] We use here the Einstein convention of summing over repeated indices.

the vortex line has a component varying inversely as the distance from the vortex line. In addition, there is a contribution to the velocity field which varies smoothly with position outside the core of the vortex, and we shall denote this by $\mathbf{u}$. We remark that at distances of order the coherence length ξ from the vortex there are 'backflow' contributions to the velocity field due to the fact that the deficit of particles in the core behaves as an obstacle, but at distances much greater than ξ from the vortex line these are negligible. Thus on the surface we consider we may treat $\mathbf{u}$ as constant.

In the frame moving with the vortex, the flow is stationary and therefore the fluid velocity satisfies the Bernoulli equation which follows from Eq. (7.20),

$$\mu + \frac{1}{2}m(\mathbf{v}-\mathbf{u})^2 = \mu_0 + \frac{1}{2}mu^2, \tag{9.50}$$

where μ_0 is the chemical potential far from the vortex core.

The change in the chemical potential is

$$\delta\mu = \mu - \mu_0 = \frac{1}{2}m[u^2 - (\mathbf{v}-\mathbf{u})^2] = m\mathbf{v}\cdot\mathbf{u} - \frac{1}{2}mv^2. \tag{9.51}$$

The change in pressure is related to the change in the chemical potential by the thermodynamic relationship valid at zero temperature $dp = nd\mu$. To calculate the momentum flux across a contour moving with the fluid, we transform the momentum flux density tensor to the frame in which the vortex is stationary, where the fluid velocity is $\mathbf{v} - \mathbf{u}$. The change in the momentum flux density tensor to first order in $\mathbf{u}$ is

$$\delta\Pi_{ij} = nm(\mathbf{v}\cdot\mathbf{u}\delta_{ij} - u_i v_j - v_i u_j). \tag{9.52}$$

The total momentum transported out over the boundary is given by integrating the scalar product of the momentum flux density tensor and the element of surface area of the boundary. The force acting on the vortex is the negative of this quantity. For definiteness, we take $\mathbf{u}$ to be in the x direction, $\mathbf{u} = (u_x, 0, 0)$. Per unit length of the vortex, the components of the force are then given by

$$\mathcal{F}_x = u_x \oint nm(v_x e_x + v_y e_y) dl, \tag{9.53}$$

and

$$\mathcal{F}_y = u_x \oint nm(v_y e_x - v_x e_y) dl, \tag{9.54}$$

where dl is the line element of the contour in the xy plane, and $\hat{\mathbf{e}}$ is the unit vector in the direction of the outward normal to the surface. The integrand on the right hand side of Eq. (9.53) is the net flux of mass across the contour, which must vanish. Thus the force in the x direction vanishes. This is an example of d'Alembert's paradox, that in potential flow past a body there is no drag force. Since the line element on the contour is given by $d\mathbf{l} = dl(-e_y, e_x)$, the integral in Eq. (9.54) is the circulation around the contour,

$$\kappa_z = \oint (v_y e_x - v_x e_y) dl = \oint \mathbf{v}\cdot d\mathbf{l}. \tag{9.55}$$

The force in the z direction vanishes by symmetry, and therefore if one regards the circulation as a vector in the direction of the vortex line, the total force per unit length of the vortex may be written as

$$\mathcal{F} = nm\boldsymbol{\kappa} \times \mathbf{u}, \tag{9.56}$$

where $\boldsymbol{\kappa}$ is the circulation of the vortex, its direction being that of the vortex line. This force is usually referred to as the Magnus force. Note that the force is independent of conditions inside the surface considered. In particular, the result holds even if a solid object, such as the wing of an aircraft, is present there; this force provides the lift responsible for flight.

From the result (9.56) one can draw the important conclusion that in steady flow a vortex moves with the local fluid velocity if there are no other forces acting on the vortex. If this were not so, there would be unbalanced forces. If the vortex is subjected to other forces, they must exactly cancel the Magnus force. We draw attention to the fact that the 'carrying vorticity with the fluid' effect is not present in the equations of motion we derived in Sec. 7.1 , where it was assumed that $\boldsymbol{\nabla} \times \mathbf{v} = 0$, and therefore there was no vorticity. To include these effects requires a careful treatment of the regions near the vortex lines, and these give rise to terms like the $\mathbf{v} \times (\boldsymbol{\nabla} \times \mathbf{v})$ one in Eq. (7.25), which was derived from the Euler equation for an ideal fluid, without assuming that the flow was irrotational.

9.5 The weakly-interacting Bose gas under rotation

The weakly-interacting Bose gas, whose equilibrium properties and excitation spectrum were considered in Chapters 6 and 8, provides an instructive model for studies of rotation. Consider N identical bosons in a harmonic-oscillator potential which is axially symmetric about the z axis. The ground state of the system has been considered before and, provided the interaction energy is small compared with the oscillator quantum of energy, the energy of the state may be estimated by perturbation theory, and is given in Eq. (6.20). Properties of vibrational modes were considered in Sec. 8.2.1. Let us now turn to states with non-zero angular momentum about the z axis. Our discussion closely follows Ref. [20]. The difference between this problem and the one considered earlier is that the oscillator frequencies for the x and y directions are the same, and therefore the modes for oscillations in the two directions are mixed by the interaction. To solve this problem we work in terms of states with definite angular momentum about the z axis, rather than states with definite numbers of quanta of motion in the x and y directions. The wave function for the motion in the z direction plays no role in our discussion, so we shall suppress it and consider a purely two-dimensional problem. The energy levels of an isotropic two-dimensional oscillator are given by

$$E = (n_x + n_y + 1)\hbar\omega_\perp, \tag{9.57}$$

where n_x and n_y are the numbers of quanta for the two directions, and $\omega_\perp$ is the oscillation frequency in the xy plane. Energy levels with more than a single quantum are degenerate, and one may construct combinations of states with the same energy that have simple properties under rotations about the z axis. The spectrum may be expressed in terms of the angular momentum quantum number m of the state and the number n_ρ of radial nodes of the wave function as

$$E = (|m| + n_\rho + 1)\hbar\omega_\perp. \tag{9.58}$$

The lowest-energy single-particle state with angular momentum $m\hbar$ has an excitation energy $|m|\hbar\omega_\perp$ relative to the ground state. This state corresponds classically to a circular orbit in the xy plane. For a given angular momentum L, the lowest-energy states of the many-body system in the absence of interactions are obtained by populating states with no radial nodes, and with the angular momentum of the particle having the same sense as L. For definiteness, it is convenient to take L to be positive. If one denotes the number of atoms in a single-particle state with angular momentum $m\hbar$ and no radial nodes by N_m, the angular momentum is given by

$$L = \hbar \sum_{m \geq 0} m N_m \tag{9.59}$$

and the energy by

$$E_L - E_0 = L\omega_\perp. \tag{9.60}$$

For $L = \hbar$, the state is unique, since it has $N - 1$ particles in the ground state of the oscillator, and one particle in the state with one unit of angular momentum. We shall denote this state by $|0^{N-1}1^1\rangle$. For higher values of the angular momentum the lowest-energy state is degenerate. For example, for $L = 2\hbar$, the state $|0^{N-1}2^1\rangle$, which has $N-1$ particles in the ground state and one in the state with $m = 2$, is degenerate with the state $|0^{N-2}1^2\rangle$ with $N - 2$ particles in the ground state and two in the state with $m = 1$. The degeneracy increases rapidly with L.

Let us now investigate the effect of interactions on the energies of states with small angular momentum. The simplest states to consider are ones in which $N-1$ particles are in the ground state, and one particle is in the state with angular momentum $m\hbar$. The expectation value of the energy of the state is (cf. Eq. (8.58))

$$E_m = E_0 + \hbar|m|\omega_\perp - (N-1)(\langle 00|U|00\rangle - 2\langle 0m|U|0m\rangle). \tag{9.61}$$

The fact that the interaction mixes the state $|0^{N-1}m^1\rangle$ with ones such as $|0^{N-|m|}1^m\rangle$ does not affect the contribution to the energy of order N (see Problem 9.5).

The matrix elements of the interaction may be calculated straightforwardly since the wave function of the state with angular momentum $m\hbar$ is proportional to $\rho^{|m|}e^{im\varphi}$. One finds

$$\langle 0m|U|0m\rangle = \frac{1}{2^{|m|}}\langle 00|U|00\rangle. \tag{9.62}$$

The excitation energies are therefore given by

$$E_m = E_0 + \hbar|m|\omega_\perp - N\langle 00|U|00\rangle(1 - 1/2^{|m|-1}), \tag{9.63}$$

where in the interaction term we have replaced $N - 1$ by N. Notice that the interaction energy is unchanged for $|m| = 1$. This is due to the fact that the angular momentum of the excited state considered is associated purely with the centre-of-mass motion and, consequently, all correlations between atoms are precisely the same as in the ground state. This effect is analogous to that for vibrational motion along one of the coordinate axes, which we considered in Chapter 7. Now let us consider higher multipole excitations.

Repulsive interactions

For repulsive interactions the excitation energy is reduced compared to that for non-interacting particles. The energy cost per unit angular momentum is given by

$$\frac{E_m - E_0}{|m|\hbar} = \omega_\perp - N\langle 00|U|00\rangle\frac{(1 - 1/2^{|m|-1})}{|m|\hbar}, \tag{9.64}$$

and it is therefore lowest for quadrupole ($|m| = 2$) and octupole ($|m| = 3$) modes, and in this approximation the value is the same for these two sorts of excitation [20].

To determine the least energetic state for a non-zero value of the angular momentum per particle one must take into account interactions between the elementary excitations considered above, and one finds that at low angular momentum the lowest-energy state is achieved by exciting quadrupole modes [10]. The change in the interaction energy is negative because the single-particle wave function in an excited state is more spread out than in the ground state, and therefore the expectation value of the interaction energy is less. The results here fit nicely with the ones deduced for the Thomas–Fermi regime (Sec. 9.3.1), where for small total angular momentum ($L \ll N\hbar$) the lowest-energy excitations per unit angular momentum were found to be surface waves with angular momentum of order $(R/a_{\rm osc})^{4/3}\hbar$. In the weak-coupling limit, R and $a_{\rm osc}$ are comparable in magnitude, and therefore one would expect the most favourable angular momentum of excitations to be of order unity, as we found by explicit calculation.

Attractive interactions

For attractive interactions the situation is different: generally the energy required to excite an atom from the ground state is greater than the free-particle value because some of the attractive energy is lost due to particles being further apart. An exception is the centre-of-mass excitation, for which there is no interaction contribution to the excitation energy. Thus the most energetically economical way to add angular momentum to the cloud is to excite motion of the centre of mass, without disturbing the internal correlations of the ground state [21]. This result is general, even though we have demonstrated it only for weak interactions and small angular momenta. Angular momentum can be added partly to the centre-of-mass motion, which leaves the interaction energy unchanged, or may be put into internal excitations, which will increase the interaction energy. Thus, exciting the centre-of-mass motion is the most energetically favourable way of giving angular momentum to a cloud of atoms with attractive interactions.

From the above example we see that the behaviour under rotation of a cloud of atoms with attractive interactions is qualitatively different from that for repulsive ones: the system undergoes a phase transition when the coupling passes through zero. This may be illustrated by considering the model discussed above for a state with more than one unit of angular momentum. As we have shown above, in the absence of interactions the ground state is degenerate. For weak interactions the eigenvalues are obtained by diag-

9.5 The weakly-interacting Bose gas under rotation

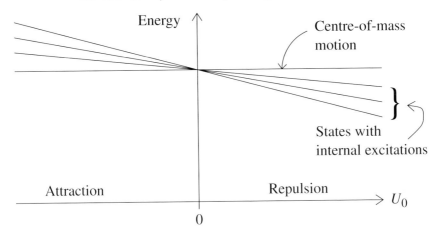

Fig. 9.3. Schematic plot of energies of many-atom eigenstates with a given non-zero angular momentum as a function of interaction strength for weak coupling between atoms.

onalizing the interaction Hamiltonian in the subspace of degenerate states. All the matrix elements are proportional to the strength of the potential, and therefore the eigenvalues are linear functions of the strength. Thus the behaviour of the eigenvalues is as shown schematically in Fig. 9.3. From this one sees that the lowest-energy state changes character as the strength of the interaction passes through zero. The calculations above have shown that the state corresponding to exciting only the centre-of-mass degree of freedom is the lowest-energy one for attractive interactions, and thus, for repulsive interactions, it has the highest energy among the states in the manifold.

Problems

PROBLEM 9.1 Calculate the condensate wave function for a vortex in a uniform Bose gas in the Thomas–Fermi approximation, in which one neglects the term in the energy containing radial gradients of the wave function. Show that the condensate wave function vanishes for $\rho \leq |\ell|\xi$ where ℓ is the number of quanta of circulation of the vortex, and ξ is the coherence length far from the vortex. Show that to logarithmic accuracy the vortex energy per unit length is $\pi\hbar^2(n/m)\ell^2 \ln(R/|\ell|\xi)$, n being the density far from the vortex. Compare this result with the exact one for a vortex line with a single quantum of circulation. When the radial gradient terms are included in the calculation, particles tunnel into the centrifugal barrier at small radii, yielding a condensate wave function which for small ρ has the $\rho^{|\ell|}$ behaviour familiar for free particles.

PROBLEM 9.2 Calculate the vortex energy from (9.17) using the trial solution given by (9.19), and show that it is a minimum for $\alpha = 2$.

PROBLEM 9.3 Consider two parallel vortices with circulations ℓ_1 and ℓ_2 separated by a distance d within a cylindrical container of radius R. The mass density of the medium is uniform, $\rho = nm$. Show that the interaction energy per unit length of the vortices is given in terms of their velocity fields $\mathbf{v}_1$ and $\mathbf{v}_2$ as

$$\epsilon_{\text{int}} = \rho \int d^2\mathbf{r}\, \mathbf{v}_1 \cdot \mathbf{v}_2 = \frac{2\pi \ell_1 \ell_2 \hbar^2 n}{m} \ln \frac{R}{d},$$

where the final expression holds for $R \gg d$ and $d \gg \xi$.

PROBLEM 9.4 Calculate the energy of a rectilinear vortex in a uniform fluid contained in a cylinder of radius R when the vortex is coaxial with the cylinder and at a distance b from the axis of the cylinder. [Hint: The problem is equivalent to that of evaluating the capacitance of two parallel cylindrical conductors of radii R and ξ whose axes are displaced with respect to each other by an amount b. See, e.g., Ref. [22].]

PROBLEM 9.5 Consider the state $|0^{N-1}2^1\rangle$, in which one particle is in the state $\rho^{|m|}e^{im\varphi}e^{-\rho^2/2a_{\text{osc}}^2}$ for $m = 2$ which is a solution of the Schrödinger equation for the isotropic harmonic oscillator in two dimensions. The interaction can induce transitions to the state with two atoms in the $m = 1$ state and the remainder in the $m = 0$ one. Show that the shift of the energy of the state $|0^{N-1}2^1\rangle$ due to this process is independent of N and of order $\langle 00|U|00\rangle$ when the interaction is weak.

References

[1] For an extensive review of the effects of rotation on superfluid liquid ^{4}He, see R. J. Donnelly, *Quantized Vortices in Liquid He II*, (Cambridge University Press, Cambridge, 1991).

[2] A. L. Fetter and A. A. Svidzinsky, *J. Phys.: Condens. Matter* **13**, 135 (2001).

[3] L. Onsager, *Nuovo Cimento* **6**, Suppl. 2, 249 (1949).

[4] R. P. Feynman, in *Progress in Low Temperature Physics*, Vol. 1, ed. C. J. Gorter, (North-Holland, Amsterdam, 1955), Chapter 2.

[5] V. L. Ginzburg and L. P. Pitaevskii, *Zh. Eksp. Teor. Fiz.* **34**, 1240 (1958) [*Sov. Phys.–JETP* **7**, 858 (1958)].

[6] A. L. Fetter, in *Lectures in Theoretical Physics*, eds. K. T. Mahanthappa and W. E. Brittin (Gordon and Breach, N.Y., 1969), Vol. XIB, p. 351.

[7] E. Lundh, C. J. Pethick, and H. Smith, *Phys. Rev. A* **55**, 2126 (1997).

[8] F. Dalfovo and S. Stringari, *Phys. Rev. A* **53**, 2477 (1996).

[9] D. A. Butts and D. S. Rokhsar, *Nature* **397**, 327 (1999).

[10] G. M. Kavoulakis, B. Mottelson, and C. J. Pethick, *Phys. Rev. A* **62**, 063605 (2000).

[11] L. D. Landau and E. M. Lifshitz, *Mechanics*, Third edition, (Pergamon, Oxford, 1976), §39.

[12] V. K. Tkachenko, *Zh. Eksp. Teor. Fiz.* **49**, 1875 (1965) [*Sov. Phys.-JETP* **22**, 1282 (1966)].

[13] M. R. Matthews, B. P. Anderson, P. C. Haljan, D. S. Hall, C. E. Wieman, and E. A. Cornell, *Phys. Rev. Lett.* **83**, 2498 (1999).

[14] J. E. Williams and M. J. Holland, *Nature* **401**, 568 (1999).

[15] B. P. Anderson, P. C. Haljan, C. E. Wieman, and E. A. Cornell, *Phys. Rev. Lett.* **85**, 2857 (2000).

[16] K. W. Madison, F. Chevy, W. Wohlleben, and J. Dalibard, *Phys. Rev. Lett.* **84**, 806 (2000).

[17] F. Chevy, K. W. Madison, and J. Dalibard, *Phys. Rev. Lett.* **85**, 2223 (2000).

[18] J. R. Abo-Shaeer, C. Raman, J. M. Vogels, and W. Ketterle, *Science* **292**, 476 (2001).

[19] L. D. Landau and E. M. Lifshitz, *Fluid Mechanics*, Second edition, (Pergamon, Oxford, 1987).

[20] B. Mottelson, *Phys. Rev. Lett.* **83**, 2695 (1999).

[21] N. K. Wilkin, J. M. F. Gunn, and R. A. Smith, *Phys. Rev. Lett.* **80**, 2265 (1998).

[22] L. D. Landau and E. M. Lifshitz, *Electrodynamics of Continuous Media*, Second edition, (Pergamon, Oxford, 1984), §3, p. 16.

10
Superfluidity

The phenomena of superfluidity and superconductivity are intimately connected with the existence of a condensate, a macroscopically occupied quantum state. Such condensates occur in a variety of different physical systems, as described in Chapter 1. The foundation for the description of superfluidity is a picture of the system as being comprised of a condensate and elementary excitations. In Chapter 8 we have seen how physical properties such as the energy and the density of a Bose–Einstein condensed system may be expressed in terms of a contribution from the condensate, plus one from the elementary excitations, and in this chapter we shall consider further developments of this basic idea to other situations. As a first application, we determine the critical velocity for creation of an excitation in a homogeneous system (Sec. 10.1). Following that, we show how to express the momentum density in terms of the velocity of the condensate and the distribution function for excitations. This provides the basis for a two-component description, the two components being the condensate and the thermal excitations (Sec. 10.2). In the past, this framework has proved to be very effective in describing the properties of superfluids and superconductors, and in Sec. 10.3 we apply it to dynamical processes.

To describe the state of a superfluid, one must specify the condensate velocity, in addition to the variables needed to characterize the state of an ordinary fluid. As a consequence, the collective behaviour of a superfluid is richer than that of an ordinary one. Collective modes are most simply examined when excitations collide frequently enough that they are in local thermodynamic equilibrium. Under these conditions the excitations may be regarded as a fluid, and a hydrodynamic description is possible. This is referred to as the two-fluid model. As an illustration, we show in Sec. 10.4 that, as a consequence of the additional macroscopic variable, there are two sound-like modes, so-called first and second sound, rather than the single sound mode in an ordinary fluid.

When collisions are so infrequent that local thermodynamic equilibrium is not established, the dynamics of the excitations must be treated microscopically. In Sec. 10.5 we illustrate this approach by considering the damping and frequency shift of low-frequency collective modes of the condensate due to coupling to thermal excitations.

10.1 The Landau criterion

Consider a uniform Bose–Einstein condensed liquid in its ground state. Imagine that a heavy obstacle moves through the liquid at a constant velocity v, and let us ask at what velocity it becomes possible for excitations to be created. In the reference frame in which the fluid is at rest, the obstacle exerts a time-dependent potential on the particles in the fluid. To simplify the analysis it is convenient to work in the frame in which the obstacle is at rest. If the energy of a system in one frame of reference is E, and its momentum is $\mathbf{p}$, the energy in a frame moving with velocity $\mathbf{v}$ is given by the standard result for Galilean transformations,

$$E(\mathbf{v}) = E - \mathbf{p}\cdot\mathbf{v} + \frac{1}{2}Mv^2, \tag{10.1}$$

where $M = Nm$ is the total mass of the system. Thus in the frame moving with the obstacle the energy of the ground state, which has momentum zero in the original frame, is

$$E(\mathbf{v}) = E_0 + \frac{1}{2}Nmv^2, \tag{10.2}$$

E_0 being the ground-state energy in the frame where the fluid is at rest.

Consider now the state with a single excitation of momentum $\mathbf{p}$. In the original frame the energy is

$$E_{\text{ex}} = E_0 + \epsilon_p, \tag{10.3}$$

and therefore the energy in the frame moving with the obstacle is

$$E_{\text{ex}}(\mathbf{v}) = E_0 + \epsilon_p - \mathbf{p}\cdot\mathbf{v} + \frac{1}{2}Nmv^2. \tag{10.4}$$

Consequently, the energy needed to create an excitation in the frame moving with the obstacle is $\epsilon_p - \mathbf{p}\cdot\mathbf{v}$, the difference between (10.4) and (10.2). In the frame of the obstacle, the potential produced by it is static, and therefore the obstacle is unable to transfer energy to the fluid. Thus at a velocity

$$v = \frac{\epsilon_p}{p}, \tag{10.5}$$

that is, when the phase velocity of the excitation is equal to the velocity of the fluid relative to the obstacle, it becomes possible kinematically for the obstacle to create an excitation with momentum parallel to **v**. For higher velocities, excitations whose momenta make an angle $\cos^{-1}(\epsilon_p/pv)$ with the velocity vector **v** may be created. This process is the analogue of the Cherenkov effect, in which radiation is emitted by a charged particle passing through a material at a speed in excess of the phase velocity of light in the medium. The minimum velocity at which it is possible to create excitations is given by

$$v_c = \min\left(\frac{\epsilon_p}{p}\right), \tag{10.6}$$

which is referred to as the *Landau critical velocity*. For velocities less than the minimum value of ϵ_p/p, it is impossible to create excitations. There is consequently no mechanism for degrading the motion of the condensate, and the liquid will exhibit superfluidity.

Equation (10.6) shows that the excitations created at the lowest velocity are those with the lowest phase velocity. In a uniform, Bose–Einstein condensed gas, for which the spectrum is given by the Bogoliubov expression (8.29), the phase velocity may be written as $s[1 + (p/2ms)^2]^{1/2}$. Therefore the lowest critical velocity is the sound speed, and the corresponding excitations are long-wavelength sound waves. This situation should be contrasted with that in liquid ^{4}He. If the relevant excitations were those given by the standard phonon-roton dispersion relation (Fig. 1.1), which is not convex everywhere, the lowest critical velocity would correspond to creation of rotons, which have finite wavelengths. However, the critical velocity observed in experiments is generally lower than this, an effect which may be accounted for by the creation of vortex rings, which have a phase velocity lower than that of phonons and rotons. We remark in passing, that for the non-interacting Bose gas, the critical velocity is zero, since the phase velocity of free particles, $p/2m$, vanishes for $p = 0$. For this reason Landau argued that the picture of liquid ^{4}He as an ideal Bose gas was inadequate for explaining superfluidity. As the Bogoliubov expression (7.31) for the excitation spectrum of the dilute gas shows, interactions are crucial for obtaining a non-zero critical velocity. In a recent experiment a critical velocity for the onset of a pressure gradient in a Bose–Einstein condensed gas has been measured by stirring the condensate with a laser beam

[1], but a detailed theoretical understanding of the results remains to be found.

Another way of deriving the Landau criterion is to work in the frame in which the fluid is at rest and the obstacle is moving. Let us denote the potential at point $\mathbf{r}$ due to the obstacle by $V(\mathbf{r} - \mathbf{R}(t))$, where $\mathbf{R}(t)$ is the position of some reference point in the obstacle. For uniform motion with velocity $\mathbf{v}$, the position of the obstacle is given by $\mathbf{R}(t) = \mathbf{R}(0) + \mathbf{v}t$, and therefore the potential will be of the form $V(\mathbf{r} - \mathbf{v}t - \mathbf{R}(0))$. From this result one can see that the Fourier component of the potential with wave vector $\mathbf{q}$ has frequency $\mathbf{q} \cdot \mathbf{v}$. Quantum-mechanically this means that the potential can transfer momentum $\hbar \mathbf{q}$ to the liquid only if it transfers energy $\hbar \mathbf{q} \cdot \mathbf{v}$. The condition for energy and momentum conservation in the creation of an excitation leads immediately to the criterion (10.6).

In the discussion above we considered a fluid with no excitations present initially. However the argument may be generalized to arbitrary initial states, and the critical velocity is given by the same expression as before, except that the excitation energy to be used is the one appropriate to the initial situation, allowing for interactions with other excitations.

10.2 The two-component picture

The description of superfluids and superconductors in terms of two interpenetrating components, one associated with the condensate and the other with the excitations, is conceptually very fruitful. In an ordinary fluid, only the component corresponding to the excitations is present and, consequently, that component is referred to as the *normal component*. That associated with the condensate is referred to as the *superfluid component*. The two components do not correspond to physically distinguishable species, as they would in a mixture of two different kinds of atoms. The expression (8.35) is an example of the two-component description for the particle density. We now develop the two-component picture by considering flow in a uniform system and calculating the momentum carried by excitations in a homogeneous gas.

10.2.1 Momentum carried by excitations

In the ground state of a gas the condensate is stationary, and the total momentum is zero. Let us now imagine that excitations are added without changing the velocity of the condensate. The total momentum per unit volume is thus equal to that carried by the excitations,

$$\mathbf{j}_{\text{ex}} = \int \frac{d\mathbf{p}}{(2\pi\hbar)^3} \mathbf{p} f_{\mathbf{p}}. \tag{10.7}$$

Now let us perform a Galilean transformation to a reference frame moving with a velocity $-\mathbf{v}_s$, in which the condensate is therefore moving with velocity $\mathbf{v}_s$. Under the Galilean transformation the total momentum changes by an amount $Nm\mathbf{v}_s$, and therefore the total momentum density in the new frame is

$$\mathbf{j} = \rho\mathbf{v}_s + \mathbf{j}_{\text{ex}} = \rho\mathbf{v}_s + \int \frac{d\mathbf{p}}{(2\pi\hbar)^3}\mathbf{p}f_\mathbf{p}, \tag{10.8}$$

where $\rho = nm$ is the mass density. For a system invariant under Galilean transformations, the momentum density is equal to the mass current density, which enters the equation for mass conservation.

10.2.2 Normal fluid density

The expression (10.8) for the momentum density forms the basis for the introduction of the concept of the normal density. We consider a system in equilibrium at finite temperature and ask how much momentum is carried by the excitations. We denote the velocity of the condensate by $\mathbf{v}_s$. The gas of excitations is assumed to be in equilibrium, and its velocity is denoted by $\mathbf{v}_n$. From the results of Sec. 10.1 one can see that the energy of an excitation in the original frame is $\epsilon_p + \mathbf{p}\cdot\mathbf{v}_s$. The distribution function is that for excitations in equilibrium in the frame moving with velocity $\mathbf{v}_n$, and therefore the energy that enters the Bose distribution function is the excitation energy appropriate to this frame. This is the energy in the original frame, shifted by an amount $-\mathbf{p}\cdot\mathbf{v}_n$, and therefore the equilibrium distribution function is

$$f_\mathbf{p} = \frac{1}{\exp\{[\epsilon_p - \mathbf{p}\cdot(\mathbf{v}_n - \mathbf{v}_s)]/kT\} - 1}. \tag{10.9}$$

By inserting this expression into Eq. (10.8) one finds the momentum density of the excitations to be

$$\mathbf{j}_{\text{ex}} = \int \frac{d\mathbf{p}}{(2\pi\hbar)^3}\mathbf{p}f_\mathbf{p} = \rho_n(|\mathbf{v}_n - \mathbf{v}_s|)(\mathbf{v}_n - \mathbf{v}_s), \tag{10.10}$$

where

$$\rho_n(v) = \int \frac{d\mathbf{p}}{(2\pi\hbar)^3}\frac{\mathbf{p}\cdot\mathbf{v}}{v^2}\frac{1}{\exp[(\epsilon_p - \mathbf{p}\cdot\mathbf{v})/kT] - 1}. \tag{10.11}$$

For small velocities one finds

$$\rho_n = \int \frac{d\mathbf{p}}{(2\pi\hbar)^3}(\mathbf{p}\cdot\hat{\mathbf{v}})^2\left(-\frac{\partial f_p^0}{\partial \epsilon_p}\right) = \int \frac{d\mathbf{p}}{(2\pi\hbar)^3}\frac{p^2}{3}\left(-\frac{\partial f_p^0}{\partial \epsilon_p}\right), \tag{10.12}$$

where $f_p^0 = [\exp(\epsilon_p/kT) - 1]^{-1}$. The temperature-dependent quantity ρ_n is referred to as the density of the normal fluid, or simply the *normal density*. For a dilute Bose gas, the spectrum of elementary excitations is the Bogoliubov one, Eq. (8.29). At temperatures low compared with $T_* = ms^2/k$, Eq. (8.64), the dominant excitations are phonons, for which $\epsilon_p \simeq sp$. Substitution of this expression into Eq. (10.12) gives

$$\rho_n = \frac{2\pi^2}{45} \frac{(kT)^4}{\hbar^3 s^5}. \tag{10.13}$$

At a temperature T, the typical wave number of a thermal phonon is $\sim kT/\hbar s$, and therefore the density of excitations is of order $(kT/\hbar s)^3$, the volume of a sphere in wave number space having a radius equal to the thermal wave number. Thus a thermal excitation behaves as though it has a mass $\sim kT/s^2$, which is much less than the atomic mass if $T \ll T_*$. As the temperature approaches T_*, the mass becomes of order the atomic mass. In the other limiting case, $T \gg T_*$, the energy of an excitation is approximately the free-particle energy $p^2/2m$. If one integrates by parts the expression (10.12) for the normal density, one finds

$$\rho_n = m n_{\text{ex}}, \tag{10.14}$$

where $n_{\text{ex}} = n - n_0$ is the number density of particles not in the condensate, given by

$$n_{\text{ex}} = n \left(\frac{T}{T_c}\right)^{3/2}. \tag{10.15}$$

The simple result for $T \gg T_*$ is a consequence of the fact that the excitations are essentially free particles, apart from the Hartree–Fock mean field, and therefore, irrespective of their momenta, they each contribute one unit to the particle number, and the particle mass m to the normal density. Note that the density of the normal component is not in general proportional to n_{ex} (see Problem 10.3).

According to (10.8) the total momentum density is then obtained by adding $\rho \mathbf{v}_s$ to $\mathbf{j}_{\text{ex}}$,

$$\mathbf{j} = \rho_n (\mathbf{v}_n - \mathbf{v}_s) + \rho \mathbf{v}_s. \tag{10.16}$$

If we define the density of the superfluid component or the *superfluid density* as the difference between the total and normal densities,

$$\rho_s = \rho - \rho_n, \tag{10.17}$$

the momentum density may be expressed as

$$\mathbf{j} = \rho_s \mathbf{v}_s + \rho_n \mathbf{v}_n, \tag{10.18}$$

which has the same form as for two interpenetrating fluids. There are however important differences between this result and the corresponding one for a fluid containing two distinct species. The density of the normal component is defined in terms of the response of the momentum density to the velocity difference $\mathbf{v}_n - \mathbf{v}_s$. It therefore depends both on temperature and on $\mathbf{v}_n - \mathbf{v}_s$.

10.3 Dynamical processes

In the previous section we considered steady-state phenomena in uniform systems. In this section we treat dynamical phenomena, taking into account spatial non-uniformity. The most basic description of dynamical processes is in terms of the eigenstates of the complete system or, more generally, in terms of the density matrix. However, for many purposes this approach is both cumbersome and more detailed than is necessary. Here we shall assume that spatial variations are slow on typical microscopic length scales. It is then possible to use the semi-classical description, in which the state of the excitations is specified in terms of their positions $\mathbf{r}_i$ and momenta $\mathbf{p}_i$ or equivalently their distribution function $f_\mathbf{p}(\mathbf{r})$, as we did earlier in the description of equilibrium properties in Sec. 2.3. The superfluid is characterized by the condensate wave function $\psi(\mathbf{r}, t)$.

The condensate wave function is specified by its magnitude and its phase, $\psi = |\psi| e^{i\phi}$. The condensate density equals $|\psi|^2$, while the superfluid velocity is given in terms of ϕ by the relation $\mathbf{v}_s = \hbar \boldsymbol{\nabla} \phi / m$. For many purposes it is convenient to eliminate the magnitude of the condensate wave function in favour of the local density $n(\mathbf{r}, t)$, and use this and the phase $\phi(\mathbf{r}, t)$ of the condensate wave function as the two independent variables. One advantage of working with the total density, rather than the density $|\psi|^2$ of the condensate, is that collisions between excitations alter the condensate density locally, but not the total density. Another is that, as we shall see, $\phi(\mathbf{r}, t)$ and the total density $n(\mathbf{r}, t)$ are canonical variables. In Sec. 7.1 we showed how a pure condensate is described in these terms, and we now generalize these considerations to take into account excitations.

Consider first a uniform system. Its state is specified by the number of particles N, the occupation numbers $N_\mathbf{p}$ for excitations, and the condensate velocity $\mathbf{v}_s$. The condensate wave function is the matrix element of the annihilation operator between the original state and the state with the same number of excitations in all states, but one fewer particles. Thus the phase of the condensate wave function varies as

$$\frac{d\phi}{dt} = -\frac{E(N, \{N_\mathbf{p}\}, \mathbf{v}_s) - E(N-1, \{N_\mathbf{p}\}, \mathbf{v}_s)}{\hbar} \simeq -\frac{1}{\hbar}\frac{\partial E}{\partial N}, \qquad (10.19)$$

where $\{N_\mathbf{p}\}$ indicates the set of occupation numbers for all momentum states. In generalizing this result to non-uniform systems we shall confine ourselves to situations where the energy density $\mathcal{E}(\mathbf{r})$ is a local function of the particle density $n(\mathbf{r})$, the condensate velocity $\mathbf{v}_s$, and the distribution function for excitations, $f_\mathbf{p}(\mathbf{r})$. The rate of change of the phase locally is therefore given by generalizing (10.19) to spatially varying situations,

$$\frac{\partial \phi(\mathbf{r},t)}{\partial t} = -\frac{1}{\hbar}\frac{\delta E}{\delta n(\mathbf{r})} = -\frac{1}{\hbar}\frac{\partial \mathcal{E}}{\partial n}. \tag{10.20}$$

Formally this result, which at zero temperature is identical with (7.22), is an expression of the fact that the density n and the phase ϕ times $\hbar$ are canonically conjugate variables.

For slow variations the form of the Hamiltonian density may be determined from Galilean invariance, since the local energy density is well approximated by that for a uniform system. It is convenient to separate the energy density of the system in the frame in which the superfluid is at rest into a part due to the external potential $V(\mathbf{r})$ and a part $\mathcal{E}(n,\{f_\mathbf{p}\})$ coming from the internal energy of the system. Examples of this energy functional are the expressions for the ground-state energy density within Bogoliubov theory obtained from Eq. (8.43), and for the Hartree–Fock energy density which one obtains from Eq. (8.73). In the frame in which the superfluid has velocity $\mathbf{v}_s$, the energy density may be found from the standard expression (10.1) for the total energy, and is

$$\mathcal{E}_{tot} = \mathcal{E}(n,\{f_\mathbf{p}\}) + \mathbf{j}_{ex}\cdot\mathbf{v}_s + \frac{1}{2}\rho v_s^2 + V(\mathbf{r})n. \tag{10.21}$$

In the long-wavelength approximation we have adopted, the energy density depends locally on the density, and therefore the equation of motion for the phase is simply

$$\hbar\frac{\partial \phi}{\partial t} = -(\mu_{int} + V + \frac{1}{2}mv_s^2). \tag{10.22}$$

In this equation $\mu_{int} = \partial\mathcal{E}(n,\{f_\mathbf{p}\})/\partial n$ is the contribution to the chemical potential due to the internal energy, and the two other terms are due to the external potential and the kinetic energy associated with the flow. From this relationship one immediately finds the equation of motion for the superfluid velocity

$$m\frac{\partial \mathbf{v}_s}{\partial t} = -\boldsymbol{\nabla}(\mu_{int} + V + \frac{1}{2}mv_s^2). \tag{10.23}$$

This result has the same form as Eq. (7.20), but the quantum pressure term proportional to spatial derivatives of the density is absent because it has been neglected in the long-wavelength approximation made here. However, it is important to note that the chemical potential μ_{int} in Eq. (10.23) contains the effects of excitations, unlike the one in Eq. (7.20).

We next consider variations of the density. As we have noted already, the number of particles in the condensate is not conserved when excitations are present, but the total number of particles is. For a Galilean-invariant system, the total current density of particles is equal to the momentum density divided by the particle mass, and the momentum density has already been calculated, Eq. (10.8). Therefore the condition for particle number conservation is

$$\frac{\partial n}{\partial t} + \frac{1}{m}\boldsymbol{\nabla}\cdot\mathbf{j} = 0. \qquad (10.24)$$

One may also derive this result from the Hamiltonian formalism by using the second member of the pair of equations for the canonical variables n and $\hbar\phi$ (Problem 10.4).

It is important to notice that the equations of motion for $\mathbf{v}_s$ and n both contain effects due to excitations. In Eq. (10.23) they enter through the dependence of the chemical potential on the distribution of excitations, which gives rise to a coupling between the condensate and the excitations. The energy of an excitation depends in general on the distribution of excitations and also on the total density and the superfluid velocity. The distribution function for the excitations satisfies a kinetic equation similar in form to the conventional Boltzmann equation for a dilute gas which we employ in Sec. 11.3. The important differences are that the equations governing the motion of an excitation contain effects of the interaction, while the collision term has contributions not only from the mutual scattering of excitations, as it does for a gas of particles, but also from processes in which excitations are created or annihilated due to the presence of the condensate.

It is difficult to solve the kinetic equation in general, so it is useful to exploit conservation laws, whose nature does not depend on the details of the collision term. We have already encountered the conservation law for mass (or, equivalently, particle number). The momentum and energy conservation laws are derived by multiplying the kinetic equation by the momentum and energy of an excitation, respectively, and integrating over momenta. In general, these equations have terms which take into account transfer of a physical quantity between the excitations and the condensate. However, if one adds to these equations the corresponding ones for the condensate contributions to the physical quantities, one arrives at conservation laws which do not include explicitly the transfer term. Using the Einstein convention of summation over repeated indices, we may write the condition for momentum conservation as

$$\frac{\partial j_i}{\partial t} = -\frac{\partial \Pi_{ik}}{\partial x_k} - n\frac{\partial V}{\partial x_i}, \qquad (10.25)$$

where Π_{ik} is the momentum flux density, and that for energy conservation as

$$\frac{\partial \mathcal{E}}{\partial t} = -\frac{\partial Q_k}{\partial x_k} - \frac{1}{m}j_i\frac{\partial V}{\partial x_i}, \qquad (10.26)$$

where $\mathbf{Q}$ is the energy flux density. The last terms in Eqs. (10.25) and (10.26) represent the effects of the external potential. In the following section we consider the forms of Π_{ik} and $\mathbf{Q}$ under special conditions and we shall use the conservation laws when discussing the properties of sound modes in uniform Bose gases.

10.4 First and second sound

A novel feature of a Bose–Einstein condensed system is that to describe the state of the system, it is necessary to specify the velocity of the superfluid, in addition to the variables needed to describe the excitations. This new degree of freedom gives rise to phenomena not present in conventional fluids. To illustrate this, we now consider sound-like modes in a uniform Bose gas. Let us assume that thermal excitations collide frequently enough that they are in local thermodynamic equilibrium. Under these conditions, the state of the system may be specified locally in terms of the total density of particles, the superfluid velocity $\mathbf{v}_\mathrm{s}$, the temperature T, and the velocity $\mathbf{v}_\mathrm{n}$ of the excitations. The general theory of the hydrodynamics of superfluids is well described in standard works [2].

Basic results

The mass density ρ and the mass current density $\mathbf{j}$ satisfy the conservation law

$$\frac{\partial \rho}{\partial t} + \boldsymbol{\nabla}\cdot\mathbf{j} = 0. \qquad (10.27)$$

The equation of motion for $\mathbf{j}$ involves the momentum flux density Π_{ik} according to (10.25). We shall neglect non-linear effects and friction, in which case the momentum flux density is $\Pi_{ik} = p\delta_{ik}$, where p is the pressure. In the absence of friction and external potentials the time derivative of the mass current density is thus

$$\frac{\partial \mathbf{j}}{\partial t} = -\boldsymbol{\nabla} p. \qquad (10.28)$$

By eliminating $\mathbf{j}$ we then obtain

$$\frac{\partial^2 \rho}{\partial t^2} - \nabla^2 p = 0, \tag{10.29}$$

which relates changes in the density to those in the pressure. Since, in equilibrium, the pressure depends on the temperature as well as the density, Eq. (10.29) gives us one relation between density changes and temperature changes. To determine the frequencies of modes we need a second such relation, which we now derive.

The acceleration of the superfluid is given by Eq. (10.23). In the absence of an external potential, and since the non-linear effect of the superfluid velocity is neglected, the quantity μ_{int} is just the usual chemical potential μ and therefore

$$m\frac{\partial \mathbf{v}_s}{\partial t} = -\boldsymbol{\nabla}\mu. \tag{10.30}$$

In local thermodynamic equilibrium, a small change $d\mu$ in the chemical potential is related to changes in pressure and temperature by the Gibbs–Duhem relation

$$N d\mu = V dp - S dT, \tag{10.31}$$

where S is the entropy and N the particle number. When written in terms of the mass density $\rho = Nm/V$ and the entropy $\tilde{s}$ per unit mass, defined by

$$\tilde{s} = \frac{S}{Nm}, \tag{10.32}$$

the Gibbs–Duhem relation (10.31) shows that the gradient in the chemical potential is locally related to the gradients in pressure and temperature according to the equation

$$\boldsymbol{\nabla}\mu = \frac{m}{\rho}\boldsymbol{\nabla}p - \tilde{s}m\boldsymbol{\nabla}T. \tag{10.33}$$

It then follows from (10.33), (10.30) and (10.28) together with (10.18) that

$$\frac{\partial(\mathbf{v}_n - \mathbf{v}_s)}{\partial t} = -\tilde{s}\frac{\rho}{\rho_n}\boldsymbol{\nabla}T. \tag{10.34}$$

In the absence of dissipation the entropy is conserved. Since entropy is carried by the normal component only, the conservation equation reads

$$\frac{\partial(\rho\tilde{s})}{\partial t} + \boldsymbol{\nabla}\cdot(\rho\tilde{s}\mathbf{v}_n) = 0. \tag{10.35}$$

The linearized form of this equation is

$$\tilde{s}\frac{\partial\rho}{\partial t} + \rho\frac{\partial\tilde{s}}{\partial t} + \tilde{s}\rho\boldsymbol{\nabla}\cdot\mathbf{v}_n = 0. \tag{10.36}$$

10.4 First and second sound

By using the mass conservation equation (10.27) in (10.36) we find that

$$\frac{\partial \tilde{s}}{\partial t} = \tilde{s}\frac{\rho_s}{\rho}\nabla \cdot (\mathbf{v}_s - \mathbf{v}_n). \tag{10.37}$$

After combining (10.37) with (10.34) we arrive at the equation

$$\frac{\partial^2 \tilde{s}}{\partial t^2} = \frac{\rho_s}{\rho_n}\tilde{s}^2 \nabla^2 T, \tag{10.38}$$

which relates variations in the temperature to those in the entropy per unit mass. Since the entropy per unit mass is a function of density and temperature, this equation provides the second relation between density and temperature variations.

The collective modes of the system are obtained by considering small oscillations of the density, pressure, temperature and entropy, with spatial and temporal dependence given by $\exp i(\mathbf{q}\cdot\mathbf{r} - \omega t)$. In solving (10.29) and (10.38) we choose density and temperature as the independent variables and express the small changes in pressure and entropy in terms of those in density and temperature. Denoting the latter by $\delta\rho$ and δT we obtain from (10.29) the result

$$\omega^2 \delta\rho - q^2 \left[\left(\frac{\partial p}{\partial \rho}\right)_T \delta\rho + \left(\frac{\partial p}{\partial T}\right)_\rho \delta T\right] = 0, \tag{10.39}$$

and from (10.38) that

$$\omega^2 \left[\left(\frac{\partial \tilde{s}}{\partial \rho}\right)_T \delta\rho + \left(\frac{\partial \tilde{s}}{\partial T}\right)_\rho \delta T\right] - q^2 \frac{\rho_s}{\rho_n}\tilde{s}^2 \delta T = 0. \tag{10.40}$$

In terms of the phase velocity $u = \omega/q$ of the wave, the condition for the existence of non-trivial solutions to the coupled equations (10.39) and (10.40) is a quadratic equation for u^2,

$$(u^2 - c_1^2)(u^2 - c_2^2) - u^2 c_3^2 = 0. \tag{10.41}$$

The constant c_1 is the isothermal sound speed, given by

$$c_1^2 = \left(\frac{\partial p}{\partial \rho}\right)_T, \tag{10.42}$$

while c_2, given by

$$c_2^2 = \frac{\rho_s \tilde{s}^2 T}{\rho_n \tilde{c}}, \tag{10.43}$$

is the velocity of temperature waves, if the density of the medium is held constant. Here $\tilde{c}$ denotes the specific heat at constant volume, per unit mass,

$$\tilde{c} = T \left(\frac{\partial \tilde{s}}{\partial T}\right)_\rho. \tag{10.44}$$

If one uses the Maxwell relation

$$\left(\frac{\partial p}{\partial T}\right)_\rho = \left(\frac{\partial S}{\partial V}\right)_T = -\rho^2 \left(\frac{\partial \tilde{s}}{\partial \rho}\right)_T, \tag{10.45}$$

the quantity c_3, which is a measure of the coupling between density and temperature variations, is given by

$$c_3^2 = \left(\frac{\partial p}{\partial T}\right)_\rho^2 \frac{T}{\rho \tilde{c}}. \tag{10.46}$$

The sound velocities are the solutions of Eq. (10.41), and are given explicitly by

$$u^2 = \frac{1}{2}(c_1^2 + c_2^2 + c_3^2) \pm \left[\frac{1}{4}(c_1^2 + c_2^2 + c_3^2)^2 - c_1^2 c_2^2\right]^{1/2}. \tag{10.47}$$

Thus in a Bose–Einstein condensed system there are two different sound modes, which are referred to as first and second sound, corresponding to the choice of positive and negative signs in this equation. The existence of two sound speeds, as opposed to the single one in an ordinary fluid, is a direct consequence of the new degree of freedom associated with the condensate. The combination $c_1^2 + c_3^2$ has a simple physical interpretation. From the Maxwell relation (10.45) and the identity $(\partial p/\partial T)_V = -(\partial p/\partial V)_T (\partial V/\partial T)_p$, which is derived in the same manner as (2.72), it follows that $c_1^2 + c_3^2 = (\partial p/\partial \rho)_{\tilde{s}}$, which is the square of the velocity of adiabatic sound waves.

The ideal Bose gas

To understand the character of the two modes we investigate how the sound velocities depend on temperature and the interparticle interaction. First let us consider the non-interacting gas. By 'non-interacting' we mean that the effect of interactions on thermodynamic properties may be neglected. However, interactions play an essential role because they are responsible for the collisions necessary to ensure that thermodynamic equilibrium is established locally. They are also important in another respect, because in Sec. 10.1 we argued that the critical velocity for creating excitations in an ideal Bose gas is zero. By considering the equilibrium of an ideal Bose gas one can see that there is no equilibrium state in which the velocity of the condensate is different from that of the excitations. However, if the particles interact, the excitations at long wavelengths are sound waves, and the critical velocity is non-zero. Consequently, there is a range of flow velocities for which the system is superfluid.

The pressure and the entropy density $\rho\tilde{s}$ of an ideal Bose–Einstein condensed gas depend on temperature but not on density. Therefore c_1, Eq. (10.42), vanishes, the first-sound velocity is

$$u_1 = \sqrt{c_2^2 + c_3^2}, \qquad (10.48)$$

and the second-sound velocity vanishes, while $\partial\tilde{s}/\partial\rho = -\tilde{s}/\rho$. Substituting this result into Eq. (10.45) and using Eq. (10.46) and the fact that $\rho_s + \rho_n = \rho$ (Eq. (10.17)), one finds from Eq. (10.48) that

$$u_1^2 = \frac{\tilde{s}^2}{\tilde{c}} \frac{\rho}{\rho_n} T. \qquad (10.49)$$

The specific heat and the entropy per unit mass may be found from Eqs. (2.62) and (2.64), respectively, and the normal density is obtained from Eqs. (10.14) and (10.15), and therefore the first-sound speed is given by

$$u_1^2 = \frac{5\zeta(5/2)}{3\zeta(3/2)} \frac{kT}{m}. \qquad (10.50)$$

The numerical prefactor is approximately equal to 0.856. With the use of Eq. (2.63) for the pressure, this result may also be written in the form $u_1^2 = (5/3)(p/\rho_n) = dp/d\rho_n$, which is precisely what one would expect for a 'sound' wave in the excitation gas.

Properties of sound in ideal Bose gases can be determined more directly, and this is left as an exercise (Problem 10.5). The picture that emerges is simple, since the motion of the condensate and that of the excitations are essentially uncoupled: in first sound, the density of excited particles varies, while in second sound, the density of the condensate varies. The velocity of the latter mode is zero because a change in the density of condensate atoms produces no restoring force.

The interacting Bose gas

As a second example, let us consider an interacting Bose gas in the low-temperature limit, $T \to 0$. Since the ground-state energy is $E_0 = N^2 U_0/2V$, the pressure is given by

$$p = -\frac{\partial E_0}{\partial V} = \frac{1}{2}U_0 n^2 = \frac{U_0 \rho^2}{2m^2}. \qquad (10.51)$$

According to (10.42) we then find $c_1^2 = nU_0/m$. At low temperatures, the entropy is that associated with the Bogoliubov excitations and therefore varies as T^3. This implies that c_3^2 approaches zero as T tends to zero. The constant c_2, however, tends to a finite value, given by $c_2^2 = nU_0/3m$, since the entropy and specific heat of the phonons are related by $\tilde{s} = \tilde{c}/3$, while $\tilde{c}$ may be expressed in terms of the normal density (10.13), $\tilde{c} = 3\rho_{\mathrm{n}}s^2/T\rho$. In this limit the velocity u_1 of first sound is therefore given by

$$u_1^2 = c_1^2 = \frac{nU_0}{m} = s^2, \tag{10.52}$$

while that of second sound is given by

$$u_2^2 = c_2^2 = \frac{nU_0}{3m} = \frac{s^2}{3}. \tag{10.53}$$

In this limit, first sound is a pure density modulation, and it corresponds to a long-wavelength Bogoliubov excitation in the condensate. Second sound, which has a velocity $1/\sqrt{3}$ times the first-sound velocity, corresponds to a variation in the density of excitations, with no variation in the total particle density; it is a pure temperature wave.

As a final example, we consider situations when the Hartree–Fock theory described in Sec. 8.3.1 applies. The condition for this is that $T \gg T_* = nU_0/k$. The velocity of second sound is zero when interactions are neglected, and we shall now calculate the leading corrections to this result due to interactions. In the absence of interactions, modes of the condensate are completely decoupled from those of the excitations. Since the modes are non-degenerate, the leading corrections to the mode frequencies may be estimated without taking into account the coupling between the modes. The coupling is at least of first order in the interaction and, consequently, the frequency shift due to coupling between modes must be of higher order than first in the interaction. We therefore look for the frequencies of modes of the condensate when the distribution function for the excitations does not vary. When the Hartree–Fock theory is valid, the total mass density is given by

$$\rho = mn_0 + mn_{\mathrm{ex}}, \tag{10.54}$$

and the momentum density by

$$\mathbf{j} = mn_0\mathbf{v}_{\mathrm{s}} + mn_{\mathrm{ex}}\mathbf{v}_{\mathrm{n}}. \tag{10.55}$$

When the normal component does not move, the continuity equation (10.24) becomes

$$\frac{\partial n_0}{\partial t} + \boldsymbol{\nabla}\cdot(n_0\mathbf{v}_{\mathrm{s}}) = 0. \tag{10.56}$$

According to (8.85) the chemical potential is given by

$$\mu = (n_0 + 2n_{\text{ex}})U_0, \tag{10.57}$$

and therefore the change in the chemical potential is $\delta\mu = U_0\delta n_0$, and Eq. (10.30) for the acceleration of the condensate is

$$m\frac{\partial \mathbf{v}_{\text{s}}}{\partial t} = -U_0\boldsymbol{\nabla}\delta n_0. \tag{10.58}$$

Combining Eqs. (10.56) and (10.58) and linearizing, one finds

$$\frac{\partial^2 \delta n_0}{\partial t^2} = \frac{n_0 U_0}{m}\nabla^2 \delta n_0. \tag{10.59}$$

Not surprisingly, this equation has precisely the same form as Eq. (7.30) for the modes of a pure condensate in the long-wavelength limit $q \to 0$, except that n_0 appears instead of the total density n, and the velocity of the mode is given by

$$u_2^2 = \frac{n_0 U_0}{m}. \tag{10.60}$$

When coupling between the condensate and the excitations is taken into account, the interaction between condensate particles is screened by the thermal excitations. For a repulsive interaction this reduces the effective interaction, but this effect is of higher order in U_0 than the effect we have considered.

By similar arguments one can show that the changes in the first-sound velocity due to the interaction are small provided the total interaction energy is small compared with the thermal energy.

Let us now summarize the results of our calculations. In the cases examined, the motions of the condensate and of the thermal excitations are essentially independent of each other. The condensate mode is a Bogoliubov phonon in the condensate, with velocity $u = [n_0(T)U_0/m]^{1/2}$. This is the second-sound mode at higher temperatures, and the first-sound mode for temperatures close to zero. The mode associated with the thermal excitations is first sound at high temperatures, and second sound at low temperatures. First and second sound change their character as the temperature changes because the motion of the condensate is strongly coupled to that of the thermal excitations for temperatures at which the velocities of the modes are close to each other, and this leads to an 'avoided crossing' of the two sound velocities. A more extensive discussion of first and second sound in uniform Bose gases may be found in Ref. [3].

In experiments on collective modes in dilute atomic gases, local thermodynamic equilibrium is usually not established, and therefore the calculations above cannot be applied directly to experiments sensitive to the normal component. They are, however, relevant for modes of the condensate, because these are generally only weakly coupled to the motion of the thermal excitations.

It is instructive to compare dilute Bose gases with liquid ^{4}He. At all temperatures below the lambda point T_λ, the potential energy due to interparticle interactions dominates the thermal energy. As a consequence, modes corresponding to density fluctuations have higher velocities than do modes corresponding to temperature fluctuations. The coupling between the two sorts of modes, which is governed by the quantity c_3, Eq. (10.46), is small because $(\partial p/\partial T)_V$ is small, except very close to T_λ. At low temperatures the dominant thermal excitations are phonons, and therefore u_1 is the phonon velocity and $u_2 = u_1/\sqrt{3}$ as for a dilute gas.

10.5 Interactions between excitations

In the preceding section we discussed sound modes in the hydrodynamic limit, when collisions are sufficiently frequent that matter is in local thermodynamic equilibrium. We turn now to the opposite extreme, when collisions are relatively infrequent. Mode frequencies of clouds of bosons at zero temperature were calculated in Chapter 7, and we now address the question of how thermal excitations shift the frequencies, and damp the modes. If, after performing the Bogoliubov transformation as described in Chapter 8, we retain in the Hamiltonian only terms that are at most quadratic in the creation and annihilation operators, elementary excitations have a well-defined energy and they do not decay. However, when terms with a larger number of creation and annihilation operators are taken into account (see below), modes are coupled, and this leads to damping and to frequency shifts.

We begin by describing the processes that can occur. The full Hamiltonian, when expressed in terms of creation and annihilation operators for excitations, has contributions with differing numbers of operators. Those with no more than two operators correspond to non-interacting excitations as we saw in Sec. 8.1. It is convenient to classify the more complicated terms by specifying the numbers a and b of excitations in the initial and final states, respectively, and we use the shorthand notation a–b to label the process. The next more complicated terms after the quadratic ones are those cubic in the creation and annihilation operators. These give rise to 1–2 processes (in which one excitation decays into two), 2–1 processes (in which two incoming excitations merge to produce a third), and 0–3 and 3–0 processes (in which three excitations are created or annihilated). The remaining terms are quartic in the creation and annihilation operators, and correspond to 2–2, 1–3, 3–1, 0–4, and 4–0 processes. Since the energy of an excitation is positive by definition, the 0–3, 3–0, 0–4, and 4–0 processes are forbidden by energy conservation. In a normal gas, the excitations are particles and, consequently, the only processes allowed by particle number conservation are the 2–2 ones. These correspond to the binary collisions of atoms taken into account in the kinetic theory of gases.

At zero temperature, an elementary excitation can decay into two or three other excitations. The excitations in the final states must have energies less than that of the original excitation, and consequently for a low-energy initial excitation, the final-state phase space available is very restricted, and the resulting damping is small. Likewise, for phase-space reasons, decay of a low-energy excitation into two excitations is more important than decay into three. This process is referred to as *Beliaev damping* [4].

When more than one excitation is present, modes can decay by processes other than the 1–2 and 1–3 ones. At low temperatures, the 2–1 process and the related 1–2 one are more important than those with three excitations in the initial or final state, and the damping they give rise to was first discussed in the context of plasma oscillations by Landau, and is referred to as *Landau damping*. It plays a key role in phenomena as diverse as the anomalous skin effect in metals, the damping of phonons in metals, and the damping of quarks and gluons in quark–gluon plasmas. In the context of trapped Bose gases, it was proposed as a mechanism for damping of collective modes in Ref. [5]. We shall now calculate its rate for a low-energy, long-wavelength excitation.

10.5.1 Landau damping

Consider the decay of a collective mode i due to its interaction with a thermal distribution of excitations. The 2–1 process in which an excitation i merges with a second excitation j to give a single excitation k is shown schematically in Fig. 10.1.

The rate at which quanta in the state i are annihilated by absorbing a quantum from the state j and creating one in the state k may be evaluated from Fermi's Golden Rule, and is given by

$$\left.\frac{df_i}{dt}\right|_{2-1} = -\frac{2\pi}{\hbar} \sum_{jk} |M_{ij,k}|^2 f_i f_j (1 + f_k) \delta(\epsilon_i + \epsilon_j - \epsilon_k), \tag{10.61}$$

where f denotes the distribution function for excitations, while $M_{ij,k}$ denotes the matrix element for the process. The factors f_i and f_j express the fact

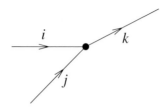

Fig. 10.1. Diagram representing the decay of a collective mode.

that the rate of the process is proportional to the numbers of incoming excitations, and the factor $1+f_k$ takes account of the fact that for bosons the scattering rate is enhanced by the presence of excitations in the final state. The first term, 1, corresponds to spontaneous emission, and the second, f_k, to induced emission. This is to be contrasted with the blocking factor $1-f_k$ that occurs for fermions.

Excitations in the state i are created by the process which is the inverse of the one above, and its rate is given by

$$\left.\frac{df_i}{dt}\right|_{1-2} = \frac{2\pi}{\hbar} \sum_{jk} |M_{ij,k}|^2 f_k (1+f_i)(1+f_j)\delta(\epsilon_i + \epsilon_j - \epsilon_k), \qquad (10.62)$$

where we have used the fact that $|M|^2$ is the same for the forward and inverse processes. The total rate of change of the distribution function is

$$\frac{df_i}{dt} = -\frac{2\pi}{\hbar} \sum_{jk} |M_{ij,k}|^2 [f_i f_j (1+f_k) - f_k(1+f_i)(1+f_j)]\delta(\epsilon_i + \epsilon_j - \epsilon_k). \qquad (10.63)$$

In equilibrium the rate vanishes, as may be verified by inserting the equilibrium Bose distribution $f_i = f_i^0$ into this expression.

When the number of quanta f_i in state i is disturbed from equilibrium, while the distribution function for other states has its equilibrium value, we may rewrite (10.63) as

$$\frac{df_i}{dt} = -2\frac{(f_i - f_i^0)}{\tau_i^{\mathrm{amp}}}, \qquad (10.64)$$

where

$$\frac{1}{\tau_i^{\mathrm{amp}}} = \frac{\pi}{\hbar} \sum_{jk} |M_{ij,k}|^2 (f_j^0 - f_k^0)\delta(\epsilon_i + \epsilon_j - \epsilon_k). \qquad (10.65)$$

The quantity τ_i^{amp} is the decay time for the amplitude of the mode, as we shall now explain. According to Eq. (10.64), the time for the excess number of quanta in the collective mode to decay by a factor e is $\tau_i^{\mathrm{amp}}/2$. In experiments, the decay of modes is often studied by exciting a mode, for example a long-wavelength collective mode, to a level far above the thermal equilibrium one, so the f_i^0 term in Eq. (10.64) may be neglected. Since the energy in the mode varies as the square of the amplitude, the decay time for the amplitude, which is the quantity generally measured experimentally, is τ_i^{amp}.

The general expression for the matrix element may be found by extracting the term proportional to $\alpha_k^\dagger \alpha_j \alpha_i$ in the Bogoliubov Hamiltonian, as was done by Pitaevskii and Stringari [6], and it is given by

$$\begin{aligned} M_{ij,k} &= 2U_0 \int d\mathbf{r} \psi(\mathbf{r})[u_i(u_j u_k^* + v_j v_k^* - v_j u_k^*) - v_i(u_j u_k^* + v_j v_k^* - u_j v_k^*)] \\ &= 2U_0 \int d\mathbf{r} \psi(\mathbf{r})[(u_i - v_i)(u_j u_k^* + v_j v_k^* - \frac{1}{2}(v_j u_k^* + u_j v_k^*)) \\ &\quad - \frac{1}{2}(u_i + v_i)(v_j u_k^* - u_j v_k^*)]. \end{aligned} \quad (10.66)$$

The latter form of the expression is useful for interpreting the physical origin of the coupling, since the combinations $u_i - v_i$ and $u_i + v_i$ are proportional to the density and velocity fields, respectively, produced by the collective mode.

To proceed further with the calculation for a trapped gas is complicated, so we shall consider a homogeneous gas. The excitations are then characterized by their momenta, and their total momentum is conserved in collisions. We denote the momenta of the excitations i, j and k by $\mathbf{q}, \mathbf{p}$ and $\mathbf{p} + \mathbf{q}$. The decay rate (10.65) is then given by

$$\frac{1}{\tau_q^{\text{amp}}} = \frac{\pi}{\hbar} \sum_\mathbf{p} |M_{\mathbf{qp,p+q}}|^2 (f_\mathbf{p}^0 - f_{\mathbf{p+q}}^0) \delta(\epsilon_\mathbf{p} + \epsilon_\mathbf{q} - \epsilon_{\mathbf{p+q}}). \quad (10.67)$$

Rather than calculating the matrix element (10.66) directly from the expressions for the coefficients u and v, we shall evaluate it for a long-wavelength collective mode using physical arguments. Provided the wavelength h/q of the collective mode is large compared with that of the thermal excitations, h/p, we may obtain the matrix elements for coupling of excitations to the collective mode by arguments similar to those used to derive the long-wavelength electron–phonon coupling in metals, or the coupling of ^{3}He quasiparticles to ^{4}He phonons in dilute solutions of ^{3}He in liquid ^{4}He [7]. When a collective mode is excited, it gives rise to oscillations in the local density and in the local superfluid velocity. For disturbances that vary sufficiently slowly in space, the coupling between the collective mode and an excitation may be determined by regarding the system locally as being spatially uniform. In the same approximation, the momentum of an excitation may be regarded as being well-defined. The energy of the excitation in the reference frame in which the superfluid is at rest is the usual result, $\epsilon_p(n_0)$, evaluated at the local condensate density, and therefore, by Galilean invariance, the energy in the frame in which the superfluid moves with velocity $\mathbf{v}_s$ is $\epsilon_p(n_0) + \mathbf{p}\cdot\mathbf{v}_s$. The modulation of the energy of the excitation due to the collective mode is therefore

$$\delta\epsilon_\mathbf{p} = \frac{\partial \epsilon_p}{\partial n_0}\delta n_0 + \mathbf{p}\cdot\mathbf{v}_s \tag{10.68}$$

to first order in δn_0.

In a sound wave, the amplitudes of the density oscillations and those of the superfluid velocity are related by the continuity equation for pure condensate motion, $\partial \delta n_0/\partial t = -\boldsymbol{\nabla}\cdot(n_0\mathbf{v}_s)$, which implies that $\epsilon_q \delta n_0 = sq\delta n_0 = n_0 \mathbf{q}\cdot\mathbf{v}_s$. Since the velocity field is longitudinal, the superfluid velocity is parallel to $\mathbf{q}$, and therefore $\mathbf{v}_s = s\hat{\mathbf{q}}\delta n_0/n_0$. Consequently, the interaction is

$$\delta\epsilon_\mathbf{p} = \left(\frac{\partial \epsilon_p}{\partial n_0} + \frac{s}{n_0}\mathbf{p}\cdot\hat{\mathbf{q}}\right)\delta n_0. \tag{10.69}$$

The final step in calculating the matrix element is to insert the expression for the density fluctuation in terms of phonon creation and annihilation operators. The density operator is (see Eq. (8.2))

$$\hat{\psi}^\dagger\hat{\psi} \simeq \psi^*\psi + \psi^*\delta\hat{\psi} + \psi\delta\hat{\psi}^\dagger. \tag{10.70}$$

The fluctuating part $\delta\hat{\psi}$ of the annihilation operator expressed in terms of phonon creation and annihilation operators is given by (see (8.51) and (7.44))

$$\delta\hat{\psi}(\mathbf{r}) = \frac{1}{V^{1/2}}\sum_\mathbf{q}(u_q\alpha_\mathbf{q} - v_q^*\alpha_{-\mathbf{q}}^\dagger)e^{i\mathbf{q}\cdot\mathbf{r}/\hbar}, \tag{10.71}$$

and therefore we may write the density fluctuation as

$$\delta n_0 = \frac{N_0^{1/2}}{V}[(u_q - v_q)\alpha_\mathbf{q} + (u_q^* - v_q^*)\alpha_{-\mathbf{q}}^\dagger]e^{i\mathbf{q}\cdot\mathbf{r}/\hbar}. \tag{10.72}$$

At long wavelengths $u_q - v_q \simeq (\epsilon_q/2\xi_q)^{1/2} \simeq (q/2ms)^{1/2}$. The $q^{1/2}$ dependence is similar to that for coupling of electrons to long-wavelength acoustic phonons in metals. The matrix element for absorbing a phonon is thus

$$M = \left(\frac{\partial \epsilon_p}{\partial n_0} + \frac{s}{n_0}\mathbf{p}\cdot\hat{\mathbf{q}}\right)\left(\frac{q}{2ms}\right)^{1/2}\frac{N_0^{1/2}}{V}. \tag{10.73}$$

To carry out the integration over the momentum of the incoming thermal excitation we make use of the fact that for a given value of p, the angle between **p** and **q** is fixed by energy conservation. Since by assumption $q \ll p$, one finds $\hat{\mathbf{p}} \cdot \hat{\mathbf{q}} = s/v_p$, where $v_p = \partial \epsilon_p / \partial p$ is the group velocity of the excitation. Thus the matrix element is given by

$$M = \left(\frac{\partial \epsilon_p}{\partial n_0} + \frac{s^2}{n_0}\frac{p}{v_p}\right)\left(\frac{q}{2ms}\right)^{1/2}\frac{N_0^{1/2}}{V}. \quad (10.74)$$

The results here are quite general, since they do not rely on any particular model for the excitation spectrum. For the Bogoliubov spectrum (8.29) the velocity is given by

$$v_p = \frac{\xi_p}{\epsilon_p}\frac{p}{m}, \quad (10.75)$$

where $\xi_p = \epsilon_p^0 + n_0 U_0$. The velocity (10.75) tends to the sound velocity for $p \ll ms$, and to the free-particle result p/m for $p \gg ms$. Thus the momenta of phonon-like thermal excitations that can be absorbed are almost collinear with the momentum of the collective mode, while those of high-energy ones, which behave essentially as free particles, are almost perpendicular. Since $\partial \epsilon_p / \partial n_0 = U_0 p^2 / 2m\epsilon_p$ and $ms^2 = n_0 U_0$, the total matrix element is

$$M = U_0 \left(\frac{p^2}{2m\epsilon_p} + \frac{\epsilon_p}{\xi_p}\right)\left(\frac{q}{2ms}\right)^{1/2}\frac{N_0^{1/2}}{V}. \quad (10.76)$$

The p-dependent factor here is 2 at large momenta, with equal contributions from the density dependence of the excitation energy and from the interaction with the superfluid velocity field. At low momenta the factor tends to zero as $3p/2ms$, the superfluid velocity field contributing twice as much as the density modulation.

The damping rate may be found by substituting the matrix element (10.76) into the expression (10.67), and using the delta function for energy conservation to perform the integral over the angle between **p** and **q**. If we further assume that $\hbar \omega_q \ll kT$, we may write

$$f_{\mathbf{p}}^0 - f_{\mathbf{p+q}}^0 \simeq -(\epsilon_{\mathbf{p+q}} - \epsilon_{\mathbf{p}})\partial f^0/\partial \epsilon_p = -\hbar \omega_q \partial f^0/\partial \epsilon_p \quad (10.77)$$

in Eq. (10.67) and the result is

$$\frac{1}{\tau_q^{\mathrm{amp}}} = \pi^{1/2}(n_0 a^3)^{1/2}\omega_q \int_0^\infty d\epsilon_p \left(-\frac{\partial f^0(\epsilon_p)}{\partial \epsilon_p}\right)\left(\frac{p^2}{2m\xi_p} + \frac{\epsilon_p^2}{\xi_p^2}\right)^2. \quad (10.78)$$

This integral may be calculated analytically in two limiting cases. First, when the temperature is low compared with $T_* = n_0 U_0/k$, the dominant excitations are phonons, and one may use the long-wavelength approximation in evaluating the matrix element and put $\epsilon_p = sp$. One then finds

$$\frac{1}{\tau_q^{\text{amp}}} = \frac{27\pi}{16}\frac{\rho_n}{\rho}\omega_q = \frac{3\pi^{9/2}}{5}(n_0 a^3)^{1/2}\frac{T^4}{T_*^4}\omega_q, \tag{10.79}$$

where ρ_n is the normal density given in Eq. (10.13). Second, at temperatures large compared with T_*, the result is

$$\frac{1}{\tau_q^{\text{amp}}} = \frac{3\pi^{3/2}}{4}(n_0 a^3)^{1/2}\frac{T}{T_*}\omega_q \approx 2.09\frac{a^{1/2}}{n_0^{1/2}\lambda_T^2}\omega_q, \tag{10.80}$$

where λ_T is the thermal de Broglie wavelength given by Eq. (1.2).

As one can see from the energy conservation condition, transitions occur if the group velocity of the thermal excitation is equal to the phase velocity of the collective mode. By 'surf-riding' on the wave, a thermal excitation may gain (or lose) energy, since the excitation experiences a force due to the interaction with the wave. Whether gain or loss is greater depends on the distribution function for excitations, but for a thermal distribution $\partial f^0/\partial \epsilon$ is negative and therefore there is a net loss of energy from the wave.

In addition to damping the collective mode, the interaction with thermal excitations also changes the frequency of the mode. The shift may be calculated from second-order perturbation theory. There are two types of intermediate states that contribute. The first are those in which there is one fewer quanta in the collective mode, one fewer excitations of momentum $\mathbf{p}$ and one more excitation with momentum $\mathbf{p}+\mathbf{q}$. This gives a contribution to the energy

$$\Delta E_{2-1} = \sum_{\mathbf{qp}}|M|^2\frac{f_\mathbf{q} f_\mathbf{p}(1+f_{\mathbf{p+q}})}{\epsilon_\mathbf{q}+\epsilon_\mathbf{p}-\epsilon_{\mathbf{p+q}}}. \tag{10.81}$$

This term corresponds to the 2–1 term in the calculation of the damping, and the distribution functions that occur here have the same origin. The second class of terms correspond to the inverse process, and they are the analogue of the 1–2 term in the damping rate. An extra quantum is created in each of the states $\mathbf{p}$ and $\mathbf{q}$, and one is destroyed in state $\mathbf{p+q}$. The energy denominator is the negative of that for the 2–1 term, and the corresponding energy shift is

$$\Delta E_{1-2} = \sum_{\mathbf{qp}}|M|^2\frac{(1+f_\mathbf{q})(1+f_\mathbf{p})f_{\mathbf{p+q}}}{\epsilon_{\mathbf{p+q}}-\epsilon_\mathbf{p}-\epsilon_\mathbf{q}}. \tag{10.82}$$

The total energy shift is obtained by adding Eqs. (10.81) and (10.82). It has a term proportional to the number of quanta in the mode $f_\mathbf{q}$, and the corresponding energy shift of a quantum is therefore given by

$$\Delta\epsilon_q = \sum_\mathbf{p} |M|^2 \left(\frac{f_\mathbf{p}(1+f_{\mathbf{p+q}})}{\epsilon_q + \epsilon_\mathbf{p} - \epsilon_\mathbf{p+q}} + \frac{f_\mathbf{p+q}(1+f_\mathbf{p})}{\epsilon_\mathbf{p+q} - \epsilon_\mathbf{p} - \epsilon_q} \right)$$

$$= V \int \frac{d\mathbf{p}}{(2\pi\hbar)^3} |M|^2 \frac{f_\mathbf{p} - f_\mathbf{p+q}}{\epsilon_q + \epsilon_\mathbf{p} - \epsilon_\mathbf{p+q}}. \quad (10.83)$$

For long-wavelength phonons the matrix element is given by (10.73), and the frequency shift $\Delta\omega_q = \Delta\epsilon_q/\hbar$ may be written in the form

$$\frac{\Delta\omega_q}{\omega_q} = -(n_0 a^3)^{1/2} F(T/T_*), \quad (10.84)$$

where $F(T/T_*)$ is a dimensionless function. The evaluation of the frequency shift at temperatures greater than T_* is the subject of Problem 10.6, and one finds that $\Delta\omega_q \propto T^{3/2}$.

Landau damping also provides the mechanism for damping of sound in liquid ^{4}He at low temperatures. For pressures less than 18 bar, the spectrum of elementary excitations for small p has the form $\epsilon_p = sp(1+\gamma p^2)$, where γ is positive, and therefore Landau damping can occur. However for higher pressures γ is negative, and the energy conservation condition cannot be satisfied. Higher-order effects give thermal excitations a finite width and when this is taken into account Landau damping can occur for phonons almost collinear with the long-wavelength collective mode.

To calculate the rate of damping in a trap, the starting point is again Eq. (10.65), where the excitations i, j, and k are those for a trapped gas. The damping rate has been calculated by Fedichev et al. [8]. Results are sensitive to details of the trapping potential, since they depend on the orbits of the excitations in the trap. For the traps that have been used in experiments, the theoretical damping rates for low-lying modes are only a factor 2–3 larger than those given by Eq. (10.80).

Problems

PROBLEM 10.1 Show that the critical velocity for simultaneous creation of two excitations can never be less than that for creation of a single excitation.

PROBLEM 10.2 Determine at temperatures much less than Δ/k, where Δ is the minimum roton energy, the normal density associated with the roton excitations in liquid ^{4}He discussed in Chapter 1.

PROBLEM 10.3 Demonstrate for the uniform Bose gas that $\rho_n \simeq m n_{\text{ex}}$ and $n_{\text{ex}} \simeq n(T/T_c)^{3/2}$ at temperatures T much higher than $T_* = ms^2/k$.

Determine the number of excitations n_ex at low temperature ($T \ll T_*$) and compare the result with ρ_n/m, Eq. (10.13).

PROBLEM 10.4 Derive the continuity equation (10.24) from the Hamiltonian equation

$$\frac{\partial n}{\partial t} = \frac{1}{\hbar}\frac{\delta \mathcal{E}}{\delta \phi}.$$

[Hint: The result

$$\frac{\delta \mathcal{E}}{\delta \phi} = -\frac{\partial}{\partial x_i}\frac{\partial \mathcal{E}}{\partial(\partial \phi/\partial x_i)}$$

that follows from Eq. (10.21) and the relation $\mathbf{v}_\text{s} = \hbar\boldsymbol{\nabla}\phi/m$ may be useful.]

PROBLEM 10.5 Use Eq. (10.29) to calculate the velocity of the hydrodynamic sound mode of a homogeneous ideal Bose gas below T_c when the condensate density is held fixed and the superfluid velocity is zero. Show that the velocity of the resulting mode, which is first sound, is given by Eq. (10.50). Demonstrate that the velocity of sound u in a homogeneous ideal Bose gas just above T_c is equal to the velocity of first sound just below T_c. Next, consider motion of the condensate and show that Eq. (10.30) for the acceleration of the superfluid immediately leads to the conclusion that modes associated with the condensate have zero frequency. [Hint: The chemical potential is constant in the condensed state.]

PROBLEM 10.6 Determine the function $F(T/T_*)$ that occurs in the expression (10.84) for the shift of the phonon frequency, and show that the shift becomes proportional to $(T/T_*)^{3/2}$ for $T \gg T_*$.

References

[1] R. Onofrio, C. Raman, J. M. Vogels, J. R. Abo-Shaeer, A. P. Chikkatur, and W. Ketterle, *Phys. Rev. Lett.* **85**, 2228 (2000).

[2] L. D. Landau and E. M. Lifshitz, *Fluid Mechanics*, Second edition, (Pergamon, New York, 1987), §139.

[3] A. Griffin and E. Zaremba, *Phys. Rev. A* **56**, 4839 (1997).

[4] S. T. Beliaev, *Zh. Eksp. Teor. Fiz.* **34**, 433 (1958) [*Sov. Phys.-JETP* **34 (7)**, 299 (1958)].

[5] W. V. Liu and W. C. Schieve, cond-mat/9702122.

[6] L. P. Pitaevskii and S. Stringari, *Phys. Lett. A* **235**, 398 (1997).

[7] G. Baym and C. J. Pethick, *Landau Fermi-liquid Theory*, (Wiley, New York, 1991), p. 149.

[8] P. O. Fedichev, G. V. Shlyapnikov, and J. T. M. Walraven, *Phys. Rev. Lett.* **80**, 2269 (1998).

11
Trapped clouds at non-zero temperature

In this chapter we consider selected topics in the theory of trapped gases at non-zero temperature when the effects of interactions are taken into account. The task is to extend the considerations of Chapters 8 and 10 to allow for the trapping potential. In Sec. 11.1 we begin by discussing energy scales, and then calculate the transition temperature and thermodynamic properties. We show that at temperatures of the order of T_c the effect of interactions on thermodynamic properties of clouds in a harmonic trap is determined by the dimensionless parameter $N^{1/6}a/\bar{a}$. Here $\bar{a}$, which is defined in Eq. (6.24), is the geometric mean of the oscillator lengths for the three principal axes of the trap. Generally this quantity is small, and therefore under many circumstances the effects of interactions are small. At low temperatures, thermodynamic properties may be evaluated in terms of the spectrum of elementary excitations of the cloud in its ground state, which we considered in Secs. 7.2, 7.3, and 8.2. At higher temperatures it is necessary to take into account thermal depletion of the condensate, and useful approximations for thermodynamic functions may be obtained using the Hartree–Fock theory as a starting point.

The remainder of the chapter is devoted to non-equilibrium phenomena. As we have seen in Secs. 10.3–10.5, two ingredients in the description of collective modes and other non-equilibrium properties of uniform gases are the two-component nature of condensed Bose systems, and collisions between excitations. For atoms in traps a crucial new feature is the inhomogeneity of the gas. This in itself would not create difficulties if collisions between excitations were sufficiently frequent that matter remained in thermodynamic equilibrium locally. However, this condition is rarely satisfied in experiments on dilute gases. In Sec. 11.2 we shall first give a qualitative discussion of collective modes in Bose–Einstein-condensed gases. To illustrate the effects of collisions we then consider the normal modes of a Bose gas above T_c in the hydrodynamic regime. This calculation is valuable for bringing out the differences between modes of a condensate and those of a normal gas. Section 11.3 contains a calculation of the damping of modes of a trapped Bose gas above T_c in the collisionless regime, a subject which is both theoretically tractable and experimentally relevant.

11.1 Equilibrium properties

To implement the finite-temperature theories of equilibrium properties of trapped gases described in Chapter 8 is generally a complicated task, since the number of particles in the condensate and the distribution function for the excitations must be determined self-consistently. We begin by estimating characteristic energy scales, and find that under many conditions the effects of interactions between non-condensed particles are small. We then investigate how the transition temperature is changed by interactions. Finally, we discuss a simple approximation that gives a good description of thermodynamic properties of trapped clouds under a wide range of conditions.

11.1.1 Energy scales

According to the Thomas–Fermi theory described in Sec. 6.2, the interaction energy per particle in a pure condensate E_{int}/N equals $2\mu/7$ where, according to (6.35), the chemical potential is given by

$$\mu = \frac{15^{2/5}}{2} \left(\frac{Na}{\bar{a}}\right)^{2/5} \hbar\bar{\omega}. \tag{11.1}$$

Using the expression (11.1) for the zero-temperature chemical potential we conclude that the interaction energy per particle corresponds to a temperature T_0 given by

$$T_0 = \frac{E_{\text{int}}}{Nk} = \frac{15^{2/5}}{7} \left(\frac{Na}{\bar{a}}\right)^{2/5} \frac{\hbar\bar{\omega}}{k}. \tag{11.2}$$

This quantity also gives a measure of the effective potential acting on a thermal excitation, and it is the same to within a numerical factor as the temperature $T_* = nU_0/k$ evaluated at the centre of the trap. When the temperature is large compared with T_0, interactions of excitations with the condensate have little effect on the properties of the thermal excitations, which consequently behave as non-interacting particles to a first approximation. The result (11.2) expressed in terms of the transition temperature

$T_{\rm c}$ for the non-interacting system, Eq. (2.20), is

$$T_0 = 0.45 \left(\frac{N^{1/6}a}{\bar{a}}\right)^{2/5} T_{\rm c}. \qquad (11.3)$$

The quantity $N^{1/6}a/\bar{a}$ is a dimensionless measure of the influence of interactions on the properties of thermal excitations. Later we shall find that it also arises in other contexts. The corresponding quantity governing the effect of interactions on the properties of the condensate itself at $T = 0$ is $Na/\bar{a}$, as we saw in Sec. 6.2. At non-zero temperatures, the corresponding parameter is $N_0 a/\bar{a}$ where N_0 is the number of atoms in the condensate, not the total number of atoms. While $Na/\bar{a}$ is large compared with unity in typical experiments, $N^{1/6}a/\bar{a}$ is less than unity, and it depends weakly on the particle number N. As a typical trap we consider the one used in the measurements of the temperature dependence of collective mode frequencies in ^{87}Rb clouds [1]. With $N = 6000$ and $\bar{\omega}/2\pi = 182$ Hz we obtain for the characteristic temperature (11.3) $T_0 = 0.11 T_{\rm c}$. For $N = 10^6$ the corresponding result is $T_0 = 0.15\, T_{\rm c}$.

At temperatures below T_0, thermodynamic quantities must be calculated with allowance for interactions between the excitations and the condensate, using, for example, the semi-classical Bogoliubov excitation spectrum (8.102). However, since T_0 is so low, it is difficult experimentally to explore this region. At temperatures above T_0, one may use the Hartree–Fock description, and interactions between excited particles may be neglected to a first approximation. This will form the basis for our discussion of thermodynamic properties in Sec. 11.1.3.

As another example, we estimate the effects of interaction at temperatures close to $T_{\rm c}$ or above, $T \gtrsim T_{\rm c}$. To a first approximation the gas may be treated as classical, and therefore the mean kinetic energy of a particle is roughly $3kT/2$. The maximum shift of the single-particle energies in Hartree–Fock theory (Sec. 8.3) is $2n(0)U_0$. The density is given approximately by the classical expression, Eq. (2.39), and at the trap centre the density is $n(0) \sim N/R_1 R_2 R_3$, where $R_i = (2kT/m\omega_i^2)^{1/2}$, Eq. (2.40), is much greater than $a_i = (\hbar/m\omega_i)^{1/2}$, when $kT \gg \hbar\omega_i$. The effects of interaction are thus small provided

$$U_0 N \ll R_1 R_2 R_3 kT. \qquad (11.4)$$

The inequality (11.4) may be written in terms of the transition temperature

T_c for the non-interacting system as

$$\left(\frac{T}{T_c}\right)^{5/2} \gg \frac{N^{1/6}a}{\bar{a}}, \tag{11.5}$$

or

$$\frac{T}{T_c} \gg \left(\frac{N^{1/6}a}{\bar{a}}\right)^{2/5}. \tag{11.6}$$

Again the factor $N^{1/6}a/\bar{a}$ appears, and because it is small, the inequalities are always satisfied at temperatures of order T_c and above. By similar arguments one can show that interactions between non-condensed particles are small if $T \gg T_0$.

11.1.2 Transition temperature

The estimates given above lead one to expect that interactions will change the transition temperature only slightly. We shall now confirm this by calculating the shift in the transition temperature to first order in the scattering length [2]. The calculation follows closely that of the shift of transition temperature due to zero-point motion (Sec. 2.5), and to leading order the effects of zero-point motion and interactions are additive.

At temperatures at and above T_c, the single-particle energy levels are given within Hartree–Fock theory by Eq. (8.100)

$$\epsilon_\mathbf{p}(\mathbf{r}) = \frac{p^2}{2m} + V(\mathbf{r}) + 2n(\mathbf{r})U_0 \tag{11.7}$$

in the semi-classical approximation. The trapping potential $V(\mathbf{r})$ is assumed to be the anisotropic harmonic-oscillator one (2.7).

When the semi-classical approximation is valid, the thermal energy kT is large compared with $\hbar\omega_i$, and the cloud of thermal particles has a spatial extent R_i in the i direction much larger than the oscillator length a_i, implying that the size of the thermal cloud greatly exceeds that of the ground-state oscillator wave function. In determining the lowest single-particle energy ϵ_0 in the presence of interactions we can therefore approximate the density $n(\mathbf{r})$ in (11.7) by its central value $n(0)$ and thus obtain

$$\epsilon_0 = \frac{3}{2}\hbar\omega_m + 2n(0)U_0, \tag{11.8}$$

where $\omega_m = (\omega_1 + \omega_2 + \omega_3)/3$ is the algebraic mean of the trap frequencies. Bose–Einstein condensation sets in when the chemical potential μ becomes equal to the lowest single-particle energy, as in the case of non-interacting particles.

The Bose–Einstein condensation temperature T_c is determined by the condition that the number of particles in excited states be equal to the total number of particles. By inserting (11.8) in the expression for the number of particles in excited states, which is given by

$$N_\text{ex} = \int d\mathbf{r} \int \frac{d\mathbf{p}}{(2\pi\hbar)^3} \frac{1}{e^{(\epsilon-\epsilon_0)/kT}-1}, \quad (11.9)$$

we obtain an equation determining the critical temperature,

$$N = \int d\mathbf{r} \int \frac{d\mathbf{p}}{(2\pi\hbar)^3} \frac{1}{e^{(\epsilon-\epsilon_0)/kT_c}-1}, \quad (11.10)$$

analogous to that for the non-interacting case discussed in Sec. 2.2. When interactions are absent and the zero-point motion is neglected, this yields the non-interacting particle result (2.20) for the transition temperature, which we denote here by T_{c0}. By expanding the right hand side of (11.10) to first order in $\Delta T_c = T_c - T_{c0}$, ϵ_0, and $n(\mathbf{r})U_0$, one finds

$$0 = \frac{\partial N}{\partial T}\Delta T_c + \frac{\partial N}{\partial \mu}\left[\frac{3}{2}\hbar\omega_\text{m} + 2n(0)U_0\right] - 2U_0\int d\mathbf{r}\, n(\mathbf{r})\frac{\partial n(\mathbf{r})}{\partial \mu}, \quad (11.11)$$

where the partial derivatives are to be evaluated for $\mu = 0$, $T = T_{c0}$, and $U_0 = 0$. The last term in Eq. (11.11) represents the change in particle number due to the interaction when the chemical potential is held fixed, and it may be written as $-(\partial E_\text{int}/\partial \mu)_T$.[1]

The partial derivatives in (11.11) were calculated in Sec. 2.4, Eqs. (2.74) and (2.75). For $\alpha = 3$ they are $\partial N/\partial T = 3N/T_c$ and $\partial N/\partial \mu = [\zeta(2)/\zeta(3)]N/kT_c$. The last term in (11.11) may be evaluated using Eq. (2.48) and it is

$$-\frac{\partial E_\text{int}}{\partial \mu} = -\frac{2\mathcal{S}}{\zeta(3)}\frac{NU_0}{\lambda_{T_c}^3 kT_c}, \quad (11.12)$$

where

$$\mathcal{S} = \sum_{n,n'=1}^{\infty} \frac{1}{n^{1/2}}\frac{1}{n'^{3/2}}\frac{1}{(n+n')^{3/2}} \approx 1.206, \quad (11.13)$$

---

[1] That the result must have this form also follows from the theorem of small increments [3]. Small changes in an external parameter change a thermodynamic potential by an amount which is independent of the particular thermodynamic potential under consideration, provided the natural variables of the thermodynamic potential are held fixed. For the potential $\Omega = E - TS - \mu N$ associated with the grand canonical ensemble, the natural variables are T and μ. It therefore follows that $(\delta\Omega)_{T,\mu} = (\delta E)_{S,N}$. The change in the energy when the interaction is turned on is the expectation value E_int of the interaction energy in the state with no interaction. Since the particle number is given by $N = -\partial\Omega/\partial\mu$, the change in the number of particles due to the interaction when μ and T are held fixed is given by $\Delta N = -(\partial E_\text{int}/\partial\mu)_T$.

and the thermal de Broglie wavelength at T_c is given by $\lambda_{T_c} = (2\pi\hbar^2/mkT_c)^{1/2}$, Eq. (1.2). After collecting the numerical factors we obtain

$$\frac{\Delta T_c}{T_c} \approx -0.68\frac{\hbar\bar{\omega}}{kT_c} - 3.43\frac{a}{\lambda_{T_c}} \approx -0.73\frac{\omega_m}{\bar{\omega}}N^{-1/3} - 1.33\frac{a}{\bar{a}}N^{1/6}. \quad (11.14)$$

That repulsive interactions reduce the transition temperature is a consequence of the fact that they lower the central density of the cloud. Such an effect does not occur for the homogeneous Bose gas, since the density in that case is uniform and independent of temperature. The influence of interactions on the transition temperature of the uniform Bose gas has been discussed recently in Ref. [4], which also contains references to earlier work. The relative change in T_c is of order $n^{1/3}a$ and positive, and it is due to critical fluctuations, not the mean-field effects considered here.

11.1.3 Thermodynamic properties

The thermodynamic properties of a non-interacting Bose gas were discussed in Chapter 2, and we now consider an interacting Bose gas in a trap. We shall assume that clouds are sufficiently large and temperatures sufficiently high that the semi-classical theory developed in Sec. 8.3.3 holds.

Low temperatures

At temperatures low enough that thermal depletion of the condensate is inappreciable, the elementary excitations in the bulk are those of the Bogoliubov theory, given in the Thomas–Fermi approximation by (8.103). For small momenta the dispersion relation is linear, $\epsilon = s(\mathbf{r})p$, with a sound velocity that depends on position through its dependence on the condensate density, $s(\mathbf{r}) = [n_0(\mathbf{r})U_0/m]^{1/2}$. Provided $kT \ll ms^2(\mathbf{r})$ the number density of phonon-like excitations is given to within a numerical constant by $\{kT/\hbar s(\mathbf{r})\}^3$, and therefore phonons contribute an amount $\sim (kT)^4/[\hbar s(\mathbf{r})]^3$ to the local energy density. Near the surface, however, the sound velocity vanishes, and there the majority of the thermal excitations are essentially free particles. For such excitations the number density varies as $(mkT/\hbar^2)^{3/2}$ (see Eq. (2.30)) and therefore, since the energy of an excitation is $\sim kT$, the energy density is proportional to $T^{5/2}$ and independent of the condensate density. The volume in which the free-particle term dominates the thermal energy density is a shell at the surface of the cloud extending to the depth at which thermal excitations become more like phonons than free particles. The density at the inner edge of this shell is therefore determined by the condition $kT \approx n_0(\mathbf{r})U_0$. Since in the Thomas–Fermi approximation the density varies linearly with distance from the surface, the thickness of the surface region where free-particle behaviour dominates the thermodynamic properties is proportional to T, and therefore the total thermal energy due to excitations in the surface varies as $T^{7/2}$. Free-particle states in the region outside the condensate cloud contribute a similar amount. The phonon-like excitations in the interior of the cloud contribute to the total energy an amount $\sim \int d\mathbf{r} T^4/n_0(\mathbf{r})^{3/2}$, where the integration is cut off at the upper limit given by $n_0(\mathbf{r})U_0 \approx kT$. This integral is also dominated by the upper limit, and again gives a contribution $\sim T^{7/2}$. Thus we conclude that the total contribution to the thermal energy at low temperatures varies as $T^{7/2}$ [5]. Evaluating the coefficient is the subject of Problem 11.1.

Let us now estimate the thermal depletion of the condensate at low temperatures. The number of particles associated with thermal excitations is given by an expression of the form (8.63), integrated over space,

$$N_{\text{ex}}(T) - N_{\text{ex}}(T=0) = \int d\mathbf{r} \int \frac{d\mathbf{p}}{(2\pi\hbar)^3} \frac{\xi_p}{\epsilon_p} f_p. \qquad (11.15)$$

The thermal phonons in the interior of the cloud each contribute an amount $ms(\mathbf{r})/p \sim ms(\mathbf{r})^2/kT$ to the depletion. The number density of excitations was estimated above, and therefore the total depletion of the condensate due to phonon-like excitations is $\sim \int d\mathbf{r} T^2/n_0(\mathbf{r})^{1/2}$. This integral converges as the surface is approached, and therefore we conclude that excitations in the interior dominate the thermal depletion. Consequently, the total thermal depletion of the condensate varies as T^2. Arguments similar to those for the energy density in the surface region show that the number of excitations in the surface region, where $kT \gtrsim n_0 U_0$, scales as $T^{5/2}$. Since the effective particle number associated with a free-particle-like excitation is essentially unity, the thermal depletion due to excitations in the surface region varies as $T^{5/2}$, and it is therefore less important than the interior contribution in the low-temperature limit.

Higher temperatures

The low-temperature expansions described above are limited to temperatures below T_0. At higher temperatures one can exploit the fact that excitations are free particles to a good approximation. Within Hartree–Fock theory excitations are particles moving in an effective potential given by (see Eq. (8.94))

$$V_{\text{eff}}(\mathbf{r}) = V(\mathbf{r}) + 2n(\mathbf{r})U_0, \qquad (11.16)$$

where $n(\mathbf{r})$ is the total density, which is the sum of the condensate density $n_0(\mathbf{r})$ and the density of excited particles, $n_{\text{ex}}(\mathbf{r})$,

$$n(\mathbf{r}) = n_0(\mathbf{r}) + n_{\text{ex}}(\mathbf{r}). \tag{11.17}$$

For $T \gg T_0$ the thermal cloud is more extended than the condensate and has a lower density than the central region of the condensate, except very near T_c. To a first approximation we may neglect the effect of interactions on the energies of excited particles and approximate $V_{\text{eff}}(\mathbf{r})$ by $V(\mathbf{r})$, since interactions have little effect on particles over most of the region in which they move.

Interactions are, however, important for the condensate. In the condensate cloud the density of thermal excitations is low, so the density profile of the condensate cloud is to a good approximation the same as for a cloud of pure condensate, as calculated in Sec. 6.2, except that the number N_0 of particles in the condensate enters, rather than the total number of particles, N. If N_0 is large enough that the Thomas–Fermi theory is valid, the chemical potential is given by Eq. (6.35),

$$\mu(T) \approx \frac{15^{2/5}}{2} \left(\frac{N_0(T) a}{\bar{a}} \right)^{2/5} \hbar \bar{\omega}. \tag{11.18}$$

This equation provides a useful starting point for estimating the influence of interactions on thermodynamic quantities such as the condensate fraction and the total energy [5].

The value of the chemical potential is crucial for determining the distribution function for excitations, which therefore depends on particle interactions even though the excitation spectrum does not. The excitations behave as if they were non-interacting particles, but with a shifted chemical potential given by (11.18).

Since the interaction enters only through the temperature-dependent chemical potential, $\mu(T)$, we can use the free-particle description of Chapter 2 to calculate the number of excited particles, $N_{\text{ex}} = N - N_0$. For particles in a three-dimensional trap, the parameter α in the density of states (2.12) equals 3. Consequently N_{ex} is given by

$$N_{\text{ex}} = C_3 \int_0^\infty d\epsilon \, \epsilon^2 \frac{1}{e^{(\epsilon - \mu)/kT} - 1}. \tag{11.19}$$

Typical particle energies are large compared with the chemical potential (11.18), and we may therefore expand this expression about its value for

$\mu = 0$, and include only the term linear in μ:

$$N_{\text{ex}}(T, \mu) \approx N_{\text{ex}}(T, \mu = 0) + \frac{\partial N_{\text{ex}}(T, \mu = 0)}{\partial \mu} \mu. \tag{11.20}$$

The first term on the right hand side is the expression for the number of excited particles for a non-interacting gas, and is given by $N_{\text{ex}}(T, \mu = 0) = N(T/T_c)^3$, Eq. (2.27). We have calculated $\partial N/\partial \mu$ for free particles in a trap at T_c before (see Eq. (2.74)), and more generally for any temperature less than T_c the result is

$$\frac{\partial N_{\text{ex}}}{\partial \mu} = \frac{\zeta(2)}{\zeta(3)} \frac{N_{\text{ex}}}{kT}. \tag{11.21}$$

We thus find

$$N_{\text{ex}} = N \left[t^3 + \frac{\zeta(2)}{\zeta(3)} t^2 \frac{\mu}{kT_c} \right], \tag{11.22}$$

where we have defined a reduced temperature $t = T/T_c$. Substituting Eq. (11.18) for $\mu(T)$ in this expression and using the fact that $N_{\text{ex}} + N_0(T) = N$ gives the result

$$N_{\text{ex}} = N \left\{ t^3 + \frac{15^{2/5} \zeta(2)}{2\zeta(3)} \left[\frac{(N - N_{\text{ex}})a}{\bar{a}} \right]^{2/5} \frac{\hbar \bar{\omega}}{kT_c} t^2 \right\}, \tag{11.23}$$

which, with the use of the result $kT_c = N^{1/3} \hbar \bar{\omega}/[\zeta(3)]^{1/3}$, Eq. (2.20), gives

$$\frac{N_{\text{ex}}}{N} \approx t^3 + 2.15 \left(\frac{N^{1/6} a}{\bar{a}} \right)^{2/5} \left(1 - \frac{N_{\text{ex}}}{N} \right)^{2/5} t^2. \tag{11.24}$$

This exhibits explicitly the interaction parameter $N^{1/6} a/\bar{a}$, and it shows that the effects of interactions on N_{ex} are small as long as $N^{1/6} a/\bar{a}$ is small, as one would expect from the qualitative arguments made in Sec. 11.1.1. It is consequently a good approximation to replace N_{ex} on the right hand side of Eq. (11.24) by its value for the non-interacting gas, and one finds

$$\frac{N_{\text{ex}}}{N} \approx t^3 + 2.15 \left(\frac{N^{1/6} a}{\bar{a}} \right)^{2/5} t^2 (1 - t^3)^{2/5}. \tag{11.25}$$

By similar methods one may derive an approximate expression for the energy. This consists of two terms. One is the contribution of the condensed cloud, which is $5N_0(T)\mu(T)/7$ by analogy with the zero-temperature case discussed in Chapter 6. The other is due to the thermal excitations and is obtained by expanding E to first order in μ/kT, as we did earlier for N_{ex} in deriving Eq. (11.22). The result is (Problem 11.2)

$$\frac{E}{NkT_c} = 3\frac{\zeta(4)}{\zeta(3)}t^4 + \frac{5+16t^3}{7}\frac{\mu(T)}{kT_c}. \tag{11.26}$$

To obtain a result analogous to Eq. (11.25), we replace $\mu(T)$ on the right hand side by (11.18), using the value $N_0 = N(1-t^3)$ appropriate to the non-interacting gas, and find

$$\frac{E}{NkT_c} = 2.70t^4 + 1.12\left(\frac{N^{1/6}a}{\bar{a}}\right)^{2/5}(1+3.20t^3)(1-t^3)^{2/5}. \tag{11.27}$$

The results (11.23) and (11.27) for the condensate depletion and the energy are in good agreement with more elaborate calculations based on the Popov approximation over most of the temperature range of interest [5].

Measurements of the ground-state occupation have been made for condensed clouds of ^{87}Rb atoms [6], and the results agree well with the predictions for the non-interacting Bose gas. Since the number of particles used in the experiments was relatively small, this is consistent with the results of the present section.

11.2 Collective modes

In this section and the following one we take up a number of topics in the theory of collective modes in traps at non-zero temperature. Our approach will be to extend the results of Chapter 7, where we discussed collective modes of a pure condensate in a trap at zero temperature, and of Chapter 10, where we considered examples of modes in homogeneous systems when both condensate and thermal excitations are present.

An important conclusion to be drawn from the calculations in Secs. 10.4 and 10.5 for uniform systems is that, under many conditions, the motion of the condensate is only weakly coupled to that of the excitations. As we saw in the calculations of first and second sound, this is untrue only if the modes of the condensate and those of the thermal excitations have velocities that are close to each other. We would expect this conclusion to hold also for traps, and therefore to a first approximation the motions of the condensate and the excitations are independent. At low temperatures there are few thermal excitations, and consequently the modes of the condensate are those described in Sec. 7.3, but with the number of particles N_0 in the condensate replacing the total number of particles N. With increasing temperature, the condensate becomes immersed in a cloud of thermal excitations, and to the extent that the thermal excitations do not participate in the motion, their only effect is to provide an extra external potential in which the condensate oscillates. However, the potential produced by the thermal cloud is of order $n_{\rm ex}U_0$. This is small compared with the potential due to the condensate $\sim n_0 U_0$, since $n_{\rm ex} \ll n_0$ except very close to $T_{\rm c}$. Consequently, we expect the modes of the condensate to have frequencies given to a first approximation by the results in Sec. 7.3. When the number of particles in the condensate is sufficiently large that the Thomas–Fermi approximation is valid, the mode frequencies depend only on the trap frequencies and are therefore independent of temperature. This result should hold irrespective of how frequently excitations collide, since there is little coupling between condensate and excitations. It is confirmed theoretically by calculations for the two-fluid model, in which the excitations are assumed to be in local thermodynamic equilibrium [7]. Experimentally, the mode frequencies exhibit some temperature dependence even under conditions when one would expect the Thomas–Fermi approximation to be valid [1], and this is a clear indication of coupling between the condensate and the thermal cloud. Sufficiently close to $T_{\rm c}$ the number of particles in the condensate will become so small that the Thomas–Fermi approximation is no longer valid, and the frequencies of the modes of the condensate will then approach the result (11.28) for free particles given below.

Now, let us consider thermal excitations. In Sec. 10.4 we studied modes under the assumption that the excitations are in local thermodynamic equilibrium, and one can extend such calculations to traps, as was done in Ref. [7]. Under most conditions realized in experiment, thermal excitations collide so infrequently that their mean free paths are long compared with the size of the cloud. In a uniform system the modes associated with excitations are not collective because interactions between excitations are weak. In traps, however, the motion of excitations can resemble a collective mode. Consider a gas in a harmonic trap at temperatures large compared with T_0. The excitations are to a first approximation free particles oscillating in the trap, and therefore mode frequencies are sums of integer multiples of the frequencies ω_i for single-particle motion in the trap,

$$\omega = \sum_{i=1}^{3} \nu_i \omega_i, \qquad (11.28)$$

where the ν_i are integers. Classically, the period for motion parallel to one of the principal axes of the trap is independent of the amplitude. Consequently, the motion of many particles may appear to be collective because, for example, after a time $\mathcal{T} = 2\pi/\omega_1$, particles have the same x coordinates as they did originally. For initial configurations with symmetry the particle distribution can return to its original form after an integral fraction of $\mathcal{T}$. When the effects of the condensate are taken into account, the motion of the excitations will be less coherent because the potential in which the excitations move is no longer harmonic. Consequently, the periods of the single-particle motion will depend on amplitude, and motions of a large number of excitations will not have a well-defined frequency.

We turn now to the damping of modes, beginning with those associated with the condensate. In Chapter 10 we calculated the rate of Landau damping of collective modes in a uniform Bose gas. Strictly speaking, the result (10.80) does not apply to collective modes in a trap, but it is interesting to compare its magnitude with the measured damping [1]. In doing this we identify the condensate density and the sound velocity with their values at the centre of the cloud, as calculated within the Thomas–Fermi approximation at $T = 0$. With $kT_* = ms^2 = n_0(0)U_0$ and the zero-temperature Thomas–Fermi result $n_0(0) = \mu/U_0$, the damping rate (10.80) may be written in the form

$$\frac{1}{\tau_q^{\text{amp}}} = 0.91\omega_q \left(\frac{N^{1/6}a}{\bar{a}}\right)^{4/5} \frac{T}{T_c}, \qquad (11.29)$$

where the transition temperature T_c is given by Eq. (2.20). When the experimental parameters $N = 6000$ and $a/\bar{a} = 0.007$ are inserted into (11.29), the theoretical values of $1/\tau^{\text{amp}}$ are in fair agreement with the measured magnitude and temperature dependence of the damping rate [1]. The linear temperature dependence of the damping rate (11.29) is a consequence of our replacing the condensate density by its zero-temperature value. At higher temperatures other processes, such as the 1–3, 3–1, and 2–2 processes described in Sec. 10.5 become important.

In the following subsection we consider modes associated with the excitations. These are damped by collisions, which also couple the excitations and the condensate. The general theory is complicated, so we illustrate the effects of collisions by considering the example of a gas above T_c, when there is no condensate. The study of modes under these conditions provides physical insight into the effects of collisions, and is also relevant experimentally, since measurements of the decay of modes above T_c are used to deduce properties of interatomic interactions. The two problems we consider are the nature of modes in the hydrodynamic regime, and the damping of modes when collisions are infrequent.

11.2.1 Hydrodynamic modes above T_c

As we have seen, for a gas in a harmonic trap in the absence of collisions, motions parallel to the axes of the trap are independent of each other, and the frequencies of normal modes are given by Eq. (11.28). Collisions couple the motions, thereby changing the character of the normal modes, damping them and changing their frequencies. One measure of the effect of collisions is the mean free path l, which is given by

$$l = \frac{1}{n\sigma}, \qquad (11.30)$$

where n is the particle density and $\sigma = 8\pi a^2$ is the total scattering cross section. The typical time τ between collisions is thus given by

$$\frac{1}{\tau} \approx n\sigma\bar{v}, \qquad (11.31)$$

where $\bar{v}$ is the average particle velocity. If the mean free path is small compared with the typical length scale of the mode, and if the collision time is small compared with the period of the mode, particles will remain in local thermodynamic equilibrium, and the properties of the mode will be governed by classical hydrodynamics. Since the wavelength of a mode is less than or of order the linear size of the system, a necessary condition for hydrodynamic behaviour is that the mean free path be small compared with the size of the cloud,

$$l \ll R. \qquad (11.32)$$

In an isotropic harmonic-oscillator trap the characteristic size of the cloud is of order $R \approx (kT/m\omega_0^2)^{1/2}$. Thus a characteristic density is of order N/R^3, and, with the use of Eq. (11.30), one finds

$$\frac{l}{R} = \frac{1}{n\sigma R} \approx \frac{R^2}{N\sigma}. \qquad (11.33)$$

To be in the hydrodynamic regime, the condition $R/l \gg 1$ must apply, which amounts to $N\sigma/R^2 = 8\pi N(a/R)^2 \gg 1$. Using the fact that $N \approx (kT_c/\hbar\omega_0)^3$, we may rewrite this condition as

$$8\pi \left(\frac{N^{1/3}a}{a_{\text{osc}}}\right)^2 \frac{T_c}{T} \gg 1. \qquad (11.34)$$

At temperatures of order T_c the importance of collisions is determined by the dimensionless quantity $N^{1/3}a/a_{\text{osc}}$, which is intermediate between $N^{1/6}a/a_{\text{osc}}$ which is a measure of the importance of interactions on the energy of the thermal cloud at T_c and Na/a_{osc} which is the corresponding quantity at $T = 0$. This result shows that more than 10^6 particles are required for local thermodynamic equilibrium to be established for typical traps. Most experiments have been carried out in the collisionless regime, but conditions approaching hydrodynamic ones have been achieved in some experiments. We note that hydrodynamics never applies in the outermost parts of trapped clouds. The density becomes lower as the distance from the centre of the cloud increases and, eventually, collisions become so rare that thermodynamic equilibrium cannot be established locally.

General formalism

We now calculate frequencies of low-lying modes in a trapped gas above T_c in the hydrodynamic limit, when collisions are so frequent that departures from local thermodynamic equilibrium may be neglected, and there is no dissipation. The calculation brings out the differences between the collective modes of ordinary gases, for which the pressure comes from the thermal motion of particles, and those of a Bose–Einstein condensate, where interactions between particles provide the pressure. The equations of (single-fluid) hydrodynamics are the continuity equation and the Euler equation. The continuity equation has its usual form,

$$\frac{\partial \rho}{\partial t} + \boldsymbol{\nabla} \cdot (\rho \mathbf{v}) = 0. \tag{11.35}$$

For simplicity, we denote the velocity of the fluid by $\mathbf{v}$, even though in a two-fluid description it corresponds to $\mathbf{v}_n$, the velocity of the normal component. The continuity equation is Eq. (10.27), with the mass current density given by the result for a single fluid, $\mathbf{j} = \rho \mathbf{v}$. The Euler equation was given in Eq. (7.24), and it amounts to the condition for conservation of momentum. It is equivalent to Eq. (10.25) if the superfluid is absent, and the momentum flux density tensor is replaced by its equilibrium value $\Pi_{ik} = p\delta_{ik} + \rho v_i v_k$ where p is the pressure. To determine the frequencies of normal modes we linearize the equations about equilibrium, treating the velocity and the deviation of the mass density from its equilibrium value $\rho_{\rm eq}$ as small. The linearized Euler equation is

$$\rho_{\rm eq} \frac{\partial \mathbf{v}}{\partial t} = -\boldsymbol{\nabla} p + \rho \mathbf{f}, \tag{11.36}$$

where $\mathbf{f}$ is the force per unit mass, given by

$$\mathbf{f} = -\frac{1}{m} \boldsymbol{\nabla} V. \tag{11.37}$$

11.2 Collective modes

The linearized continuity equation is

$$\frac{\partial \rho}{\partial t} + \boldsymbol{\nabla} \cdot (\rho_{\text{eq}} \mathbf{v}) = 0. \tag{11.38}$$

According to the Euler equation (11.36), the pressure p_{eq} and density ρ_{eq} in equilibrium must satisfy the relation

$$\boldsymbol{\nabla} p_{\text{eq}} = \rho_{\text{eq}} \mathbf{f}. \tag{11.39}$$

We now take the time derivative of (11.36), which yields

$$\rho_{\text{eq}} \frac{\partial^2 \mathbf{v}}{\partial t^2} = -\boldsymbol{\nabla} \left(\frac{\partial p}{\partial t} \right) + \frac{\partial \rho}{\partial t} \mathbf{f}. \tag{11.40}$$

As demonstrated in Sec. 11.1, the effects of particle interactions on equilibrium properties are negligible above T_c. Therefore we may treat the bosons as a non-interacting gas. Since collisions are assumed to be so frequent that matter is always in local thermodynamic equilibrium, there is no dissipation, and the entropy per unit mass is conserved as a parcel of gas moves. The equation of state of a perfect, monatomic, non-relativistic gas under adiabatic conditions is $p/\rho^{5/3} = \text{constant}$. In using this result it is important to remember that it applies to a given element of the fluid, which changes its position in time, not to a point fixed in space. If we denote the displacement of a fluid element from its equilibrium position by $\boldsymbol{\xi}$, the condition is thus

$$\frac{p(\mathbf{r} + \boldsymbol{\xi})}{\rho(\mathbf{r} + \boldsymbol{\xi})^{5/3}} = \frac{p_{\text{eq}}(\mathbf{r})}{\rho_{\text{eq}}(\mathbf{r})^{5/3}}, \tag{11.41}$$

or

$$p(\mathbf{r}) = p_{\text{eq}}(\mathbf{r} - \boldsymbol{\xi}) \left[\frac{\rho(\mathbf{r})}{\rho_{\text{eq}}(\mathbf{r} - \boldsymbol{\xi})} \right]^{5/3}. \tag{11.42}$$

Small changes δp in the pressure are therefore related to $\boldsymbol{\xi}$ and small changes $\delta \rho$ in the density by the expression

$$\delta p = \frac{5}{3} \frac{p_{\text{eq}}}{\rho_{\text{eq}}} (\delta \rho + \boldsymbol{\xi} \cdot \boldsymbol{\nabla} \rho_{\text{eq}}) - \boldsymbol{\xi} \cdot \boldsymbol{\nabla} p_{\text{eq}}. \tag{11.43}$$

By taking the time derivative of (11.43) and using (11.38) together with (11.39) and $\mathbf{v} = \dot{\boldsymbol{\xi}}$ we obtain

$$\frac{\partial p}{\partial t} = -\frac{5}{3} p_{\text{eq}} \boldsymbol{\nabla} \cdot \mathbf{v} - \rho_{\text{eq}} \mathbf{f} \cdot \mathbf{v}. \tag{11.44}$$

Inserting this equation into Eq. (11.40) we find

$$\rho_{\text{eq}} \frac{\partial^2 \mathbf{v}}{\partial t^2} = \frac{5}{3} \boldsymbol{\nabla}(p_{\text{eq}} \boldsymbol{\nabla} \cdot \mathbf{v}) + \boldsymbol{\nabla}(\rho_{\text{eq}} \mathbf{f} \cdot \mathbf{v}) - \mathbf{f} \boldsymbol{\nabla} \cdot (\rho_{\text{eq}} \mathbf{v}). \tag{11.45}$$

From Eq. (11.39) it follows that $\mathbf{f}$ and $\boldsymbol{\nabla}\rho_{\text{eq}}$ are parallel, and therefore $(\mathbf{f}\cdot\mathbf{v})\boldsymbol{\nabla}\rho_{\text{eq}} = \mathbf{f}(\mathbf{v}\cdot\boldsymbol{\nabla})\rho_{\text{eq}}$. With this result and Eq. (11.39) for the pressure gradient, we may rewrite Eq. (11.45) in the convenient form

$$\frac{\partial^2 \mathbf{v}}{\partial t^2} = \frac{5}{3}\frac{p_{\text{eq}}}{\rho_{\text{eq}}}\boldsymbol{\nabla}(\boldsymbol{\nabla}\cdot\mathbf{v}) + \boldsymbol{\nabla}(\mathbf{f}\cdot\mathbf{v}) + \frac{2}{3}\mathbf{f}(\boldsymbol{\nabla}\cdot\mathbf{v}). \tag{11.46}$$

This is the general equation of motion satisfied by the velocity field $\mathbf{v}$. In the context of Bose gases the result was first derived from kinetic theory [8].

In the absence of a confining potential ($\mathbf{f} = 0$), the equation has longitudinal waves as solutions. These are ordinary sound waves, and have a velocity $(5p_{\text{eq}}/3\rho_{\text{eq}})^{1/2}$. For temperatures high compared with T_{c}, the velocity becomes $(5kT/3m)^{1/2}$, the familiar result for the adiabatic sound velocity of a classical monatomic gas. For $T = T_{\text{c}}$, the velocity agrees with that of first sound, Eq. (10.50). For transverse disturbances ($\boldsymbol{\nabla}\cdot\mathbf{v} = 0$) there is no restoring force and, consequently, these modes have zero frequency. When dissipation is included they become purely decaying modes.

Low-frequency modes

Let us now consider an anisotropic harmonic trap with a potential given by Eq. (2.7). The force per unit mass, Eq. (11.37), is given by

$$\mathbf{f} = -(\omega_1^2 x, \omega_2^2 y, \omega_3^2 z). \tag{11.47}$$

We shall look for normal modes of the form

$$\mathbf{v} = (ax, by, cz), \tag{11.48}$$

where the coefficients a, b, and c depend on time as $e^{-i\omega t}$. The motion corresponds to homologous expansion and contraction of the cloud, with a scaling factor that may depend on the coordinate axis considered. We note that $\boldsymbol{\nabla}\times\mathbf{v} = 0$, and therefore the flow is irrotational. Since the divergence of the velocity field is constant in space ($\boldsymbol{\nabla}\cdot\mathbf{v} = a+b+c$), Eq. (11.46) becomes

$$-\omega^2 \mathbf{v} = \boldsymbol{\nabla}(\mathbf{f}\cdot\mathbf{v}) + \frac{2}{3}\mathbf{f}(\boldsymbol{\nabla}\cdot\mathbf{v}). \tag{11.49}$$

For a velocity field of the form (11.48), this equation contains only terms linear in x, y or z. Setting the coefficients of each of these equal to zero, we obtain the following three coupled homogeneous equations for a, b and c,

$$(-\omega^2 + \frac{8}{3}\omega_1^2)a + \frac{2}{3}\omega_2^2 b + \frac{2}{3}\omega_3^2 c = 0, \tag{11.50}$$

$$(-\omega^2 + \frac{8}{3}\omega_2^2)b + \frac{2}{3}\omega_3^2 c + \frac{2}{3}\omega_1^2 a = 0, \tag{11.51}$$

and

$$(-\omega^2 + \frac{8}{3}\omega_3^2)c + \frac{2}{3}\omega_1^2 a + \frac{2}{3}\omega_2^2 b = 0. \tag{11.52}$$

For an isotropic oscillator, $\omega_1 = \omega_2 = \omega_3 = \omega_0$, there are two eigenfrequencies, given by

$$\omega^2 = 4\omega_0^2 \text{ and } \omega^2 = 2\omega_0^2. \tag{11.53}$$

The $2\omega_0$ oscillation corresponds to $a = b = c$. The velocity is thus proportional to $\mathbf{r}$. Since the radial velocity has the same sign everywhere, the mode is a breathing mode analogous to that for a condensate discussed below Eq. (7.73) and in Sec. 7.3.3. The density oscillation in the breathing mode is independent of angle, corresponding to the spherical harmonic Y_{lm} with $l = m = 0$, and the mode frequency is the same as in the absence of collisions. There are two degenerate modes with frequency $\sqrt{2}\omega_0$, and a possible choice for two orthogonal mode functions is $a = b = -c/2$ and $a = -b, c = 0$. These modes have angular symmetry corresponding to $l = 2$. From general principles, one would expect there to be $2l + 1 = 5$ degenerate modes having $l = 2$. The other three have velocity fields proportional to $\nabla(xy) = (y, x, 0)$ and the two other expressions obtained from this by cyclic permutation, and they are identical with the scissors modes in the condensate considered in Sec. 7.3.2.

For anisotropic traps one obtains from Eqs. (11.50)–(11.52) a cubic equation for ω^2 with, in general, three different roots. There are also three transverse scissors modes of the type $\mathbf{v} \propto \nabla(xy)$ with different frequencies given by $\omega^2 = \omega_1^2 + \omega_2^2$ and the corresponding expressions obtained by cyclic permutation.

For a trap with axial symmetry, the force constants in the xy plane are equal and differ from that in the z direction, $\omega_1 = \omega_2 = \omega_0$, $\omega_3 = \lambda\omega_0$. Due to the axial symmetry, the mode with $a = -b, c = 0$ and frequency $\omega = \sqrt{2}\omega_0$ found for an isotropic trap is still present, since in it there is no motion in the z direction. For the two other frequencies one finds

$$\omega^2 = \omega_0^2 \left(\frac{5}{3} + \frac{4}{3}\lambda^2 \pm \frac{1}{3}\sqrt{25 - 32\lambda^2 + 16\lambda^4} \right), \tag{11.54}$$

which are plotted in Fig. 11.1.

The method we have used here can also be applied to calculate the frequencies of modes of a trapped Bose–Einstein condensate when the number of particles is sufficiently large that the Thomas–Fermi approximation may be used. The only difference is that the equation of state for a zero-temperature condensate must be used instead of that for a thermal gas. One may derive an equation for the velocity field analogous to Eq. (11.46), and calculate

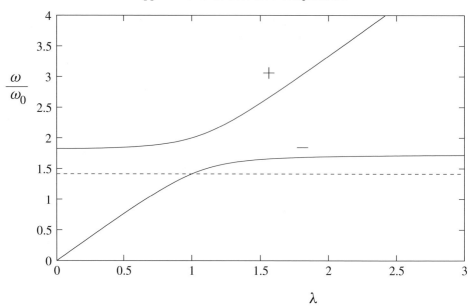

Fig. 11.1. Frequencies of low-lying hydrodynamic modes for an axially-symmetric harmonic trap, as a function of the anisotropy parameter λ. The full lines correspond to Eq. (11.54), while the dotted line ($\omega = \sqrt{2}\omega_0$) corresponds to the mode with $a = -b, c = 0$, which is degenerate with the xy scissors mode.

the frequencies of low-lying modes corresponding to a homologous scaling of the density distribution. The results agree with those derived in Sec. 7.3 by considering the density distribution (see Problem 11.3).

The results of this subsection show that the frequencies of modes for trapped gases depend on the equation of state, since the results obtained for a thermal gas differ from those for a pure condensate. For a thermal gas the mode frequencies also depend on collisions, since the modes in the hydrodynamic limit differ from those in the collisionless limit, which are given by Eq. (11.28). This demonstrates how properties of collective modes may be used as a diagnostic tool for probing the state of a gas.

11.3 Collisional relaxation above T_c

As we indicated above, in most experiments on oscillations in trapped alkali gases collisions are so infrequent that the gas is not in local thermodynamic equilibrium, so now we consider the opposite limit, when collisions are rare. The modes have frequencies given to a first approximation by Eq. (11.28), and collisions damp the modes. Such dissipation processes in trapped Bose gases have been investigated experimentally in a variety of ways. One is to measure the damping of collective oscillations. The atomic cloud is excited at the frequency of the normal mode of interest by means of a weak external perturbation, and the damping of the mode is then extracted from the measured time dependence of the oscillation amplitude in the absence of the perturbation [9]. Another way is to study the relaxation of temperature anisotropies in a gas in a harmonic trap [10, 11]. In a harmonic trap the motion of a free particle is separable since the Hamiltonian may be written as a sum of independent terms corresponding to the motions parallel to each of the axes of the trap. Thus if an atomic cloud is prepared in a state in which the average particle energy for motion parallel to an axis of the trap is not the same for all axes, these energies do not depend on time in the absence of collisions. Such a state corresponds to a particle distribution with different effective temperatures along the various axes of the trap. However, because of collisions, temperature anisotropies decay in time. The dimensions of the cloud along the principal axes of the trap depend on the corresponding temperatures, and therefore anisotropy of the temperature may be monitored by observing how the shape of the cloud depends on time.

Most experiments on collisional relaxation have been carried out in the collisionless regime, where the mean free path is large compared with the size of the cloud. In the following we shall therefore consider the collisionless regime in some detail, and comment only briefly on the hydrodynamic and intermediate regimes. For simplicity, we limit the discussion to temperatures above the transition temperature.

We have already estimated the mean free time τ for collisions, Eq. (11.31). This sets the characteristic timescale for the decay of modes, but the decay time generally differs from τ by a significant numerical factor which depends on the mode in question. The reason for this difference is that collisions conserve particle number, total momentum, and total energy. As a consequence, collisions have no effect on certain parts of the distribution function.

The Boltzmann equation

For temperatures T greater than T_c, the thermal energy kT is large compared with the separation between the energy eigenvalues in the harmonic-oscillator potential, and consequently the semi-classical Boltzmann equation provides an accurate starting point for the calculation of relaxation rates. In addition, we may neglect the mean-field interactions with other atoms, since even when T is as low as the transition temperature, the energy nU_0 is typically no more than a few per cent of kT. The resulting Boltzmann equation for the distribution function $f_\mathbf{p}(\mathbf{r}, t)$ is

$$\frac{\partial f}{\partial t} + \frac{\mathbf{p}}{m} \cdot \boldsymbol{\nabla} f - \boldsymbol{\nabla} V \cdot \boldsymbol{\nabla}_\mathbf{p} f = -I[f], \tag{11.55}$$

where V is the external potential and $I[f]$ the collision term. In the spirit of the semi-classical approach adopted here, we replace the collision term by the expression for a bulk gas,

$$\begin{aligned} I[f] &= \int \frac{d\mathbf{p}_1}{(2\pi\hbar)^3} \int d\Omega \, \frac{d\sigma}{d\Omega} |\mathbf{v} - \mathbf{v}_1| \\ &\quad \times [f f_1 (1 + f')(1 + f_1') - (1 + f)(1 + f_1) f' f_1']. \end{aligned} \tag{11.56}$$

In the above expression we have introduced the differential cross section $d\sigma/d\Omega$, where Ω is the solid angle for the direction of the relative momentum $\mathbf{p}' - \mathbf{p}_1'$. In general, the differential cross section depends on the relative velocity $|\mathbf{v} - \mathbf{v}_1|$ of the two incoming particles, as well as on the angle of the relative velocities of the colliding particles before and after the collision, but for low-energy particles it tends to a constant, as we saw in Sec. 5.2. Because collisions are essentially local, the spatial arguments of all the distribution functions in the collision integral are the same. The first term on the second line of Eq. (11.56) is the out-scattering term, in which particles with momenta $\mathbf{p}$ and $\mathbf{p}_1$ are scattered to states with momenta $\mathbf{p}'$ and $\mathbf{p}_1'$, while the second term is the in-scattering term, corresponding to the inverse process. The two sorts of processes, as well as the enhancement factors $1 + f$ due to the Bose statistics, are familiar from the calculation of Landau damping in Sec. 10.5.1.

Since the atoms are bosons, their distribution function f^0 in equilibrium is

$$f_\mathbf{p}^0(\mathbf{r}) = \frac{1}{e^{(p^2/2m + V - \mu)/kT} - 1}, \tag{11.57}$$

where μ is the chemical potential. To investigate small deviations from equilibrium, we write $f = f^0 + \delta f$, where f^0 is the equilibrium distribution function and δf is the deviation of f from equilibrium. In equilibrium the net collision rate vanishes, because $f^0 f_1^0 (1 + f^{0\prime})(1 + f_1^{0\prime}) = (1 + f^0)(1 + f_1^0) f^{0\prime} f_1^{0\prime}$ as a consequence of energy conservation. Linearizing Eq. (11.55) and introducing the definition

$$\delta f = f^0 (1 + f^0) \Phi, \tag{11.58}$$

we find

$$\frac{\partial \delta f}{\partial t} + \frac{\mathbf{p}}{m} \cdot \boldsymbol{\nabla} \delta f - \boldsymbol{\nabla} V \cdot \boldsymbol{\nabla}_\mathbf{p} \delta f = -I[\delta f], \tag{11.59}$$

where the linearized collision integral is

$$I[\delta f] = \int \frac{d\mathbf{p}_1}{(2\pi\hbar)^3} \int d\Omega \frac{d\sigma}{d\Omega} |\mathbf{v}-\mathbf{v}_1|(\Phi+\Phi_1-\Phi'-\Phi'_1) \\ \times f^0 f^0_1 (1+f^{0'})(1+f^{0'}_1). \quad (11.60)$$

For the calculations that follow it is convenient to separate out a factor $f^0(1+f^0)$ and define the operator

$$\hat{\Gamma}[\Phi] = \frac{I[\delta f]}{f^0(1+f^0)}. \quad (11.61)$$

The conservation laws imply that the collision integral (11.60) vanishes for certain forms of Φ. The simplest one, which reflects conservation of particle number, is $\Phi = a(\mathbf{r})$, where $a(\mathbf{r})$ is any function that does not depend on the momentum. Since momentum is conserved in collisions, the collision integral (11.60) also vanishes for $\Phi = \mathbf{b}(\mathbf{r}) \cdot \mathbf{p}$ where $\mathbf{b}(\mathbf{r})$ is a vector independent of $\mathbf{p}$. Finally, since collisions are local in space, they conserve the kinetic energy of particles. This implies the vanishing of the collision integral when $\Phi = c(\mathbf{r})p^2$, for any function $c(\mathbf{r})$ that is independent of momentum. These *collision invariants* will play an important role in the calculations of damping described below. To understand their significance we consider the distribution function for particles in equilibrium in a frame moving with a velocity $\mathbf{v}$. This is

$$f_\mathbf{p}(\mathbf{r}) = \frac{1}{e^{(p^2/2m+V-\mathbf{p}\cdot\mathbf{v}-\mu)/kT} - 1}. \quad (11.62)$$

Since the derivative of the equilibrium distribution function with respect to the particle energy $\epsilon = p^2/2m$ is given by

$$\frac{\partial f^0}{\partial \epsilon} = -\frac{f^0(1+f^0)}{kT}, \quad (11.63)$$

the change in f for small values of $\mathbf{v}$ and for small changes in the chemical potential and the temperature is

$$\delta f = \left[\delta\mu + \mathbf{p}\cdot\mathbf{v} + (\frac{p^2}{2m}+V-\mu)\frac{\delta T}{T}\right]\frac{f^0(1+f^0)}{kT}. \quad (11.64)$$

Thus the deviation function Φ, Eq. (11.58), for local thermodynamic equilibrium is a sum of collision invariants.

The general approach for finding the frequencies and damping rates of modes is to evaluate the eigenvalues of the linearized Boltzmann equation. When collisions are infrequent, the modes are very similar to those in the absence of collisions. One can use this idea to develop a systematic method for determining properties of modes. We illustrate the approach by finding approximate solutions to the problems of relaxation of temperature anisotropies and the damping of oscillations.

11.3.1 Relaxation of temperature anisotropies

Consider a cloud of atoms in an anisotropic harmonic trap and imagine that the distribution function is disturbed from its equilibrium form by making the temperature T_z associated with motion in the z direction different from that for motion in the x and y directions, $T_\perp$. The distribution function may thus be written

$$f_\mathbf{p}(\mathbf{r}) = \left[\exp\left(\frac{p_z^2/2m + V_z}{kT_z} + \frac{p_\perp^2/2m + V_\perp}{kT_\perp} - \frac{\mu}{kT}\right) - 1\right]^{-1}, \quad (11.65)$$

where $V_z = m\omega_3^2 z^2/2$, $V_\perp = m(\omega_1^2 x^2 + \omega_2^2 y^2)/2$, and $p_\perp^2 = p_x^2 + p_y^2$. As usual, the temperature T and the chemical potential are chosen to ensure that the total energy and the total number of particles have their actual values. For small temperature anisotropies, the deviation function has the form

$$\Phi = \frac{p_z^2/2m + V_z}{kT^2}\delta T_z + \frac{p_\perp^2/2m + V_\perp}{kT^2}\delta T_\perp, \quad (11.66)$$

where $\delta T_z = T_z - T$ and $\delta T_\perp = T_\perp - T$. From the condition that the total energy correspond to the temperature T it follows that $\delta T_z = -2\delta T_\perp$, the factor of two reflecting the two transverse degrees of freedom. We therefore obtain from Eq. (11.66) that

$$\Phi = \left(\frac{3p_z^2 - p^2}{4m} + V_z - \frac{1}{2}V_\perp\right)\frac{\delta T_z}{kT^2}. \quad (11.67)$$

The distribution function corresponding to this is a static solution of the collisionless Boltzmann equation. This result follows from the fact that Φ is a function of the energies associated with the z and transverse motions, which are separately conserved. It may be confirmed by explicit calculation.

When the collision time for a particle is long compared with the periods of all oscillations in the trap we expect that the solution of the Boltzmann equation will be similar to Eq. (11.67). We therefore adopt as an ansatz the form

$$\delta f = f^0(1 + f^0)\Phi_T g(t), \quad (11.68)$$

where

$$\Phi_T = p_z^2 - \frac{p^2}{3} + \frac{4m}{3}\left(V_z - \frac{1}{2}V_\perp\right), \quad (11.69)$$

11.3 Collisional relaxation above T_c

and $g(t)$ describes the relaxation of the distribution function towards its equilibrium value. We insert (11.68) in the Boltzmann equation (11.59), multiply by Φ_T, and integrate over coordinates and momenta. The result is

$$<\Phi_T^2>\frac{\partial g}{\partial t}=-<\Phi_T\hat{\Gamma}[\Phi_T]>g(t), \tag{11.70}$$

where $<\cdots>$ denotes multiplication by $f^0(1+f^0)$ and integration over both coordinate space and momentum space. The integral operator $\hat{\Gamma}$ is defined in Eq. (11.61).

The solution of (11.70) is thus

$$g(t)=g(0)e^{-\Gamma_T t}, \tag{11.71}$$

where the relaxation rate is

$$\Gamma_T=\frac{<\Phi_T\hat{\Gamma}[\Phi_T]>}{<\Phi_T^2>}. \tag{11.72}$$

The physical content of this equation is that the numerator is, apart from factors, the rate of entropy generation times T, and therefore gives the dissipation. The denominator is essentially the excess free energy associated with the temperature anisotropy. The rate of decay is therefore the ratio of these two quantities. The expression thus has a form similar to that for the decay rate of modes in the hydrodynamic regime [12]. The source of dissipation in both the hydrodynamic and collisionless regimes is collisions between atoms, but in the hydrodynamic regime the effect of collisions may be expressed in terms of the shear and bulk viscosities and the thermal conductivity.

If one replaces the collision integral by a naive approximation for it, $\hat{\Gamma}[\Phi]=-\Phi/\tau(\mathbf{r})$, the damping rate is $\Gamma=<\Phi_T^2/\tau(\mathbf{r})>/<\Phi_T^2>$. However, this is a poor approximation, since the potential energy terms in Φ_T are collision invariants, and therefore the collision integral gives zero when acting on them. Thus we may write

$$\hat{\Gamma}[\Phi_T]=\hat{\Gamma}[p_z^2-p^2/3]. \tag{11.73}$$

Since the collision integral is symmetric ($<A\hat{\Gamma}[B]>=<B\hat{\Gamma}[A]>$), the only term in the collision integral which survives is $<(p_z^2-p^2/3)\hat{\Gamma}[p_z^2-p^2/3]>$. After multiplication and division by $<(p_z^2-p^2/3)^2>$ the decay rate may be written as

$$\Gamma_T=\frac{<(p_z^2-p^2/3)\hat{\Gamma}[p_z^2-p^2/3]>}{<(p_z^2-p^2/3)^2>}\frac{<(p_z^2-p^2/3)^2>}{<\Phi_T^2>}. \tag{11.74}$$

In this equation, the first factor is an average collision rate for a distribution function with $\Phi \propto p_z^2 - p^2/3$, and the second factor is the ratio of the excess free energy in the part of Φ_T varying as $p_z^2 - p^2/3$, and the total excess free energy. An explicit evaluation shows that the second factor is 1/2. This reflects the fact that in a harmonic trap, the kinetic and potential energies are equal. However, collisions relax directly only the contributions to Φ_T that are anisotropic in momentum space, but not those that are anisotropic in coordinate space. The time to relax the excess free energy in the mode is therefore twice as long as it would have been if there were no potential energy contributions to Φ_T.

The first factor in Eq. (11.74) gives the decay rate for a deviation function $\Phi \propto p_z^2 - p^2/3$, which is spatially homogeneous and is proportional to the Legendre polynomial of degree $l = 2$ in momentum space. Since the collision integral is invariant under rotations in momentum space, the decay rate is the same for all disturbances corresponding to $l = 2$ in momentum space which have the same dependence on p. In particular, it is the same as for $\Phi \propto p_x v_y$. Disturbances of this form arise when calculating the shear viscosity η of a uniform gas in the hydrodynamic regime. To a very good approximation, the solution to the Boltzmann equation for a shear flow in which the fluid velocity is in the x direction and varies in the y direction is proportional to $p_x p_y$. One may define a viscous relaxation time τ_η by the equation

$$\eta = \tau_\eta \int \frac{d\mathbf{p}}{(2\pi\hbar)^3} \left(\frac{p_x p_y}{m}\right)^2 \left(-\frac{\partial f^0}{\partial \epsilon}\right). \tag{11.75}$$

This definition implies that for a classical gas $\eta = nkT\tau_\eta$. In the simplest variational approximation the viscous relaxation time is given by [13]

$$\frac{1}{\tau_\eta(\mathbf{r})} = \frac{<p_x p_y \hat{\Gamma}[p_x p_y]>_p}{<(p_x p_y)^2>_p} = \frac{<(p_z^2 - p^2/3)\hat{\Gamma}[p_z^2 - p^2/3]>_p}{<(p_z^2 - p^2/3)^2>_p}, \tag{11.76}$$

where the subscript p on $<\cdots>$ indicates that only the integral over momentum space is to be performed. The difference between the approximate result (11.76) and the result of more exact calculations is small, and we shall neglect it. For classical gases with energy-independent s-wave interactions it is less than two per cent.

The final result for the damping rate of temperature anisotropies, Eq. (11.74), may therefore be rewritten as

$$\Gamma_T = \frac{1}{2} \frac{<p^4 \tau_\eta^{-1}(\mathbf{r})>}{<p^4>}, \tag{11.77}$$

since the angular integrals in momentum space can be factored out.

The classical limit

Let us now consider the classical limit. In the averages, the quantity $f^0(1+f^0)$ reduces to the Boltzmann distribution. The momentum and space integrals then factorize, and one finds

$$\Gamma_T = \frac{1}{2\tau_{\text{av}}}, \tag{11.78}$$

where

$$\frac{1}{\tau_{\text{av}}} = \frac{\int d\mathbf{r}\, n(\mathbf{r}) \tau_\eta^{-1}(\mathbf{r})}{\int d\mathbf{r}\, n(\mathbf{r})}, \tag{11.79}$$

which is the density-weighted average of $1/\tau_\eta$ over the volume of the cloud.

We have previously estimated the relaxation rate, Eq. (11.31), and we shall now be more quantitative. In a uniform classical gas, the average collision rate $1/\tau$ is obtained by averaging the scattering rate for particles of a particular momentum over the distribution function,

$$\frac{1}{\tau} = \frac{\int d\mathbf{p}\, d\mathbf{p}_1 \sigma(|\mathbf{v}-\mathbf{v}_1|)|\mathbf{v}-\mathbf{v}_1| f_{\mathbf{p}}^0 f_{\mathbf{p}_1}^0}{(2\pi\hbar)^3 \int d\mathbf{p}\, f_{\mathbf{p}}^0}. \tag{11.80}$$

Here $\mathbf{v}-\mathbf{v}_1$ is the relative velocity of the two colliding particles, while $\sigma(|\mathbf{v}-\mathbf{v}_1|)$ denotes the total scattering cross section. For energy-independent s-wave scattering the total scattering cross section is a constant, equal to σ. The integrals over $\mathbf{p}$ and $\mathbf{p}_1$ in (11.80) are then carried out by introducing centre-of-mass and relative coordinates and using $f_{\mathbf{p}}^0 \propto \exp(-p^2/2mkT)$. The result is that the average collision rate in a homogeneous classical gas is given by

$$\frac{1}{\tau} = n\sigma\sqrt{2}\,\bar{v}, \tag{11.81}$$

where $\bar{v}$ is the mean thermal velocity of a particle (cf. (4.96)),

$$\bar{v} = \left(\frac{8kT}{\pi m}\right)^{1/2}. \tag{11.82}$$

Equation (11.81) differs from the earlier estimate (11.31) by a factor of $\sqrt{2}$ because it is the *relative* velocity that enters the scattering rate, not just the velocity of the particle.

Finally, we must relate the viscous relaxation rate to the average collision rate. These differ because the viscous relaxation time determines how rapidly an anisotropy in momentum space having $l=2$ symmetry relaxes. This depends not only on the total collision rate but also on how effectively collisions reduce momentum anisotropies. To take an extreme example, if the scattering cross section vanished except for a range of scattering angles close to zero, collisions would be ineffective in relaxing anisotropies in momentum space because the momenta of the two final particles would be essentially the same as those of the two initial particles. For an isotropic and energy-independent cross section the differences between the two relaxation times are less marked, and for a classical gas one finds [13]

$$\frac{1}{\tau_\eta} = \frac{4}{5\tau}, \tag{11.83}$$

where τ is given by (11.81). Thus collisions are on average only 80% effective in relaxing momentum anisotropies of the form $p_z^2 - p^2/3$ or similar ones with $l = 2$ symmetry.

In a harmonic trap the averaged relaxation rate $1/\tau_{\rm av}$ in the classical limit is therefore obtained by combining (11.79), (11.81) and (11.83). The spatial integral is carried out using the fact that for a harmonic potential

$$\frac{\int d\mathbf{r}\, n^2(\mathbf{r})}{\int d\mathbf{r}\, n(\mathbf{r})} = \frac{n_{\rm cl}(0)}{2^{3/2}}, \tag{11.84}$$

where $n_{\rm cl}(0) = N\bar{\omega}^3 [m/2\pi kT]^{3/2}$ is the central density (cf. (2.39) and (2.40)). We conclude that the rate (11.78) is

$$\Gamma_T = \frac{1}{5} n_{\rm cl}(0) \sigma \bar{v}, \tag{11.85}$$

where $\sigma = 8\pi a^2$ for identical bosons.

The calculation described above can be extended to take into account the effects of quantum degeneracy at temperatures above the transition temperature. The leading high-temperature correction to the classical result (11.85) is given by [14]

$$\Gamma_T \simeq \frac{1}{5} n_{\rm cl}(0) \sigma \bar{v} \left[1 + \frac{3\zeta(3)}{16} \frac{T_c^3}{T^3} \right], \tag{11.86}$$

while at lower temperatures the rate must be calculated numerically.

Throughout our discussion we have assumed that the deviation function is given by Eq. (11.67). Actually, the long-lived mode does not have precisely this form, and one can systematically improve the trial function by exploiting a variational principle essentially identical to that used to find the ground-state energy in quantum mechanics. This always reduces the decay rate, and in such a calculation for the classical limit the decay rate was found to be 7% less than the estimate (11.85) [14].

11.3.2 Damping of oscillations

The above approach can also be used to investigate the damping of oscillations. Consider an anisotropic harmonic trap with uniaxial symmetry, $\omega_1 = \omega_2 \neq \omega_3$. To begin with we study an oscillation corresponding to an extension of the cloud along the z axis. In the absence of collisions, the motions in the x, y, and z directions decouple. If the cloud is initially at rest but is more extended in the z direction than it would be in equilibrium, it will begin to contract. The kinetic energy of particles for motion in the z direction will increase. Later the contraction will halt, the thermal kinetic energy will be a maximum, and the cloud will begin to expand again towards its original configuration. Physically we would expect the deviation function to have terms in p_z^2, corresponding to a modulation of the temperature for motion in the z direction, a function of z to allow for density changes, and a term of the form p_z times a function of z, which corresponds to a z-dependent mean particle velocity. The combination $p_z + im\omega_3 z$ is the classical equivalent of a raising operator in quantum mechanics, and it depends on time as $e^{-i\omega_3 t}$. Consequently, we expect that a deviation from equilibrium proportional to $(p_z + im\omega_3 z)^2$ will give rise to an oscillatory mode at the frequency $2\omega_3$. We therefore use the deviation function[2]

$$\Phi = \Phi_{\mathrm{osc}} = C(p_z + im\omega_3 z)^2 e^{-i2\omega_3 t}, \tag{11.87}$$

where C is an arbitrary constant. That this is a solution of the collisionless Boltzmann equation may be verified by inserting it into (11.59) with $I = 0$. We note that the trial function we used for temperature relaxation may be written as a sum of terms such as $(p_z + im\omega_3 z)(p_z - im\omega_3 z) \propto E_z$ which, as we saw, have frequency zero in the absence of collisions.

We now consider the effect of collisions, and we look for a solution of the form

$$\Phi = \Phi_{\mathrm{osc}} e^{-i2\omega_3 t} g(t), \tag{11.88}$$

where $g(t)$ again describes the relaxation of the distribution function. We insert (11.88) in the Boltzmann equation (11.59) and use the fact that

$$I[(p_z + im\omega_3 z)^2] = I[p_z^2] = I[p_z^2 - p^2/3], \tag{11.89}$$

because of the existence of the collision invariants discussed above. Multiplying by Φ_{osc}^*, integrating over coordinates and momenta, and solving the

---

[2] In this section we choose to work with a complex deviation function. We could equally well have worked with a real function, with terms depending on time as $\cos 2\omega_3 t$ and $\sin 2\omega_3 t$. The dissipation rate then depends on time, but its average agrees with the result obtained using a complex deviation function.

resulting differential equation for g one finds

$$g(t) = e^{-\Gamma_{\text{osc}} t}, \qquad (11.90)$$

where

$$\Gamma_{\text{osc}} = \frac{<\Phi_{\text{osc}}^* \hat{\Gamma}[\Phi_{\text{osc}}]>}{<|\Phi_{\text{osc}}|^2>} = \frac{<(p_z^2 - p^2/3)\hat{\Gamma}[p_z^2 - p^2/3]>}{<(p_z^2 + (m\omega_3 z)^2)^2>}$$

$$= \frac{<(p_z^2 - p^2/3)\hat{\Gamma}[p_z^2 - p^2/3]>}{<(p_z^2 - p^2/3)^2>} \frac{<(p_z^2 - p^2/3)^2>}{<(p_z^2 + (m\omega_3 z)^2)^2>}. \qquad (11.91)$$

This expression has essentially the same form as that for the decay rate of temperature anisotropies. The second factor, reflecting the fraction of the free energy in the form of velocity anisotropies, may be evaluated directly, and is equal to 1/6. Following the same path as before, we find for the decay rate

$$\Gamma_{\text{osc}} = \frac{1}{6} \frac{<p^4 \tau_\eta^{-1}(\mathbf{r})>}{<p^4>}, \qquad (11.92)$$

which reduces in the classical limit to

$$\Gamma_{\text{osc}} = \frac{1}{15} n_{\text{cl}}(0) \sigma \bar{v}. \qquad (11.93)$$

Let us now consider modes in the xy plane. We shall again restrict ourselves to modes having a frequency equal to twice the oscillator frequency, in this case the transverse one $\omega_\perp$. In the absence of interactions the modes corresponding to the deviation functions $(p_x + im\omega_\perp x)^2$ and $(p_y + im\omega_\perp y)^2$ are degenerate. To calculate the damping of the modes when there are collisions, one must use degenerate perturbation theory. Alternatively, one may use physical arguments to determine the form of the modes. The combinations of the two mode functions that have simple transformation properties under rotations about the z axis are $(p_x + im\omega_\perp x)^2 \pm (p_y + im\omega_\perp y)^2$. The plus sign corresponds to a mode which is rotationally invariant, and the minus sign to a quadrupolar mode. Because of the different rotational symmetries of the two modes, they are not mixed by collisions. The damping of the modes may be calculated by the same methods as before, and the result differs from the earlier result only through the factor that gives the fraction of the free energy in the mode that is due to the $l = 2$ anisotropies in momentum space. For the rotationally invariant mode the factor is 1/12, while for the quadrupole mode it is 1/4, which are to be compared with the factor 1/6 for the oscillation in the z direction [14].

The relaxation rates given above may be compared directly with experiment. It is thereby possible to obtain information about interactions between atoms, because uncertainties in the theory are small. The results also provide a useful theoretical testing ground for approximate treatments of the collision integral.

The hydrodynamic and intermediate regimes

The calculations above of damping rates were made for the collisionless regime. However, in some experiments collision frequencies are comparable with the lowest of the trap frequencies. Let us therefore turn to the opposite limit in which the oscillation frequency ω is much less than a typical collision rate. When the hydrodynamic equations apply, the attenuation of modes, e.g., of the type (11.48), may be calculated using the standard expression for the rate of loss of mechanical energy [12]. One finds that the damping rate $1/\tau$ is proportional to an integral of the shear viscosity $\eta(\mathbf{r})$ over coordinate space, $1/\tau \propto \int d\mathbf{r} \eta(\mathbf{r})$. Since the viscosity of a classical gas is independent of density, the damping rate would diverge if one integrated over all of space. This difficulty is due to the fact that a necessary condition for hydrodynamics to apply is that the mean free path be small compared with the length scale over which the density varies. In the outer region of the cloud, the density, and hence also the collision rate, are low, and at some point the mean free path becomes so long that the conditions for the hydrodynamic regime are violated. Consequently, the dissipation there must be calculated from kinetic theory. Strictly speaking, a hydrodynamic limit does not exist for clouds confined by a trap, since the conditions for hydrodynamic behaviour are always violated in the outer region. An approximate solution to this problem is obtained by introducing a cut-off in the hydrodynamic formula [15].

There are no experimental data for conditions when $\omega_3 \tau \ll 1$, but there are experiments on the damping of oscillations for the intermediate regime when $\omega_3 \tau$ is close to unity [9]. A good semi-quantitative description of the intermediate regime may be obtained by interpolating between the hydrodynamic and collisionless limits. A simple expression for ω^2 that gives the correct frequencies ω_C in the collisionless limit and ω_H in the hydrodynamic regime and has a form typical of relaxation processes is

$$\omega^2 = \omega_C^2 - \frac{\omega_C^2 - \omega_H^2}{1 - i\omega\tilde{\tau}}, \tag{11.94}$$

where $\tilde{\tau}$ is a suitably chosen relaxation time which is taken to be independent of frequency. In the collisionless limit the leading contribution to the damping rate of the mode is $(1 - \omega_H^2/\omega_C^2)/2\tilde{\tau}$, while in the hydrodynamic limit the damping is not given correctly because of the difficulties at the edge of the cloud, described above. The form (11.94) predicts a definite relationship between the frequency of the mode and its damping which is in good agreement with experiments in the intermediate regime [9].

Problems

PROBLEM 11.1 Calculate the thermal contribution to the energy of a cloud of N bosons in an isotropic harmonic trap with frequency ω_0 at temperatures low compared with T_0. You may assume that $Na/a_{\rm osc} \gg 1$.

PROBLEM 11.2 Verify the result (11.27) for the energy of a trapped cloud at non-zero temperature and sketch the temperature dependence of the associated specific heat. Use the same approximation to calculate the energy of the cloud after the trapping potential is suddenly turned off. This quantity is referred to as the *release energy* $E_{\rm rel}$. It is equal to the kinetic energy of the atoms after the cloud has expanded so much that the interaction energy is negligible.

PROBLEM 11.3 Consider linear oscillations of a Bose–Einstein condensate in the Thomas–Fermi approximation. Show that the Euler equation and the equation of continuity lead to the following equation for the velocity field,

$$\frac{\partial^2 \mathbf{v}}{\partial t^2} = 2\frac{p_{\rm eq}}{\rho_{\rm eq}}\boldsymbol{\nabla}(\boldsymbol{\nabla}\cdot\mathbf{v}) + \boldsymbol{\nabla}(\mathbf{f}\cdot\mathbf{v}) + \mathbf{f}(\boldsymbol{\nabla}\cdot\mathbf{v}),$$

where $p_{\rm eq}/\rho_{\rm eq} = n(\mathbf{r})U_0/2m$. This result is equivalent to Eq. (7.62), which is expressed in terms of the density rather than the velocity. It is the analogue of Eq. (11.46) for a normal gas in the hydrodynamic regime. Consider now a condensate in an anisotropic harmonic trap. Show that there are solutions to the equation having the form $\mathbf{v} = (ax, by, cz)$, and calculate their frequencies for a trap with axial symmetry. Compare them with those for the corresponding hydrodynamic modes of a gas above T_c.

PROBLEM 11.4 Show that for a gas above T_c in an isotropic trap $V = m\omega_0^2 r^2/2$ there exist hydrodynamic modes with a velocity field of the form $\mathbf{v}(\mathbf{r}) \propto \boldsymbol{\nabla}[r^l Y_{lm}(\theta, \phi)]$, and evaluate their frequencies. Compare the results with those for a pure condensate in the Thomas–Fermi limit which were considered in Sec. 7.3.1. Determine the spatial dependence of the associated density fluctuations in the classical limit.

PROBLEM 11.5 Determine for the anisotropic harmonic-oscillator trap

(2.7) the frequencies of hydrodynamic modes of a Bose gas above $T_{\rm c}$ with a velocity field given by $\mathbf{v}(\mathbf{r}) \propto \boldsymbol{\nabla}(xy)$. Compare the result with that for a condensate in the Thomas–Fermi limit. Show that a local velocity having the above form corresponds in a kinetic description to a deviation function $\Phi \propto yp_x + xp_y$ if, initially, the local density and temperature are not disturbed from equilibrium. Calculate how a deviation function of this form develops in time when there are no collisions, and compare the result with that for the hydrodynamic limit. [Hint: Express the deviation function in terms of $p_x \pm im\omega_1 x$, etc., which have a simple time dependence.]

References

[1] D. S. Jin, M. R. Matthews, J. R. Ensher, C. E. Wieman, and E. A. Cornell, *Phys. Rev. Lett.* **78**, 764 (1997).

[2] S. Giorgini, L. P. Pitaevskii, and S. Stringari, *Phys. Rev. A* **54**, 4633 (1996).

[3] L. D. Landau and E. M. Lifshitz, *Statistical Physics*, Part 1, Third edition, (Pergamon, New York, 1980), §15.

[4] G. Baym, J.-P. Blaizot, M. Holzmann, F. Laloë, and D. Vautherin, *Phys. Rev. Lett.* **83**, 1703 (1999).

[5] S. Giorgini, L. P. Pitaevskii, and S. Stringari, *J. Low Temp. Phys.* **109**, 309 (1997).

[6] J. R. Ensher, D. S. Jin, M. R. Matthews, C. E. Wieman, and E. A. Cornell, *Phys. Rev. Lett.* **77**, 4984 (1996).

[7] V. B. Shenoy and T.-L. Ho, *Phys. Rev. Lett.* **80**, 3895 (1998).

[8] A. Griffin, W.-C. Wu, and S. Stringari, *Phys. Rev. Lett.* **78**, 1838 (1997).

[9] M.-O. Mewes, M. R. Andrews, N. J. van Druten, D. S. Durfee, C. G. Townsend, and W. Ketterle, *Phys. Rev. Lett.* **77**, 988 (1996).

[10] C. R. Monroe, E. A. Cornell, C. A. Sackett, C. J. Myatt, and C. E. Wieman, *Phys. Rev. Lett.* **70**, 414 (1993).

[11] M. Arndt, M. B. Dahan, D. Guéry-Odelin, M. W. Reynolds, and J. Dalibard, *Phys. Rev. Lett.* **79**, 625 (1997).

[12] L. D. Landau and E. M. Lifshitz, *Fluid Mechanics*, Second edition, (Pergamon, New York, 1987), §79.

[13] H. Smith and H. H. Jensen, *Transport Phenomena* (Oxford University Press, Oxford, 1989), p. 29.

[14] G. M. Kavoulakis, C. J. Pethick, and H. Smith, *Phys. Rev. Lett.* **81**, 4036 (1998); *Phys. Rev. A* **61**, 053603 (2000).

[15] G. M. Kavoulakis, C. J. Pethick, and H. Smith, *Phys. Rev. A* **57**, 2938 (1998).

12
Mixtures and spinor condensates

In preceding chapters we have explored properties of Bose–Einstein condensates with a single macroscopically-occupied quantum state, and spin degrees of freedom of the atoms were assumed to play no role. In the present chapter we extend the theory to systems in which two or more quantum states are macroscopically occupied.

The simplest example of such a multi-component system is a mixture of two different species of bosons, for example two isotopes of the same element, or two different atoms. The theory of such systems can be developed along the same lines as that for one-component systems developed in earlier chapters, and we do this in Sec. 12.1.

Since alkali atoms have spin, it is also possible to make mixtures of the same isotope, but in different internal spin states. This was first done experimentally by the JILA group, who made a mixture of ^{87}Rb atoms in hyperfine states $F = 2, m_F = 2$ and $F = 1, m_F = -1$ [1]. Mixtures of hyperfine states of the same isotope differ from mixtures of distinct isotopes because atoms can undergo transitions between hyperfine states, while transitions that convert one isotope into another do not occur under most circumstances. Transitions between different hyperfine states can influence equilibrium properties markedly if the interaction energy per particle is comparable with or larger than the energy difference between hyperfine levels. In magnetic traps it is difficult to achieve such conditions, since the trapping potential depends on the particular hyperfine state. However, in optical traps (see Sec. 4.2.2) the potential is independent of the hyperfine state, and the dynamics of the spin can be investigated, as has been done experimentally [2, 3]. To calculate properties of a condensate with a number of hyperfine components, one may generalize the treatment for the one-component system to allow for the spinor nature of the wave function. We describe this in Sec. 12.2. While this theory is expected to be valid under a wide range of experimental conditions, we shall show in Sec. 12.3 that in the absence of a magnetic field the ground state for atoms with an antiferromagnetic interaction is very different from that predicted by the Gross–Pitaevskii theory. This has important implications for understanding Bose–Einstein condensation.

12.1 Mixtures

Let us begin by considering a mixture of two different bosonic atoms. The generalization of the Hartree wave function (6.1) to two species, labelled 1 and 2, with N_1 and N_2 particles respectively, is

$$\Psi(\mathbf{r}_1,\ldots,\mathbf{r}_{N_1};\mathbf{r}'_1,\ldots,\mathbf{r}'_{N_2}) = \prod_{i=1}^{N_1}\phi_1(\mathbf{r}_i)\prod_{j=1}^{N_2}\phi_2(\mathbf{r}'_j), \quad (12.1)$$

where the particles of species 1 are denoted by $\mathbf{r}_i$ and those of species 2 by $\mathbf{r}'_j$. The corresponding single-particle wave functions are ϕ_1 and ϕ_2. The atomic interactions generally depend on the species, and we shall denote the effective interaction for an atom of species i with one of species j by U_{ij}. For a uniform system, the interaction energy is given by the generalization of Eq. (6.6),

$$E = \frac{N_1(N_1-1)U_{11}}{2V} + \frac{N_1 N_2 U_{12}}{V} + \frac{N_2(N_2-1)U_{22}}{2V}. \quad (12.2)$$

If we introduce the condensate wave functions for the two components according to the definitions $\psi_1 = N_1^{1/2}\phi_1$ and $\psi_2 = N_2^{1/2}\phi_2$, the energy functional corresponding to Eq. (6.9) for a one-component system is

$$E = \int d\mathbf{r}\left[\frac{\hbar^2}{2m_1}|\boldsymbol{\nabla}\psi_1|^2 + V_1(\mathbf{r})|\psi_1|^2 + \frac{\hbar^2}{2m_2}|\boldsymbol{\nabla}\psi_2|^2 + V_2(\mathbf{r})|\psi_2|^2 \right.$$
$$\left. + \frac{1}{2}U_{11}|\psi_1|^4 + \frac{1}{2}U_{22}|\psi_2|^4 + U_{12}|\psi_1|^2|\psi_2|^2\right], \quad (12.3)$$

where we have neglected effects of order $1/N_1$ and $1/N_2$, which are small when N_1 and N_2 are large. Here m_i is the mass of an atom of species i, and V_i is the external potential. In a magnetic trap, the potential depends on the energy of an atom as a function of magnetic field, and therefore it varies from one hyperfine state, isotope, or atom to another. The constants U_{11}, U_{22} and $U_{12} = U_{21}$ are related to the respective scattering lengths a_{11}, a_{22} and $a_{12} = a_{21}$ by $U_{ij} = 2\pi\hbar^2 a_{ij}/m_{ij}$ ($i,j = 1,2$), where $m_{ij} = m_i m_j/(m_i + m_j)$ is the reduced mass for an atom i and an atom j.

The interaction conserves separately the numbers of atoms of the two species. To minimize the energy functional subject to the constraint that the number of atoms of each species be conserved, one therefore introduces the two chemical potentials μ_1 and μ_2. The resulting time-independent Gross–Pitaevskii equations are

$$-\frac{\hbar^2}{2m_1}\nabla^2\psi_1 + V_1(\mathbf{r})\psi_1 + U_{11}|\psi_1|^2\psi_1 + U_{12}|\psi_2|^2\psi_1 = \mu_1\psi_1, \qquad (12.4)$$

and

$$-\frac{\hbar^2}{2m_2}\nabla^2\psi_2 + V_2(\mathbf{r})\psi_2 + U_{22}|\psi_2|^2\psi_2 + U_{12}|\psi_1|^2\psi_2 = \mu_2\psi_2. \qquad (12.5)$$

These will form the basis of our analysis of equilibrium properties of mixtures.

12.1.1 Equilibrium properties

Let us first examine a homogeneous gas, where the densities $n_i = |\psi_i|^2$ of the two components are constant. For each component, the energy is minimized by choosing the phase to be independent of space, and the Gross–Pitaevskii equations become

$$\mu_1 = U_{11}n_1 + U_{12}n_2 \quad \text{and} \quad \mu_2 = U_{12}n_1 + U_{22}n_2, \qquad (12.6)$$

which relate the chemical potentials to the densities.

Stability

For the homogeneous solution to be stable, the energy must increase for deviations of the density from uniformity. We imagine that the spatial scale of the density disturbances is so large that the kinetic energy term in the energy functional plays no role. Under these conditions the total energy E may be written as

$$E = \int d\mathbf{r}\,\mathcal{E}(n_1(\mathbf{r}), n_2(\mathbf{r})), \qquad (12.7)$$

where $\mathcal{E}$ denotes the energy density as a function of the densities n_1 and n_2 of the two components. We consider the change in total energy arising from small changes δn_1 and δn_2 in the densities of the two components. The first-order variation δE must vanish, since the number of particles of each species is conserved,

$$\int d\mathbf{r}\,\delta n_i = 0, \quad i = 1, 2. \qquad (12.8)$$

The second-order variation $\delta^2 E$ is given by the quadratic form

$$\delta^2 E = \frac{1}{2}\int d\mathbf{r}\left[\frac{\partial^2 \mathcal{E}}{\partial n_1^2}(\delta n_1)^2 + \frac{\partial^2 \mathcal{E}}{\partial n_2^2}(\delta n_2)^2 + 2\frac{\partial^2 \mathcal{E}}{\partial n_1 \partial n_2}\delta n_1 \delta n_2\right]. \tag{12.9}$$

The derivative of the energy density with respect to the particle density, $\partial \mathcal{E}/\partial n_i$, is the chemical potential μ_i of species i ($i = 1, 2$). The quadratic form (12.9) is thus positive definite, provided

$$\frac{\partial \mu_1}{\partial n_1} > 0, \quad \frac{\partial \mu_2}{\partial n_2} > 0, \tag{12.10}$$

and

$$\frac{\partial \mu_1}{\partial n_1}\frac{\partial \mu_2}{\partial n_2} - \frac{\partial \mu_1}{\partial n_2}\frac{\partial \mu_2}{\partial n_1} > 0. \tag{12.11}$$

Since $\partial \mu_1/\partial n_2 = \partial^2 \mathcal{E}/\partial n_1 \partial n_2 = \partial \mu_2/\partial n_1$, the condition (12.11) implies that $(\partial \mu_1/\partial n_1)(\partial \mu_2/\partial n_2) > (\partial \mu_1/\partial n_2)^2 \geq 0$, and therefore a sufficient condition for stability is that Eq. (12.11) and one of the two conditions in (12.10) be satisfied, since the second condition in (12.10) then holds automatically. For the energy functional (12.3),

$$\mathcal{E} = \frac{1}{2}n_1^2 U_{11} + \frac{1}{2}n_2^2 U_{22} + n_1 n_2 U_{12}, \tag{12.12}$$

and therefore

$$\frac{\partial \mu_i}{\partial n_j} = U_{ij}. \tag{12.13}$$

Consequently the stability conditions (12.10) and (12.11) become

$$U_{11} > 0, \quad U_{22} > 0, \quad \text{and} \quad U_{11}U_{22} > U_{12}^2. \tag{12.14}$$

The first condition ensures stability against collapse when only the density of the first component is varied, and therefore it is equivalent to the requirement that long-wavelength sound modes in that component be stable. Similarly, the second condition ensures stable sound waves in the second component if it alone is perturbed. The final condition ensures that no disturbance in which the densities of both components are varied can lower the energy.

The stability conditions may be understood in physical terms by observing that the last of the conditions (12.14) is equivalent to the requirement that

$$U_{11} - \frac{U_{12}^2}{U_{22}} > 0, \tag{12.15}$$

since $U_{22} > 0$ by the second of the conditions (12.14). The first term is the so-called *direct interaction* between atoms of species 1, and it gives the change in energy when the density of the second species is held fixed. The term $-U_{12}^2/U_{22}$, which is referred to as the *induced interaction*, corresponds to an interaction mediated by the atoms of the second species. Such effects will be discussed in greater detail in Secs. 14.3.2 and 14.4.1 in the context of Fermi systems and mixtures of bosons and fermions. The result (12.15) states that the total effective interaction, consisting of the direct interaction U_{11} and the induced interaction, must be positive for stability. The argument may also be couched in terms of the effective interaction between two atoms of the second species. If $U_{12}^2 > U_{11}U_{22}$ and U_{12} is negative, the gas is unstable to formation of a denser state containing both components, while if U_{12} is positive, the two components will separate. One can demonstrate that the conditions (12.14) also ensure stability against large deviations from uniformity [4].

Density profiles

Now we consider the density distributions in trapped gas mixtures. We shall work in the Thomas–Fermi approximation, in which one neglects the kinetic energy terms in Eqs. (12.4) and (12.5), which then become

$$\mu_1 = V_1 + U_{11}n_1 + U_{12}n_2, \quad (12.16)$$

and

$$\mu_2 = V_2 + U_{22}n_2 + U_{12}n_1. \quad (12.17)$$

These equations may be inverted to give

$$n_1 = \frac{U_{22}(\mu_1 - V_1) - U_{12}(\mu_2 - V_2)}{U_{11}U_{22} - U_{12}^2}, \quad (12.18)$$

and

$$n_2 = \frac{U_{11}(\mu_2 - V_2) - U_{12}(\mu_1 - V_1)}{U_{11}U_{22} - U_{12}^2}. \quad (12.19)$$

The denominator is positive, since it is necessary that $U_{11}U_{22} > U_{12}^2$ for stability of bulk matter, as we saw in Eq. (12.14).

The solutions (12.18) and (12.19) make sense only if the densities n_1 and n_2 are positive. When one component is absent, the other one obeys the Gross–Pitaevskii equation for a single component. The chemical potentials μ_1 and μ_2 must be determined self-consistently from the condition that the total number of particles of a particular species is given by integrating the density distributions over space.

Let us now give a specific example. We assume that the trapping potentials are isotropic and harmonic, $V_1(\mathbf{r}) = m_1\omega_1^2 r^2/2$, and that $V_2 =$

$m_2\omega_2^2 r^2/2 = \lambda V_1$, where $\lambda = m_2\omega_2^2/m_1\omega_1^2$ is a constant. We define lengths R_1 and R_2 according to the equation

$$\mu_i = \frac{1}{2}m_i\omega_i^2 R_i^2, \quad i = 1, 2. \tag{12.20}$$

After inserting the expressions for V_i and μ_i into (12.18) and (12.19) we may write the densities as

$$n_1 = \frac{\mu_1}{U_{11}} \frac{1}{1 - U_{12}^2/U_{11}U_{22}} \left[1 - \frac{U_{12}}{U_{22}}\frac{\mu_2}{\mu_1} - \frac{r^2}{R_1^2}\left(1 - \frac{\lambda U_{12}}{U_{22}}\right)\right], \tag{12.21}$$

and

$$n_2 = \frac{\mu_2}{U_{22}} \frac{1}{1 - U_{12}^2/U_{11}U_{22}} \left[1 - \frac{U_{12}}{U_{11}}\frac{\mu_1}{\mu_2} - \frac{r^2}{R_2^2}\left(1 - \frac{U_{12}}{\lambda U_{11}}\right)\right]. \tag{12.22}$$

In the absence of interaction between the two components of the mixture ($U_{12} = 0$), the densities (12.21) and (12.22) vanish at $r = R_1$ and $r = R_2$, respectively. Quite generally, if one of the densities, say n_1, vanishes in a certain region of space, one can see from Eq. (12.17) that the density n_2 of the other component will be given by the Gross–Pitaevskii equation for that component alone, $U_{22}n_2 = \mu_2 - V_2$, and therefore the density profile in that region is given by

$$n_2 = \frac{\mu_2}{U_{22}}\left(1 - \frac{r^2}{R_2^2}\right), \tag{12.23}$$

provided r is less than R_2. This agrees with the density obtained from (12.22) by setting U_{12} equal to zero. In all cases, the density profiles are linear functions of r^2, the specific form depending on whether or not two components coexist in the region in question. For more general potentials, the density profiles are linear functions of the potentials for the two species.

The density distributions are given by (12.21) and (12.22) where the two components coexist, by (12.23) when $n_1 = 0$, and by an analogous expression for n_1 when $n_2 = 0$. The chemical potentials μ_1 and μ_2 are determined by requiring that the integrals over space of the densities of the components be equal to the total numbers of particles N_1 and N_2. Depending on the ratios of the three interaction parameters, the two components may coexist in some regions of space, but remain separated in others [5]. When the numbers of the two kinds of particles are comparable, the mixture will generally exhibit a fairly large region where the two components coexist, provided the conditions (12.14) for stability of bulk mixed phases are satisfied. If one adds a small number of, say, 2-atoms to a cloud containing a large number of 1-atoms, the 2-atoms tend to reside either at the surface or in the deep

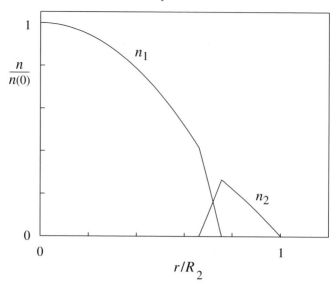

Fig. 12.1. Density profiles of a mixture of two condensates for the isotropic model discussed in the text. The values of the parameters are $\mu_1/\mu_2 = 1.5$, $\lambda = 2$, $U_{12} = 0.9\, U_{11}$, and $U_{22} = 1.08\, U_{11}$. The corresponding ratio of the numbers of particles is $N_1/N_2 = 2.4$.

interior of the cloud, depending on the ratio of the interaction parameters. In Fig. 12.1 we illustrate this by plotting the density profiles for a specific choice of parameters.

12.1.2 Collective modes

The methods used in Chapter 7 to describe the dynamical properties of a condensate with one component may be extended to mixtures. The natural generalization of the time-independent Gross–Pitaevskii equations (12.4) and (12.5) to allow for time dependence is

$$i\hbar \frac{\partial \psi_1}{\partial t} = \left[-\frac{\hbar^2}{2m_1} \nabla^2 + V_1(\mathbf{r}) + U_{11}|\psi_1|^2 + U_{12}|\psi_2|^2 \right] \psi_1, \qquad (12.24)$$

and

$$i\hbar \frac{\partial \psi_2}{\partial t} = \left[-\frac{\hbar^2}{2m_2} \nabla^2 + V_2(\mathbf{r}) + U_{22}|\psi_2|^2 + U_{12}|\psi_1|^2 \right] \psi_2. \qquad (12.25)$$

By introducing the velocities and densities of the two components one may write the time-dependent Gross–Pitaevskii equations as hydrodynamic equations, as was done for a single component in Sec. 7.1.1. As an illustration let us calculate the frequencies of normal modes of a mixture of two components. By generalizing the linearized hydrodynamic equations (7.26)–(7.28) we obtain the coupled equations

$$m_1\omega^2 \delta n_1 + \nabla \cdot (n_1 \nabla \delta \tilde{\mu}_1) = 0 \qquad (12.26)$$

and

$$m_2\omega^2 \delta n_2 + \nabla \cdot (n_2 \nabla \delta \tilde{\mu}_2) = 0, \qquad (12.27)$$

where $\delta\tilde{\mu}_i$ is obtained by linearizing the expressions

$$\tilde{\mu}_i = V_i - \frac{\hbar^2}{2m_i}\frac{1}{\sqrt{n_i}}\nabla^2\sqrt{n_i} + \sum_j U_{ij} n_j, \qquad (12.28)$$

which correspond to Eq. (7.21).

The first example we consider is a homogeneous system, and we look for travelling wave solutions proportional to $\exp(i\mathbf{q}\cdot\mathbf{r} - i\omega t)$. From (12.26) and (12.27) it follows that

$$m_i \omega^2 \delta n_i = n_i q^2 \delta \tilde{\mu}_i, \quad i = 1, 2. \qquad (12.29)$$

The change in the chemical potentials is given to first order in δn_1 and δn_2 by

$$\delta\tilde{\mu}_1 = \left(U_{11} + \frac{\hbar^2 q^2}{4m_1 n_1}\right)\delta n_1 + U_{12}\delta n_2 \qquad (12.30)$$

and

$$\delta\tilde{\mu}_2 = \left(U_{22} + \frac{\hbar^2 q^2}{4m_2 n_2}\right)\delta n_2 + U_{12}\delta n_1. \qquad (12.31)$$

The expressions (12.30) and (12.31) are now inserted into (12.29), and the frequencies are found from the consistency condition for the homogeneous equations for δn_i. The result is

$$(\hbar\omega)^2 = \frac{1}{2}(\epsilon_1^2 + \epsilon_2^2) \pm \frac{1}{2}\sqrt{(\epsilon_1^2 - \epsilon_2^2)^2 + 16\epsilon_1^0 \epsilon_2^0 n_1 n_2 U_{12}^2}. \qquad (12.32)$$

Here we have introduced the abbreviations

$$\epsilon_1^2 = 2U_{11} n_1 \epsilon_1^0 + (\epsilon_1^0)^2 \text{ and } \epsilon_2^2 = 2U_{22} n_2 \epsilon_2^0 + (\epsilon_2^0)^2, \qquad (12.33)$$

where $\epsilon_i^0 = \hbar^2 q^2 / 2m_i$ is the free-particle energy as in Chapter 7. The energies ϵ_1 and ϵ_2 are those of the Bogoliubov modes of the two components when there is no interaction between different components. This interaction, which gives rise to the term containing U_{12}, hybridizes the modes. At short wavelengths, the two mode frequencies approach the free-particle frequencies $\hbar q^2/2m_1$ and $\hbar q^2/2m_2$. As U_{12}^2 approaches $U_{11}U_{22}$, one of the two frequencies tends to zero in the long-wavelength limit $q \to 0$, signalling an instability of the system, in agreement with the conclusions we reached from static considerations, Eq. (12.14). Equation (12.32) has been used to study theoretically [6] the growth of unstable modes of a two-component system observed experimentally [7].

We next turn to particles in traps. In Sec. 7.3 we obtained the collective modes in a trap by making the Thomas–Fermi approximation, in which the contribution of the quantum pressure to $\delta\tilde{\mu}$ is neglected. The generalization of Eq. (7.58) to a two-component system is

$$\delta\tilde{\mu}_1 = U_{11}\delta n_1 + U_{12}\delta n_2 \text{ and } \delta\tilde{\mu}_2 = U_{22}\delta n_2 + U_{12}\delta n_1. \qquad (12.34)$$

In order to solve the coupled equations for δn_i ($i = 1, 2$) one must determine the equilibrium densities n_i for given choices of the interaction parameters and insert these in the equations. The situation when the two condensates coexist everywhere within the cloud is the simplest one to analyze. This is the subject of Problem 12.1, where coupled surface modes in the two components are investigated. When the two components do not overlap completely, solutions in different parts of space must be matched by imposing boundary conditions.

In summary, the properties of mixtures may be analyzed by generalizing the methods used in previous chapters for a single condensate. We have treated a mixture with two components, but it is straightforward to extend the theory to more components. We turn now to condensates where the components are atoms of the same isotope in different hyperfine states. These exhibit qualitatively new features.

12.2 Spinor condensates

In the mixtures considered above, the interaction conserves the total number of particles of each species. This is no longer true when condensation occurs in states which are different hyperfine states of the same isotope. As we noted in the previous section, overlapping condensates with atoms in two different hyperfine states have been made in magnetic traps [1], and the development of purely optical traps has made it possible to Bose–Einstein condense Na atoms in the three magnetic sublevels, corresponding to the quantum numbers $m_F = 0, \pm 1$, of the hyperfine multiplet with total spin $F = 1$ [2]. Let us for definiteness consider magnetic fields so low that the states of a single atom are eigenstates of the angular momentum, as we described in Sec. 3.2. Because of its experimental relevance and simplicity we consider atoms with $F = 1$, but the treatment may be extended to higher values of F.

In the mixtures described in Sec. 12.1, the number of particles of each component was strictly conserved. For the three hyperfine states this is no longer so, since, e.g., an atom in the $m_F = 1$ state may scatter with another in the $m_F = -1$ state to give two atoms in the $m_F = 0$ state. Let us begin by considering the interaction between atoms. Rotational invariance imposes important constraints on the number of parameters needed to characterize the interaction. Two identical bosonic atoms with $F = 1$ in an s state of the relative motion can couple to make states with total angular momentum $\mathcal{F} = 0$ or 2 units, since the possibility of unit angular momentum is ruled out by the requirement that the wave function be symmetric under exchange of the two atoms. The interaction is invariant under rotations, and therefore it is diagonal in the total angular momentum of the two atoms. We may thus write the effective interaction for low-energy collisions as $U_0 = 4\pi\hbar^2 a^{(0)}/m$ for $\mathcal{F} = 0$ and $U_2 = 4\pi\hbar^2 a^{(2)}/m$ for $\mathcal{F} = 2$, where $a^{(0)}$ and $a^{(2)}$ are the corresponding scattering lengths. For arbitrary hyperfine states in the hyperfine manifold in which there are two atoms with $F = 1$, the effective interaction may therefore be written in the form

$$U(\mathbf{r}_1 - \mathbf{r}_2) = \delta(\mathbf{r}_1 - \mathbf{r}_2)(U_0 \mathcal{P}_0 + U_2 \mathcal{P}_2), \qquad (12.35)$$

where the operators $\mathcal{P}_\mathcal{F}$ project the wave function of a pair of atoms on a state of total angular momentum $\mathcal{F}$. It is helpful to re-express this result in terms of the operators for the angular momenta of the two atoms, which we here denote by $\mathbf{S}_1$ and $\mathbf{S}_2$.[1] The eigenvalues of the scalar product $\mathbf{S}_1 \cdot \mathbf{S}_2$ are 1 when the total angular momentum quantum number $\mathcal{F}$ of the pair of atoms is 2, and -2 when $\mathcal{F} = 0$. The operator that projects onto the $\mathcal{F} = 2$ manifold is $\mathcal{P}_2 = (2 + \mathbf{S}_1 \cdot \mathbf{S}_2)/3$, and that for the $\mathcal{F} = 0$ state is $\mathcal{P}_0 = (1 - \mathbf{S}_1 \cdot \mathbf{S}_2)/3$. The strength of the contact interaction in Eq. (12.35) may therefore be written as

$$U_0 \mathcal{P}_0 + U_2 \mathcal{P}_2 = W_0 + W_2 \mathbf{S}_1 \cdot \mathbf{S}_2, \qquad (12.36)$$

where

$$W_0 = \frac{U_0 + 2U_2}{3} \quad \text{and} \quad W_2 = \frac{U_2 - U_0}{3}. \qquad (12.37)$$

Equation (12.36) has a form analogous to that for the exchange interaction between a pair of atoms when the nuclear spin is neglected, see Eq. (5.74).

---

[1] In Chapter 3 we denoted the angular momentum of an atom by $\mathbf{F}$ but, to conform with the convention used in the literature on spinor condensates, in this section we denote it by $\mathbf{S}$ even though it contains more than the contribution from the electron spin. Thus $\mathbf{S} \cdot \mathbf{S} = F(F+1)$.

For larger values of F the interaction may be written in a similar form but with additional terms containing higher powers of $\mathbf{S}_1 \cdot \mathbf{S}_2$.

In second-quantized notation, the many-body Hamiltonian for atoms with the effective interaction (12.36) is

$$\hat{H} = \int d\mathbf{r} \left(\frac{\hbar^2}{2m} \boldsymbol{\nabla}\hat{\psi}_\alpha^\dagger \cdot \boldsymbol{\nabla}\hat{\psi}_\alpha + V(\mathbf{r})\hat{\psi}_\alpha^\dagger \hat{\psi}_\alpha + g\mu_B \hat{\psi}_\alpha^\dagger \mathbf{B} \cdot \mathbf{S}_{\alpha\beta} \hat{\psi}_\beta \right.$$
$$\left. + \frac{1}{2} W_0 \hat{\psi}_\alpha^\dagger \hat{\psi}_{\alpha'}^\dagger \hat{\psi}_{\alpha'} \hat{\psi}_\alpha + \frac{1}{2} W_2 \hat{\psi}_\alpha^\dagger \hat{\psi}_{\alpha'}^\dagger \mathbf{S}_{\alpha\beta} \cdot \mathbf{S}_{\alpha'\beta'} \hat{\psi}_{\beta'} \hat{\psi}_\beta \right). \quad (12.38)$$

Here we have included the Zeeman energy to first order in the magnetic field, and g is the Landé g factor introduced in Chapter 3. Repeated indices are to be summed over, following the Einstein convention. The external potential $V(\mathbf{r})$ is assumed to be independent of the hyperfine state, as it is for an optical trap. In the following subsection we treat the Hamiltonian (12.38) in the mean-field approximation.

12.2.1 Mean-field description

A direct extension of the wave function (12.1) to three components has a definite number of particles in each of the magnetic sublevels and, consequently, it does not take into account the effect of processes such as that in which an atom with $m_F = 1$ interacts with one with $m_F = -1$ to give two atoms in the $m_F = 0$ state. It is therefore necessary to consider a more general wave function, as was done in Refs. [8] and [9]. Instead of generalizing the Hartree wave function for a single component as we did in Sec. 12.1, we imagine that all particles are condensed in a state $\phi_\alpha(\mathbf{r})$ which is a superposition of the three hyperfine substates,

$$\phi_\alpha(\mathbf{r}) = \phi_1(\mathbf{r})|1,1\rangle + \phi_0(\mathbf{r})|1,0\rangle + \phi_{-1}(\mathbf{r})|1,-1\rangle, \quad (12.39)$$

where we have used Dirac notation for the spin degrees of freedom. The wave function for the state where all particles are in the same single-particle quantum state with wave function $\phi_\alpha(\mathbf{r})$ is then written as

$$\Psi(\mathbf{r}_1, \alpha_1; \ldots; \mathbf{r}_N, \alpha_N) = \prod_{i=1}^{N} \phi_{\alpha_i}(\mathbf{r}_i), \quad (12.40)$$

where $\alpha_i = 0, \pm 1$ specifies the spin state of particle i. The state (12.40) does not have a definite number of particles in a given hyperfine state, since it has components in which all particles are in, e.g., the $m_F = 1$ state. However, the probability distribution for the number of particles in a particular hyperfine state is sharply peaked about some value. Interestingly, when a wave function of the type (12.40) is applied to the problem of mixtures of two different atoms or isotopes considered in Sec. 12.1, it leads to results which agree with those for the wave function (12.1) if the number of particles is large. To demonstrate this is the subject of Problem 12.2.

Provided we are interested only in contributions to the energy which are of order N^2, working with the wave function (12.40) is equivalent to using the Gross–Pitaevskii or mean-field prescription of treating the operator fields $\hat{\psi}$ as classical ones ψ, which we write in the form

$$\hat{\psi}_\alpha \approx \psi_\alpha = \sqrt{n(\mathbf{r})}\zeta_\alpha(\mathbf{r}), \tag{12.41}$$

where $n(\mathbf{r}) = \sum_\alpha |\psi_\alpha(\mathbf{r})|^2$ is the total density of particles in all hyperfine states, and $\zeta_\alpha(\mathbf{r})$ is a three-component spinor, normalized according to the condition

$$\zeta_\alpha^* \zeta_\alpha = 1. \tag{12.42}$$

The total energy E in the presence of a magnetic field is therefore given by

$$E - \mu N = \int d\mathbf{r} \left[\frac{\hbar^2}{2m}(\boldsymbol{\nabla}\sqrt{n})^2 + \frac{\hbar^2}{2m} n \boldsymbol{\nabla}\zeta_\alpha^* \cdot \boldsymbol{\nabla}\zeta_\alpha \right.$$
$$\left. + (V - \mu)n + g\mu_B n \mathbf{B} \cdot <\mathbf{S}> + \frac{1}{2}n^2(W_0 + W_2 <\mathbf{S}>^2) \right] \tag{12.43}$$

with

$$<\mathbf{S}> = \zeta_\alpha^* \mathbf{S}_{\alpha\beta} \zeta_\beta. \tag{12.44}$$

It is convenient to use an explicit representation of the angular momentum matrices, and we shall work with a basis of states $|F = 1, m_F\rangle$ referred to the direction of the applied magnetic field. In the representation generally used, the angular momentum matrices are written as

$$S_x = \frac{1}{\sqrt{2}}\begin{bmatrix} 0 & 1 & 0 \\ 1 & 0 & 1 \\ 0 & 1 & 0 \end{bmatrix}, \quad S_y = \frac{1}{\sqrt{2}}\begin{bmatrix} 0 & -i & 0 \\ i & 0 & -i \\ 0 & i & 0 \end{bmatrix}, \quad \text{and} \quad S_z = \begin{bmatrix} 1 & 0 & 0 \\ 0 & 0 & 0 \\ 0 & 0 & -1 \end{bmatrix}.$$

Let us examine the ground state when there is no magnetic field. Apart from the W_2 term, the energy then has essentially the same form as for a one-component system. If W_2 is negative, the energy is lowered by making the magnitude of $<\mathbf{S}>$ as large as possible. This is achieved by, e.g., putting all particles into the $m_F = 1$ state. Typographically, row vectors are preferable to column ones, so it is convenient to express some results in terms of the transpose of ζ, which we denote by $\tilde{\zeta}$, and in this notation the state is

$\tilde{\zeta} = (1, 0, 0)$. The state is ferromagnetic. Any state obtained from this one by a rotation in spin space has the same energy since the interaction is invariant under rotations. The reason for the ferromagnetic state being favoured is that, according to Eq. (12.37), W_2 is negative if the $\mathcal{F} = 0$ interaction exceeds the $\mathcal{F} = 2$ one. By putting all atoms into the $m_F = 1$ state, all pairs of atoms are in states with total angular momentum $\mathcal{F} = 2$, and there is no contribution from the $\mathcal{F} = 0$ interaction. This may be confirmed by explicit calculation, since $<\mathbf{S}>^2 = 1$ and therefore the interaction term in the energy (12.43) becomes

$$\frac{1}{2}(W_0 + W_2)n^2 = \frac{1}{2}U_2 n^2. \qquad (12.45)$$

In a trap, the density profile in the Thomas–Fermi approximation is therefore given by

$$n(\mathbf{r}) = \frac{\mu - V(\mathbf{r})}{U_2}. \qquad (12.46)$$

When $W_2 > 0$, the energy is minimized by making $<\mathbf{S}>$ as small as possible. This is achieved by, e.g., putting all particles into the $m_F = 0$ substate, since then $<\mathbf{S}> = 0$. Equivalently $\tilde{\zeta} = (0, 1, 0)$. Again, states obtained from this one by rotations in spin space have the same energy. The density profile in the Thomas–Fermi approximation is consequently obtained by replacing U_2 in (12.46) by W_0. Investigating the influence of a magnetic field is the subject of Problem 12.4.

The collective modes of spinor condensates may be obtained by linearizing the equations of motion in the deviations of ψ from its equilibrium form. It is convenient to start from the operator equations of motion, which in the absence of an external magnetic field are

$$i\hbar \frac{\partial \hat{\psi}_\alpha}{\partial t} = -\frac{\hbar^2}{2m}\nabla^2 \hat{\psi}_\alpha + [V(\mathbf{r}) - \mu]\hat{\psi}_\alpha + W_0 \hat{\psi}^\dagger_{\alpha'}\hat{\psi}_{\alpha'}\hat{\psi}_\alpha + W_2 \hat{\psi}^\dagger_{\alpha'}\mathbf{S}_{\alpha'\beta}\cdot\mathbf{S}_{\alpha\beta'}\hat{\psi}_{\beta'}\hat{\psi}_\beta. \qquad (12.47)$$

By writing $\hat{\psi}_\alpha = \psi_\alpha + \delta\hat{\psi}_\alpha$ and linearizing (12.47) in $\delta\hat{\psi}_\alpha$ about the particular equilibrium state (12.41) one obtains three coupled, linear equations for $\delta\hat{\psi}_\alpha$. The form of these equations depends on the sign of W_2. For particles with an antiferromagnetic interaction ($W_2 > 0$) in a harmonic trap one finds in the Thomas–Fermi approximation that the frequencies of density modes are independent of the interaction parameters and are exactly the same as those obtained in Chapter 7 for a single-component condensate, while the frequencies of the spin–wave modes are related to those of the density modes by a factor $(W_2/W_0)^{1/2}$. For a ferromagnetic interaction, the frequencies of the density modes remain the same, while the low-energy spin–wave modes

12.2.2 Beyond the mean-field approximation

A surprising discovery is that the ground state of the Hamiltonian (12.38) in zero magnetic field is completely different from the mean-field solution just described [11, 12]. Consider a homogeneous spin-1 Bose gas in a magnetic field, with interaction parameters W_0 (> 0) and W_2. The parameter W_2 is negative for a ferromagnetic interaction and positive for an antiferromagnetic interaction. The field operators $\hat{\psi}_\alpha(\mathbf{r})$ may be expanded in a plane-wave basis,

$$\hat{\psi}_\alpha(\mathbf{r}) = \frac{1}{V^{1/2}} \sum_{\mathbf{k}} a_{\mathbf{k},\alpha} e^{i\mathbf{k}\cdot\mathbf{r}}. \tag{12.48}$$

As long as we may neglect depletion of the condensate, it is sufficient to retain the single term associated with $\mathbf{k} = 0$,

$$\hat{\psi}_\alpha = \frac{1}{V^{1/2}} a_\alpha, \tag{12.49}$$

where we denote $a_{\mathbf{k}=0,\alpha}$ by a_α. The corresponding part of the Hamiltonian (12.38) is

$$\hat{H}_0 = g\mu_B a_\alpha^\dagger \mathbf{B}\cdot\mathbf{S}_{\alpha\beta} a_\beta + \frac{1}{2V} W_0 a_\alpha^\dagger a_{\alpha'}^\dagger a_{\alpha'} a_\alpha + \frac{1}{2V} W_2 a_\alpha^\dagger a_{\alpha'}^\dagger \mathbf{S}_{\alpha\beta}\cdot\mathbf{S}_{\alpha'\beta'} a_{\beta'} a_\beta. \tag{12.50}$$

To find the eigenstates of the Hamiltonian we introduce operators

$$\hat{N} = a_\alpha^\dagger a_\alpha \tag{12.51}$$

for the total number of particles and

$$\hat{\mathbf{S}} = a_\alpha^\dagger \mathbf{S}_{\alpha\beta} a_\beta \tag{12.52}$$

for the total spin. The operator defined by (12.52) is the total spin of the system written in the language of second quantization. Consequently, its components satisfy the usual angular momentum commutation relations. Verifying this explicitly is left as an exercise, Problem 12.3.

The spin–spin interaction term in Eq. (12.50) would be proportional to $\hat{\mathbf{S}}^2$, were it not for the fact that the order of the creation and annihilation operators matters. In the earlier part of the book we have generally ignored differences between the orders of operators and replaced operators by c numbers. However, the energy differences between the mean-field state and the true ground state is so small that here we must be more careful than usual. The Bose commutation relations yield

$$a_\alpha^\dagger a_{\alpha'}^\dagger a_{\beta'} a_\beta = a_\alpha^\dagger a_\beta a_{\alpha'}^\dagger a_{\beta'} - \delta_{\alpha'\beta} a_\alpha^\dagger a_{\beta'}. \tag{12.53}$$

The second term on the right hand side of (12.53) produces a term in the Hamiltonian proportional to $\mathbf{S}_{\alpha\beta} \cdot \mathbf{S}_{\beta\beta'} a_\alpha^\dagger a_{\beta'}$. The matrix $\mathbf{S}_{\alpha\beta} \cdot \mathbf{S}_{\beta\beta'}$ is diagonal and equal to $F(F+1)\delta_{\alpha\beta'}$ for spin F. Since we are considering $F=1$, the Hamiltonian (12.50) may be rewritten as

$$\hat{H}_0 = g\mu_{\rm B} \mathbf{B} \cdot \hat{\mathbf{S}} + \frac{W_0}{2V}(\hat{N}^2 - \hat{N}) + \frac{W_2}{2V}(\hat{\mathbf{S}}^2 - 2\hat{N}). \tag{12.54}$$

Let us now consider the nature of the ground state of the Hamiltonian (12.54), taking the number of particles N to be even. The allowed values of the quantum number S for the total spin are then $0, 2, \ldots, N$. For a ferromagnetic interaction ($W_2 < 0$), the ground state has the maximal spin $S = N$. This is the state in which the spins of all particles are maximally aligned in the direction of the magnetic field, a result which agrees with the conclusions from mean-field theory.

We now consider an antiferromagnetic interaction ($W_2 > 0$). In zero magnetic field the ground state is a singlet, with total spin $S = 0$. The ground-state energy E_0 is seen to be

$$E_0 = N(N-1)\frac{W_0}{2V} - N\frac{W_2}{V}, \tag{12.55}$$

which differs from the mean-field result $N(N-1)W_0/2V$ by the term NW_2/V. The factor $N(N-1)/2$ in (12.55) is the number of ways of making pairs of bosons. Within the mean-field Gross–Pitaevskii approach we have usually approximated this by $N^2/2$, but we do not do so here since the energy differences we are calculating are of order N.

An alternative representation of the ground state is obtained by introducing the operator $\hat{a}_x$ that destroys a particle in a state whose angular momentum component in the x direction is zero and the corresponding operators $\hat{a}_y$ and $\hat{a}_z$ for the y and z directions. In terms of the operators that destroy atoms in states with given values of m_F these operators are given by

$$a_x = \frac{1}{\sqrt{2}}(a_{-1} - a_1), \quad a_y = \frac{1}{i\sqrt{2}}(a_{-1} + a_1), \quad \text{and} \quad a_z = a_0. \tag{12.56}$$

The combination

$$\hat{A} = a_x^2 + a_y^2 + a_z^2 = a_0^2 - 2a_{-1}a_1 \qquad (12.57)$$

is rotationally invariant. When $\hat{A}^\dagger$ acts on the vacuum it creates a pair of particles with total angular momentum zero. Thus the state $(\hat{A}^\dagger)^{N/2}|0\rangle$ is a singlet, and it is in fact unique. Physically it corresponds to a Bose–Einstein condensate of $N/2$ composite bosons, each of which is made up by coupling two of the original spin-1 bosons to spin zero. This state may thus be regarded as a Bose–Einstein condensate of pairs of bosons, and it is very different from the simple picture of a Bose–Einstein condensate of spinless particles described earlier. In Sec. 13.5 we shall discuss the properties of this singlet ground state further in relation to the criterion for Bose–Einstein condensation.

The physical origin of the lowering of the energy is that the interaction acting on the Hartree wave function $(a_0^\dagger)^N|0\rangle$, which is the mean-field ground state, can scatter particles to the $m_F = \pm 1$ states. The correlations in the singlet ground state are delicate, since they give rise to a lowering of the energy of order $1/N$ compared with the total energy of a state described by a wave function of the product form (12.40).

The singlet state is the ground state when the interaction is antiferromagnetic, $W_2 > 0$, and the magnetic field is zero. It differs in a number of interesting ways from a condensate in which all atoms occupy the same state. One of these is that the mean number of particles in each of the three hyperfine states is the same, and equal to $N/3$, while the fluctuations are enormous, of order N. In the presence of a magnetic field the singlet state is no longer the ground state, and the fluctuations are cut down drastically [12]. Because of the small energy difference between the singlet state and the state predicted by mean-field theory, which becomes a good approximation even for very small magnetic fields, detecting the large fluctuations associated with the singlet state is a challenging experimental problem.

Problems

PROBLEM 12.1 Consider a mixture of two components, each with particle mass m, trapped in the isotropic potentials $V_1(r) = m\omega_1^2 r^2/2$ and $V_2(r) = m\omega_2^2 r^2/2$, with $\lambda = \omega_2^2/\omega_1^2 = 2$. The ratio N_1/N_2 of the particle numbers N_1 and N_2 is such that $R_1 = R_2 = R$. Determine the equilibrium densities (12.21) and (12.22) in terms of the interaction parameters, trap frequencies and particle numbers. In the absence of coupling ($U_{12} = 0$), surface modes with density oscillations proportional to $r^l Y_{ll}(\theta, \phi) \propto (x + iy)^l$

have frequencies given by $\omega^2 = l\omega_i^2$, with $i = 1, 2$ (cf. Sec. 7.3.1 and 7.3.2). Calculate the frequencies of the corresponding surface modes when $U_{12} \neq 0$, and investigate the limit $U_{12}^2 \to U_{11}U_{22}$.

PROBLEM 12.2 Consider a wave function of the form (12.40) for a mixture of two different spinless bosonic isotopes,

$$\Psi(\mathbf{r}_1, \alpha_1; \ldots; \mathbf{r}_N, \alpha_N) = \prod_{i=1}^{N} \phi_{\alpha_i}(\mathbf{r}_i),$$

where the label $\alpha = 1, 2$ now refers to the species. Show that for the choice $\phi_\alpha(\mathbf{r}) = (N_\alpha/N)^{1/2}\chi_\alpha(\mathbf{r})$, where χ satisfies the normalization condition $\int d\mathbf{r}|\chi_\alpha(\mathbf{r})|^2 = 1$, the average number of particles of species α is N_α. Calculate the root-mean-square fluctuations in the particle numbers in this state, and show that it is small for large N_α. Calculate the expectation value of the energy of the state and show that it agrees to order N^2 with the result of using the wave function (12.1).

PROBLEM 12.3 Show that the operators $\hat{S}_z$ and $\hat{S}_\pm = \hat{S}_x \pm i\hat{S}_y$ defined in Eq. (12.52) are given by

$$\hat{S}_+ = \sqrt{2}(a_1^\dagger a_0 + a_0^\dagger a_{-1}), \quad \hat{S}_- = (\hat{S}_+)^\dagger, \quad \hat{S}_z = (a_1^\dagger a_1 - a_{-1}^\dagger a_{-1}),$$

where the operators a_1, a_0, and a_{-1} annihilate particles in states with $m = 1$, $m = 0$, and $m = -1$. Verify that these operators satisfy the commutation relations $[\hat{S}_+, \hat{S}_-] = 2\hat{S}_z$ and $[\hat{S}_z, \hat{S}_\pm] = \pm\hat{S}_\pm$ for angular momentum operators.

PROBLEM 12.4 Determine from (12.54) for an antiferromagnetic interaction the lowest-energy state in the presence of a magnetic field and compare the result with the mean-field solution.

References

[1] C. J. Myatt, E. A. Burt, R. W. Ghrist, E. A. Cornell, and C. E. Wieman, *Phys. Rev. Lett.* **78**, 586 (1997).

[2] D. M. Stamper-Kurn, M. R. Andrews, A. P. Chikkatur, S. Inouye, H.-J. Miesner, J. Stenger, and W. Ketterle, *Phys. Rev. Lett.* **80**, 2027 (1998).

[3] J. Stenger, S. Inouye, D. M. Stamper-Kurn, H.-J. Miesner, A. P. Chikkatur, and W. Ketterle, *Nature* **396**, 345 (1998).

[4] P. Ao and S. T. Chui, *Phys. Rev. A* **58**, 4836 (1998).

[5] T.-L. Ho and V. B. Shenoy, *Phys. Rev. Lett.* **77**, 3276 (1996).

[6] P. Ao and S. T. Chui, *J. Phys. B* **33**, 535 (2000).

[7] D. S. Hall, M. R. Matthews, J. R. Ensher, C. E. Wieman, and E. A. Cornell, *Phys. Rev. Lett.* **81**, 1539 (1998).

[8] T.-L. Ho, *Phys. Rev. Lett.* **81**, 742 (1998).

[9] T. Ohmi and K. Machida, *J. Phys. Soc. Jpn.* **67**, 1822 (1998).

[10] D. Vollhardt and P. Wölfle, *The Superfluid Phases of ^{3}He*, (Taylor and Francis, London, 1990).

[11] C. K. Law, H. Pu, and N. P. Bigelow, *Phys. Rev. Lett.* **81**, 5257 (1998).

[12] T.-L. Ho and S. K. Yip, *Phys. Rev. Lett.* **84**, 4031 (2000).

13

Interference and correlations

Bose–Einstein condensates of particles behave in many ways like coherent radiation fields, and the realization of Bose–Einstein condensation in dilute gases has opened up the experimental study of many aspects of interactions between coherent matter waves. In addition, the existence of these dilute trapped quantum gases has prompted a re-examination of a number of theoretical issues. This field is a vast one, and in this chapter we shall touch briefly on selected topics.

In Sec. 13.1 we describe the classic interference experiment, in which two clouds of atoms are allowed to expand and overlap. Rather surprisingly, an interference pattern is produced even though initially the two clouds are completely isolated. We shall analyse the reasons for this effect. The marked decrease in density fluctuations in a Bose gas when it undergoes Bose–Einstein condensation is demonstrated in Sec. 13.2. Gaseous Bose–Einstein condensates can be manipulated by lasers, and this has made possible the study of coherent matter wave optics. We describe applications of these techniques to observe solitons, Bragg scattering, and non-linear mixing of matter waves in Sec. 13.3. The atom laser and amplification of matter waves is taken up in Sec. 13.4. How to characterize Bose–Einstein condensation microscopically is the subject of Sec. 13.5, where we also consider fragmented condensates.

13.1 Interference of two condensates

One of the striking manifestations of the wave nature of Bose–Einstein condensates is the observation of an interference pattern when two condensed and initially separated clouds are allowed to overlap [1]. An example is shown in Fig. 13.1. The study of the evolution of two expanding clouds has given a deeper understanding of the conditions under which interference can

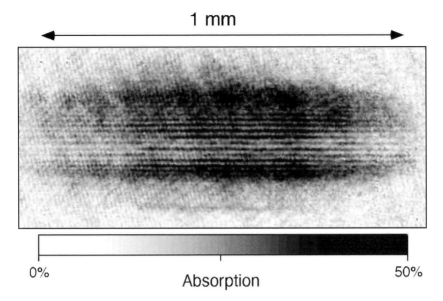

Fig. 13.1. Interference pattern formed by two overlapping clouds of sodium atoms. (From Ref. [1]).

arise. We begin by considering two clouds whose relative phase is locked, and then compare the result with that for two clouds with a fixed number of particles in each. Quite remarkably, an interference pattern appears even if the relative phase of the two clouds is not locked, as demonstrated in Refs. [2] and [3]. The presentation here follows unpublished work by G. Baym, A. J. Leggett, and C. J. Pethick.

13.1.1 Phase-locked sources

When the phase difference between two radio stations transmitting at the same frequency is fixed, the intensity of the signal at any point exhibits an interference pattern that depends on the position of the receiver. Since in a quantum-mechanical description electromagnetic waves are composed of photons, the interference pattern arises as a result of interference between photons from the two transmitters.

For two Bose–Einstein-condensed clouds a similar result holds if the relative phase of the two clouds is locked. Let us imagine that particles in the two clouds, which are assumed not to overlap initially, are described by single-particle wave functions $\psi_1(\mathbf{r}, t)$ and $\psi_2(\mathbf{r}, t)$, where the labels 1 and 2 refer to the two clouds.[1] If there is coherence between the clouds, the state may be described by a single condensate wave function, which must be of the form

$$\psi(\mathbf{r}, t) = \sqrt{N_1}\psi_1(\mathbf{r}, t) + \sqrt{N_2}\psi_2(\mathbf{r}, t), \tag{13.1}$$

where N_1 and N_2 denote the expectation values of the numbers of particles in the two clouds. For electromagnetic radiation the analogous result is that the electromagnetic field at any point is the sum of the fields produced by the two sources separately. Upon expansion the two condensates overlap and interfere, and if the effects of particle interactions in the overlap region can be neglected, the particle density at any point is given by

$$\begin{aligned} n(\mathbf{r}, t) &= |\psi(\mathbf{r}, t)|^2 = |\sqrt{N_1}\psi_1(\mathbf{r}, t) + \sqrt{N_2}\psi_2(\mathbf{r}, t)|^2 \\ &= N_1|\psi_1(\mathbf{r}, t)|^2 + N_2|\psi_2(\mathbf{r}, t)|^2 + 2\sqrt{N_1 N_2}\,\mathrm{Re}[\psi_1(\mathbf{r}, t)\psi_2^*(\mathbf{r}, t)], \end{aligned} \tag{13.2}$$

where Re denotes the real part. As a consequence of the last term in this expression, the density displays an interference pattern due to the spatial dependence of the phases of the wave functions for the individual clouds. For example, if the clouds are initially Gaussian wave packets of width R_0 centred on the points $\mathbf{r} = \pm \mathbf{d}/2$ and if the effects of particle interactions and external potentials can be neglected, the wave functions are

$$\psi_1 = \frac{e^{i\phi_1}}{(\pi R_t^2)^{3/4}} \exp\left[-\frac{(\mathbf{r} - \mathbf{d}/2)^2(1 + i\hbar t/mR_0^2)}{2R_t^2}\right] \tag{13.3}$$

and

$$\psi_2 = \frac{e^{i\phi_2}}{(\pi R_t^2)^{3/4}} \exp\left[-\frac{(\mathbf{r} + \mathbf{d}/2)^2(1 + i\hbar t/mR_0^2)}{2R_t^2}\right], \tag{13.4}$$

where m is the particle mass. Here ϕ_1 and ϕ_2 are the initial phases of the two condensates, and the width R_t of a packet at time t is given by

$$R_t^2 = R_0^2 + \left(\frac{\hbar t}{mR_0}\right)^2. \tag{13.5}$$

The results (13.3) and (13.4) are solutions of the Schrödinger equation for free particles, and correspond to the problem described in Sec. 7.5 when particle interactions may be neglected.

The interference term in Eq. (13.2) thus varies as

$$2\sqrt{N_1 N_2}\,\mathrm{Re}[\psi_1(\mathbf{r}, t)\psi_2^*(\mathbf{r}, t)] \sim A\cos\left(\frac{\hbar}{m}\frac{\mathbf{r}\cdot\mathbf{d}}{R_0^2 R_t^2}t + \phi_1 - \phi_2\right), \tag{13.6}$$

---

[1] In this chapter we denote the single-particle wave functions by ψ_1 and ψ_2 rather than by ϕ, as we did in Chapter 6.

where the prefactor A depends slowly on the spatial coordinates. Lines of constant phase are therefore perpendicular to the vector between the centres of the two clouds. The positions of the maxima depend on the relative phase of the two condensates, and if we take $\mathbf{d}$ to lie in the z direction, the distance between maxima is

$$\Delta z = 2\pi \frac{mR_t^2 R_0^2}{\hbar t d}. \qquad (13.7)$$

If the expansion time t is sufficiently large that the cloud has expanded to a size much greater than its original one, the radius is given approximately by $R_t \simeq \hbar t/mR_0$, and therefore the distance between maxima is given by the expression

$$\Delta z \simeq \frac{2\pi \hbar t}{md}. \qquad (13.8)$$

Physically this is the de Broglie wavelength associated with the momentum of a free particle that would travel the distance between the centres of the two clouds in the expansion time.

In the experiment [1] interference patterns were indeed observed. The expression (13.6) predicts that maxima and minima of the interference pattern will be planes perpendicular to the original separation of the two clouds, which is precisely what is seen in the experiments, as shown in Fig. 13.1. However, the existence of an interference pattern does not provide evidence for the phase coherence of the two clouds, since interference effects occur even if the two clouds are completely decoupled before they expand and overlap. Before explaining this we describe the above calculation in terms of many-particle states.

Phase states

The description in terms of the wave function of the condensed state is equivalent to the Hartree approximation, where all particles are in the same single-particle state. It is convenient to introduce states in which the phase difference between the single-particle wave functions for the two clouds has a definite value ϕ:

$$\psi_\phi(\mathbf{r}) = \frac{1}{\sqrt{2}}[\psi_1(\mathbf{r})e^{i\phi/2} + \psi_2(\mathbf{r})e^{-i\phi/2}]. \qquad (13.9)$$

The state used in the discussion above corresponds to the choice $\phi = \phi_1 - \phi_2$. The overall phase of the wave function plays no role, so we have put it equal to zero. The many-particle state with N particles in the state ψ_ϕ may be written as

$$|\phi, N\rangle = \frac{1}{(2^N N!)^{1/2}}(a_1^\dagger e^{i\phi/2} + a_2^\dagger e^{-i\phi/2})^N|0\rangle, \qquad (13.10)$$

where the operators $a_1^\dagger$ and $a_2^\dagger$ create particles in states for the clouds 1 and 2,

$$a_i^\dagger = \int d\mathbf{r}\, \psi_i(\mathbf{r})\hat\psi^\dagger(\mathbf{r}). \qquad (13.11)$$

If we neglect interactions between particles in one cloud with those in the other during the evolution of the state, the wave function at time t is given by

$$|\phi, N, t\rangle = \frac{1}{\sqrt{N!}}\left[\int d\mathbf{r}\,\psi_\phi(\mathbf{r},t)\hat\psi^\dagger(\mathbf{r})\right]^N|0\rangle, \qquad (13.12)$$

where the time evolution of the wave functions is to be calculated for each cloud separately.

The phase states form an overcomplete set. The overlap between two single-particle states with different phases is given by

$$\langle \phi', N=1|\phi, N=1\rangle = \int d\mathbf{r}\,\psi_{\phi'}^*(\mathbf{r},t)\psi_\phi(\mathbf{r},t). \qquad (13.13)$$

The integrand is

$$\begin{aligned}\psi_{\phi'}^*(\mathbf{r},t)\psi_\phi(\mathbf{r},t) &= \frac{1}{2}|\psi_1(\mathbf{r},t)|^2 e^{i(\phi-\phi')/2} + \frac{1}{2}|\psi_2(\mathbf{r},t)|^2 e^{-i(\phi-\phi')/2} \\ &\quad + \mathrm{Re}[\psi_1(\mathbf{r},t)\psi_2^*(\mathbf{r},t)e^{i(\phi+\phi')/2}] \\ &= \frac{1}{2}[|\psi_1(\mathbf{r},t)|^2 + |\psi_2(\mathbf{r},t)|^2]\cos[(\phi-\phi')/2] \\ &\quad + \frac{i}{2}[|\psi_1(\mathbf{r},t)|^2 - |\psi_2(\mathbf{r},t)|^2]\sin[(\phi-\phi')/2] \\ &\quad + \mathrm{Re}[\psi_1(\mathbf{r},t)\psi_2^*(\mathbf{r},t)e^{i(\phi+\phi')/2}]. \end{aligned} \qquad (13.14)$$

The overlap integral is obtained by integrating this expression over space. The last term in the integrand varies rapidly because of the spatial dependence of the phases of ψ_1 and ψ_2, and therefore it gives essentially zero on integration. The $\sin[(\phi-\phi')/2]$ term vanishes because of the normalization of the two states. Thus one finds for the overlap integral between two single-particle states

$$\langle \phi', N=1|\phi, N=1\rangle = \cos[(\phi-\phi')/2]. \qquad (13.15)$$

This has a maximum for $\phi=\phi'$, but there is significant overlap for $|\phi-\phi'|\sim \pi/2$.

13.1 Interference of two condensates

The overlap integral for N-particle phase states is the product of N such factors, and is therefore given by

$$\langle \phi', N | \phi, N \rangle = \cos^N[(\phi - \phi')/2]. \tag{13.16}$$

As N increases the overlap falls off more and more rapidly as $\phi - \phi'$ departs from zero. By using the identity $\lim_{N \to \infty}(1 - x/N)^N = e^{-x}$ one sees that the overlap integral falls off as $\exp[-N(\phi - \phi')^2/8]$ for phase differences of order $1/N^{1/2}$. Thus, while the overlap integral for a single particle depends smoothly on $\phi - \phi'$, that for a large number of particles becomes very small if the phase difference is greater than $\sim N^{-1/2}$. Consequently, states whose phases differ by more than $\sim N^{-1/2}$ are essentially orthogonal. This property will be very important in the next section, where we consider the relationship between phase states and states with a definite number of particles in each cloud initially.

As an example, let us evaluate the expectation value of the density operator in a phase state. Since the many-particle state (13.12) is properly normalized, the removal of a particle from it gives the normalized single-particle wave function ψ_ϕ times the usual factor $\sqrt{N}$,

$$\hat{\psi}(\mathbf{r})|\phi, N, t\rangle = \sqrt{N}\psi_\phi(\mathbf{r}, t)|\phi, N - 1, t\rangle. \tag{13.17}$$

The expectation value of the particle density $n(\mathbf{r})$ in the state $|\phi, N, t\rangle$ is therefore given by the product of (13.17) and its Hermitian conjugate,

$$n(\mathbf{r}, t) = \langle \phi, N, t | \hat{\psi}^\dagger(\mathbf{r})\hat{\psi}(\mathbf{r}) | \phi, N, t \rangle = \frac{N}{2}|\psi_1(\mathbf{r}, t)e^{i\phi/2} + \psi_2(\mathbf{r}, t)e^{-i\phi/2}|^2, \tag{13.18}$$

which contains an interference term

$$2\sqrt{N_1 N_2}\,\text{Re}[\psi_1(\mathbf{r}, t)\psi_2^*(\mathbf{r}, t)e^{i\phi}] \sim A\cos\left(\frac{\hbar}{m}\frac{\mathbf{r} \cdot \mathbf{d}}{R_0^2 R_t^2}t + \phi\right) \tag{13.19}$$

like that in Eq. (13.6).

13.1.2 Clouds with definite particle number

Let us now consider an initial state in which the numbers of particles N_1 and N_2 in each of the two clouds are fixed. The corresponding state vector is

$$|N_1, N_2\rangle = \frac{1}{\sqrt{N_1! N_2!}}(a_1^\dagger)^{N_1}(a_2^\dagger)^{N_2}|0\rangle, \tag{13.20}$$

which is referred to as a *Fock state*.

To begin with, we again consider the particle density. We calculate the expectation value of the density operator by using an identity analogous to (13.17),

$$\hat{\psi}(\mathbf{r})|N_1, N_2, t\rangle = \sqrt{N_1}\psi_1(\mathbf{r},t)|N_1-1, N_2, t\rangle + \sqrt{N_2}\psi_2(\mathbf{r},t)|N_1, N_2-1, t\rangle. \tag{13.21}$$

In the state $|N_1, N_2, t\rangle$ the expectation value of the density is therefore obtained by multiplying (13.21) by its Hermitian conjugate,

$$n(\mathbf{r}) = \langle N_1, N_2, t|\hat{\psi}^\dagger(\mathbf{r})\hat{\psi}(\mathbf{r})|N_1, N_2, t\rangle = N_1|\psi_1(\mathbf{r},t)|^2 + N_2|\psi_2(\mathbf{r},t)|^2, \tag{13.22}$$

which has no interference terms. However, this does not mean that there are no interference effects for a Fock state. The experimental situation differs from the one we have just treated in two respects. First, experiments on interference between two expanding clouds are of the 'one-shot' type: two clouds are prepared and allowed to expand, and the positions of atoms are then observed after a delay. However, according to the usual interpretation of quantum mechanics, a quantum-mechanical expectation value of an operator gives the *average* value for the corresponding physical quantity when the experiment is repeated many times. A second difference is that many particles are detected in experiments on condensates. Many-particle properties can exhibit interference effects even when single-particle properties do not, the most famous example being the correlation of intensities first discovered by Hanbury Brown and Twiss for electromagnetic radiation [4].

As an example of how interference effects appear even for states with a given number of particles in each cloud initially, we calculate the two-particle correlation function, which gives the amplitude for destroying particles at points $\mathbf{r}$ and $\mathbf{r}'$, and then creating them again at the same points. The correlation function is evaluated as before by expressing $\hat{\psi}(\mathbf{r})\hat{\psi}(\mathbf{r}')|N_1, N_2, t\rangle$ in terms of Fock states. This gives a linear combination of the orthogonal states $|N_1-2, N_2, t\rangle$, $|N_1-1, N_2-1, t\rangle$ and $|N_1, N_2-2, t\rangle$, and the resulting correlation function is

$$\begin{aligned}\langle N_1, N_2, t|\hat{\psi}^\dagger(\mathbf{r})\hat{\psi}^\dagger(\mathbf{r}')\hat{\psi}(\mathbf{r}')\hat{\psi}(\mathbf{r})|N_1, N_2, t\rangle = \\ [N_1|\psi_1(\mathbf{r},t)|^2 + N_2|\psi_2(\mathbf{r},t)|^2][N_1|\psi_1(\mathbf{r}',t)|^2 + N_2|\psi_2(\mathbf{r}',t)|^2] \\ - N_1|\psi_1(\mathbf{r},t)|^2|\psi_1(\mathbf{r}',t)|^2 - N_2|\psi_2(\mathbf{r},t)|^2|\psi_2(\mathbf{r}',t)|^2 \\ + 2N_1N_2\,\mathrm{Re}[\psi_1^*(\mathbf{r}',t)\psi_1(\mathbf{r},t)\psi_2^*(\mathbf{r},t)\psi_2(\mathbf{r}',t)].\end{aligned} \tag{13.23}$$

The correlation found by Hanbury Brown and Twiss is expressed in the last term in this expression. This result demonstrates that coherence between sources is not a prerequisite for interference effects.

As a simple example, let us assume that the amplitudes of the two wave

13.1 Interference of two condensates

functions are the same, but that their phases may be different:

$$\psi_i(\mathbf{r},t) = \psi_0 e^{i\phi_i(\mathbf{r},t)}. \tag{13.24}$$

The result (13.23) then becomes

$$\langle N_1, N_2, t | \hat{\psi}^\dagger(\mathbf{r})\hat{\psi}^\dagger(\mathbf{r}')\hat{\psi}(\mathbf{r}')\hat{\psi}(\mathbf{r}) | N_1, N_2, t \rangle =$$
$$N(N-1)|\psi_0|^4 + 2N_1 N_2 |\psi_0|^4 \cos[\Delta(\mathbf{r},t) - \Delta(\mathbf{r}',t)], \tag{13.25}$$

where $N = N_1 + N_2$ and

$$\Delta(\mathbf{r},t) = \phi_1(\mathbf{r},t) - \phi_2(\mathbf{r},t). \tag{13.26}$$

Let us now consider the expectation value of a more general operator $\mathcal{O}$ in a Fock state [3]. In order to relate results for this state to those for a phase state we expand the Fock state in terms of the phase states (13.9). In a phase state the component of the wave function having N_1 particles in cloud 1 and N_2 particles in cloud 2 has a phase factor $(N_1 - N_2)\phi/2$. To project this component out we multiply the phase state (13.9) by $e^{-i(N_1-N_2)\phi/2}$ and integrate over ϕ:

$$|N_1, N_2\rangle \propto \int_0^{2\pi} \frac{d\phi}{2\pi} e^{-i(N_1-N_2)\phi/2} |\phi, N\rangle. \tag{13.27}$$

From the normalization condition for a Fock state, $\langle N_1, N_2 | N_1, N_2 \rangle = 1$, one may calculate the constant of proportionality in the relation (13.27), and for N large and even, we may write the Fock state with equal numbers of particles in the two wells as

$$|N/2, N/2\rangle = \left(\frac{\pi N}{2}\right)^{1/4} \int_0^{2\pi} \frac{d\phi}{2\pi} |\phi, N\rangle. \tag{13.28}$$

Let us now express the expectation value of an operator $\mathcal{O}$ in the Fock state in terms of the phase states. This gives

$$\langle N/2, N/2 | \mathcal{O} | N/2, N/2 \rangle = \left(\frac{\pi N}{2}\right)^{1/2} \int_0^{2\pi} \frac{d\phi}{2\pi} \int_0^{2\pi} \frac{d\phi'}{2\pi} \langle \phi', N | \mathcal{O} | \phi, N \rangle. \tag{13.29}$$

The matrix element (13.29) may be written as

$$\langle N/2, N/2 | \mathcal{O} | N/2, N/2 \rangle = \int_0^{2\pi} \frac{d\phi}{2\pi} O(\phi) \tag{13.30}$$

where

$$O(\phi) = \left(\frac{\pi N}{2}\right)^{1/2} \int_0^{2\pi} \frac{d\phi'}{2\pi} \langle \phi', N | \mathcal{O} | \phi, N \rangle. \tag{13.31}$$

For many physically important operators the matrix elements $\langle \phi', N | \mathcal{O} | \phi, N \rangle$ between phase states have the property that they, like the overlap integral (13.16), are essentially diagonal in ϕ. As an example, we consider the operator $\mathcal{O}$ to be the correlation function for k particles being detected at different points $\mathbf{r}_1, \ldots, \mathbf{r}_k$:

$$\mathcal{O} = \prod_{i=1}^{k} \hat{\psi}^\dagger(\mathbf{r}_i) \hat{\psi}(\mathbf{r}_i). \tag{13.32}$$

The matrix element is given by

$$\langle \phi', N | \prod_{i=1}^{k} \hat{\psi}^\dagger(\mathbf{r}_i) \hat{\psi}(\mathbf{r}_i) | \phi, N \rangle = N^k \prod_{i=1}^{k} \psi_{\phi'}^*(\mathbf{r}_i) \psi_\phi(\mathbf{r}_i) \prod_{j=k+1}^{N} \int d\mathbf{r}_j \psi_{\phi'}^*(\mathbf{r}_j) \psi_\phi(\mathbf{r}_j)$$

$$= N^k \cos^{N-k}[(\phi - \phi')/2] \prod_{i=1}^{k} \psi_{\phi'}^*(\mathbf{r}_i) \psi_\phi(\mathbf{r}_i). \tag{13.33}$$

Apart from the factor N^k, this matrix element is identical with the overlap integral, except that the k coordinates of the particles that are detected must not be integrated over. The product of single-particle wave functions is given by Eq. (13.14), and therefore for small phase differences the matrix element (13.33) falls off rapidly when $|\phi - \phi'|$ increases from zero, just as does the integrated result that gives the overlap integral for many-particle states.

This result is remarkable, because it implies that, when the number of particles N is large, operators of the type we have considered are almost diagonal in the phase. Note that this property does not hold for all operators. For example, the non-local operator

$$\mathcal{O} = \prod_{i=1}^{N} \int d\mathbf{r}_i \hat{\psi}^\dagger(\mathbf{r}_i + \mathbf{a}) \hat{\psi}(\mathbf{r}_i) \tag{13.34}$$

that translates all particles by a constant amount $\mathbf{a}$ will generally not have a maximum for zero phase difference.

We now return to operators of the type given in Eq. (13.32). What we shall now argue is that when the positions of a large number of particles are measured in a single-shot experiment, their distribution will correspond to that for a phase state. Imagine that a particle is detected at point $\mathbf{r}_1$. The action of the annihilation operator on a Fock state is given by

$$\hat{\psi}(\mathbf{r}_1) | N \rangle = \int_0^{2\pi} \frac{d\phi}{2\pi} \psi_\phi(\mathbf{r}_1, t) | \phi, N-1 \rangle. \tag{13.35}$$

The phase states which are weighted most heavily in this expression are those for which $|\psi_\phi(\mathbf{r}_1, t)|$ has the largest value. Let us denote this value by ϕ_0. When just one particle is detected, the distribution in phase of the states is very broad, and the width in ϕ is comparable to π. If the wave functions ψ_1 and ψ_2 are equal in the overlap region, the amplitude varies as $\cos\{[\Delta(\mathbf{r}_1, t) + \phi]/2\}$. When a second particle is detected, there is a higher probability that it too will have a distribution corresponding to the phase states which are weighted most strongly in the state (13.35). Thus it will tend to be at points at which the modulus of the single-particle phase states with $\phi \approx \phi_0$ is largest. When more particles are detected, the distribution in phase becomes narrower and narrower, in essentially the same way as it did in our discussion of the overlap integral. Thus the greater the number of particles detected, the smaller the spread in phase of the final state, even though the initial state had a completely random distribution of phases. In a given shot of the experiment, the phase of the state becomes better and better defined as the number of detected particles increases. However, the particular value of the phase about which the components of the state are centred is random from one shot to another. Detailed calculations that exhibit this effect quantitatively may be found in Refs. [2] and [3].

A close parallel is provided by the Stern–Gerlach experiment. Atoms from an unpolarized source pass through an inhomogeneous magnetic field. Subsequent detection of the atoms shows that they always have projections of the magnetic moment along the magnetic field that correspond to eigenvalues of the component of the magnetic moment operator in that direction. Similarly, measurement of a many-particle correlation function in the interference experiment will always give a result corresponding to an eigenvalue of the corresponding operator. The eigenfunctions of the many-particle correlation operators are well localized in phase, and therefore the correlation function measured in a single-shot experiment corresponds with high probability to a state made up of phase states with a narrow distribution in phase.

We therefore conclude that the distribution of particles detected in a one-shot experiment in which two initially isolated clouds with definite numbers of particles are allowed to overlap corresponds to the distribution for a state with two clouds having a well-defined phase difference. To be able experimentally to demonstrate the difference between the theoretical results for a Fock state and those for two clouds with a definite phase difference, it is necessary to be able to control the relative phase of two coupled clouds, and to measure accurately the positions of the maxima in the interference pattern.

The above results are important because they demonstrate that many results for Fock states are essentially the same as those for a phase state. In calculations it is common to work with states that correspond to phase states and to talk about a broken gauge symmetry of the condensate. In the above example the gauge angle is the relative phase of the two clouds. However, the above discussion demonstrates that, for the class of operators considered, results for Fock states are equivalent to those for phase states when the number of particles is large. The discussion of the excitations of the uniform Bose gas in Sec. 8.1 is another example that illustrates this point. The traditional approach to the microscopic theory is to assume that the expectation value of the particle creation and annihilation operators have non-zero values. This corresponds to using phase states. However, the discussion at the end of Sec. 8.1 showed that essentially the same results are obtained by working with states having a definite particle number. In the context of quantum optics, Mølmer arrived at similar conclusions regarding the equivalence of results for phase states and Fock states [5].

13.2 Density correlations in Bose gases

In the previous section we considered the properties of states in which there were no thermal excitations, and we turn now to non-zero temperatures. There are striking differences between density correlations in a pure Bose–Einstein condensate and in a thermal gas. An example of this effect has already been discussed in Sec. 8.3, where we showed that the interaction energy per particle in a homogeneous thermal gas above the Bose–Einstein transition temperature is twice its value in a pure condensate of the same density. To understand this result, we express it in terms of the two-particle correlation function used in the discussion of the interference experiment, and we write the interaction energy as

$$E_{\text{int}} = \frac{U_0 V}{2} <\hat{\psi}^\dagger(\mathbf{r})\hat{\psi}^\dagger(\mathbf{r})\hat{\psi}(\mathbf{r})\hat{\psi}(\mathbf{r})>. \tag{13.36}$$

Here $<\cdots>$ denotes an expectation value in the state under consideration.[2] The correlation function is a measure of the probability that two atoms are at the same point in space. Its relationship to the density–density correlation

---

[2] The correlation functions considered in this section are ones which contain the effects of the interparticle interactions only in a mean-field sense, as in the Hartree and Hartree–Fock approaches. The true correlation functions have a short-distance structure that reflects the behaviour of the wave function for a pair of interacting particles and, for alkali atoms, for which there are many nodes in the two-body zero-energy wave function, this gives rise to rapid oscillations of correlation functions at short distances.

13.2 Density correlations in Bose gases

function may be brought out by using the commutation relations for creation and annihilation operators to write

$$<\hat{\psi}^\dagger(\mathbf{r})\hat{\psi}^\dagger(\mathbf{r}')\hat{\psi}(\mathbf{r}')\hat{\psi}(\mathbf{r})> = \\ <\hat{\psi}^\dagger(\mathbf{r})\hat{\psi}(\mathbf{r})\hat{\psi}^\dagger(\mathbf{r}')\hat{\psi}(\mathbf{r}')> - <\hat{\psi}^\dagger(\mathbf{r})\hat{\psi}(\mathbf{r})>\delta(\mathbf{r}-\mathbf{r}'). \quad (13.37)$$

This relation expresses the fact that the pair correlation function is obtained from the instantaneous density–density correlation function (the first term on the right hand side) by removing the contribution from correlations of one atom with itself (the second term on the right hand side).

As a dimensionless measure of correlations between different atoms we introduce the *pair distribution function* $g_2(\mathbf{r}, \mathbf{r}')$. For a spatially uniform system this is defined by the equation

$$g_2(\mathbf{r}, \mathbf{r}') = \frac{V^2}{N(N-1)}<\hat{\psi}^\dagger(\mathbf{r})\hat{\psi}^\dagger(\mathbf{r}')\hat{\psi}(\mathbf{r}')\hat{\psi}(\mathbf{r})>, \quad (13.38)$$

where V is the volume of the system. The normalization factor reflects the fact that the integral of the pair distribution function over both coordinates is $N(N-1)$. For large $|\mathbf{r}-\mathbf{r}'|$, correlations are negligible and therefore $<\hat{\psi}^\dagger(\mathbf{r})\hat{\psi}^\dagger(\mathbf{r}')\hat{\psi}(\mathbf{r}')\hat{\psi}(\mathbf{r})> \to (N/V)^2$ where N/V is the average density. Consequently, g_2 tends to 1 to within terms of order $1/N$, which may be neglected in the following discussion since N is large.

Thus, from the difference by a factor of two in the interaction energies of a condensate and of a gas above T_c we conclude that for small spatial separations the pair distribution function in the non-condensed state is *twice* that for a pure condensate of the same density. This result may be understood by examining the expectation value in the two situations. For a pure condensate,

$$<\hat{\psi}^\dagger(\mathbf{r})\hat{\psi}^\dagger(\mathbf{r})\hat{\psi}(\mathbf{r})\hat{\psi}(\mathbf{r})> = N(N-1)|\phi(\mathbf{r})|^4 = \frac{N(N-1)}{V^2} \simeq n_0^2 = n^2. \quad (13.39)$$

For a gas above T_c one may use the Hartree–Fock approximation or neglect interactions altogether, and we therefore expand the field operators in terms of creation and annihilation operators for atoms in plane-wave states. By the methods used in Sec. 8.3.1 one finds that

$$<\hat{\psi}^\dagger(\mathbf{r})\hat{\psi}^\dagger(\mathbf{r})\hat{\psi}(\mathbf{r})\hat{\psi}(\mathbf{r})> = 2n^2, \quad (13.40)$$

where we have neglected terms of relative order $1/N$. The factor of 2 is due to there being both direct and exchange contributions for a thermal gas, as we found in the calculations of the energy in Sec. 8.3.1.

More generally, if both a condensate and thermal excitations are present, one finds in the Hartree–Fock approximation that

$$<\hat{\psi}^\dagger(\mathbf{r})\hat{\psi}^\dagger(\mathbf{r})\hat{\psi}(\mathbf{r})\hat{\psi}(\mathbf{r})> = n_0^2 + 4n_0 n_\text{ex} + 2n_\text{ex}^2. \tag{13.41}$$

For temperatures below $T_0 \sim nU_0/k$ the correlations of thermal excitations must be calculated using Bogoliubov theory rather than the Hartree–Fock approximation, but at such temperatures the contributions of thermal excitations to the pair distribution function are small.

Many-particle distribution functions are also affected by Bose–Einstein condensation. As an example, we consider the rate of three-body processes discussed in Sec. 5.4.1 [6]. At low temperatures it is a good approximation to ignore the dependence of the rate of the process on the energies of the particles, and therefore the rate is proportional to the probability of finding three bosons at essentially the same point, that is

$$g_3(0) = <(\hat{\psi}^\dagger(\mathbf{r}))^3 (\hat{\psi}(\mathbf{r}))^3>. \tag{13.42}$$

This may be evaluated in a similar way to the density–density correlation function, and for the condensed state it is

$$g_3(0) = n_0^3 = n^3, \tag{13.43}$$

while above T_c it is

$$g_3(0) = 6n^3. \tag{13.44}$$

The factor 6 here is 3!, the number of ways of pairing up the three creation operators with the three annihilation operators in $g_3(0)$. The calculation of $g_3(0)$ for the more general situation when both condensate and thermal particles are present is left as an exercise (Problem 13.1). The results (13.43) and (13.44) show that, for a given density, the rate of three-body processes in the gas above T_c is 6 times that in a pure condensate. This reflects the strong suppression of correlations in the condensed phase. Such a reduction in the rate of three-body recombination was found in experiments on ^{87}Rb [7].

For the general correlation function $g_\nu = <(\hat{\psi}^\dagger(\mathbf{r}))^\nu (\hat{\psi}(\mathbf{r}))^\nu>$, the differences between results for the condensed and non-condensed states are a factor $\nu!$, and therefore even more dramatic for large values of ν.

13.3 Coherent matter wave optics

A wide range of experiments has been carried out on Bose–Einstein condensates to explore properties of coherent matter waves. Forces on atoms due to laser light described in Sec. 4.2 are an essential ingredient in most of them, and in this section we describe briefly a number of applications. One is phase imprinting. By applying a laser pulse whose intensity varies over the atomic cloud, one may modulate the phase of the condensate, and this effect has been used to generate solitons. A second effect is Bragg diffraction of matter waves by a 'lattice' made by two overlapping laser beams. The final example is non-linear mixing of matter waves.

Phase imprinting

To understand how the phase of a condensate may be altered, consider illuminating a condensate with a pulse of radiation whose intensity varies in space. As described in Chapters 3 and 4, the radiation shifts the energy of an atom, and thereby causes the phase of the wave function to advance at a rate different from that in the absence of the pulse. It follows from the equation of motion for the phase, Eq. (7.19), that the extra contribution $\Delta\phi(\mathbf{r},t)$ to the phase of the wave function satisfies the equation

$$\frac{\partial \Delta\phi(\mathbf{r},t)}{\partial t} = -\frac{V(\mathbf{r},t)}{\hbar}, \tag{13.45}$$

where $V(\mathbf{r},t)$ is the energy shift produced by the radiation field, Eq. (4.30). In writing this equation we have neglected the contribution due to the velocity of the atoms, and this is a good approximation provided the duration τ of the pulse is sufficiently short, and provided that the velocity is sufficiently small initially. The additional phase is then given by

$$\Delta\phi(\mathbf{r}) = -\frac{1}{\hbar}\int_0^\tau dt V(\mathbf{r},t), \tag{13.46}$$

where we have assumed that the pulse starts at $t=0$. Since the potential energy of an atom is proportional to the intensity of the radiation, spatial variation of the intensity of the radiation field makes the phase of the wave function depend on space. Phase imprinting does not require a condensate, and it works also for atoms that are sufficiently cold that their thermal motion plays a negligible role.

Phase imprinting has been used to create solitons in condensates. As we have seen in Sec. 7.6 the non-linear Gross–Pitaevskii equation has solutions in which a disturbance moves at constant speed u without changing its form, see (7.151). The phase difference across the soliton is given by Eq. (7.154) and most of the phase change occurs over a distance ξ_u, given by (7.152), from the centre of the soliton. Such an abrupt change in the phase of the condensate wave function may be produced by illuminating an initially uniform condensate with a pulse of laser light whose intensity is spatially uniform for, say, $x > 0$ and zero otherwise. The phase of the condensate wave function for $x > 0$ will be changed relative to that for $x < 0$, and the magnitude of the phase difference may be adjusted to any required value by an appropriate choice of the properties of the pulse. The distance over which the change in phase occurs is determined by how sharply the illumination can be cut off spatially. Immediately after applying the pulse the behaviour of the phase is therefore similar to that in the soliton whose centre is on the plane $x = 0$. However the density is essentially uniform, without the dip present in the soliton solution. The velocity distribution immediately following the laser pulse is proportional to the gradient of the phase imprinted. This leads to a density disturbance that moves in the positive x direction if the energy shift is negative. In addition, there is a soliton that moves towards negative values of x. Solitons have been generated by the above methods in experiments at the National Institute of Standards and Technology (NIST) [8] and in Hannover [9].

Our discussion in the previous section was appropriate for radiation fields produced by propagating electromagnetic waves, and one may ask what happens if a standing wave is applied. If the intensity of the wave varies as $\sin^2 qz$, this gives an energy shift of the form $\sin^2 qz$ according to Eq. (3.50). For short pulses, this will impose on the condensate a phase variation proportional to $\sin^2 qz$ and therefore a velocity variation equal to $(\hbar/m)\partial\phi/\partial z \propto \sin 2qz$. With time this velocity field will produce density variations of the same wavelength. For longer pulses, a density wave is created with an amplitude that oscillates as a function of the duration of the pulse (Problem 13.3). These density fluctuations behave as a grating, and may thus be detected by diffraction of light, as has been done experimentally [10].

Bragg diffraction of matter waves

The phenomenon of Bragg diffraction of electromagnetic radiation is well known and widely applied in determining crystal structures. We now consider its analogue for particles. Imagine projecting a Bose–Einstein condensate at a periodic potential produced by a standing electromagnetic wave. This gives rise to a diffracted wave with the same wave number as that of the particles in the original cloud, and its direction of propagation is determined by the usual Bragg condition. In practice, it is more convenient to adopt a method that corresponds to performing a Galilean transformation on the above process. One applies to a cloud of condensate initially at rest a potential having the form of a travelling wave produced by superimposing two laser beams with wave vectors $\mathbf{q}_1$ and $\mathbf{q}_2$ and frequencies ω_1 and ω_2.

The amplitudes of the electric fields in the two beams are proportional to $\cos(\mathbf{q}_1\cdot\mathbf{r} - \omega_1 t)$ and $\cos(\mathbf{q}_2\cdot\mathbf{r} - \omega_2 t)$, where we have omitted arbitrary phases. The effective potential acting on an atom may be calculated by second-order perturbation theory, just as we did in the discussion of Sisyphus cooling in Chapter 4. It contains the product of the amplitudes of the two waves and therefore has components with frequencies and wave vectors given by

$$\omega = \omega_1 \pm \omega_2 \quad \text{and} \quad \mathbf{q} = \mathbf{q}_1 \pm \mathbf{q}_2. \tag{13.47}$$

The term proportional to $\cos[(\mathbf{q}_1 - \mathbf{q}_2)\cdot\mathbf{r} - (\omega_1 - \omega_2)t]$ describes a Raman process in which one photon is emitted and one of another frequency absorbed. It has the important property that its frequency and wave vector may be tuned by appropriate choice of the frequencies and directions of propagation of the beams. When the frequencies of the two beams are the same, the potential is static, as we described in detail in Chapter 4, while the time-dependent potential has a wave vector $\mathbf{q}_1 - \mathbf{q}_2$ and moves at a velocity $(\omega_1 - \omega_2)/|\mathbf{q}_1 - \mathbf{q}_2|$. This can act as a diffraction grating for matter waves. The term with the sum of frequencies was considered in the study of two-photon absorption in Sec. 8.4. It is of very high frequency, and is not of interest here.

Let us, for simplicity, consider a non-interacting Bose gas. Bragg diffraction of a condensate with momentum zero will be kinematically possible if energy and momentum can be conserved. This requires that the energy difference between the photons in the two beams must be equal to the energy of an atom with momentum $\hbar(\mathbf{q}_1 - \mathbf{q}_2)$, or

$$\hbar(\omega_1 - \omega_2) = \frac{\hbar^2(\mathbf{q}_1 - \mathbf{q}_2)^2}{2m}. \tag{13.48}$$

This result is equivalent to the usual Bragg diffraction condition for a static grating. When interactions between atoms are taken into account, the free-particle dispersion relation must be replaced by the Bogoliubov one $\epsilon_\mathbf{q}$, Eq. (7.31) for $\mathbf{q} = \mathbf{q}_1 - \mathbf{q}_2$. This type of Bragg spectroscopy has been used to measure the structure factor of a Bose–Einstein condensate in the phonon regime [11].

By the above method it is possible to scatter a large fraction of a condensate into a different momentum state. Initially the two components of the condensate overlap in space, but as time elapses, they will separate as a consequence of their different velocities. This technique was used in the experiment on non-linear mixing of matter waves that we now describe.

Four-wave mixing

By combining laser beams with different frequencies and different directions it is possible to make condensed clouds with components having a number of different momenta. For example, by making two gratings with wave vectors $\mathbf{q}_a$ and $\mathbf{q}_b$ one can make a cloud with components having momenta zero (the original cloud), $\pm\hbar\mathbf{q}_a$ and $\pm\hbar\mathbf{q}_b$. As a consequence of the interaction between particles, which is due to the term $U_0|\psi(\mathbf{r})|^4/2$ in the expression for the energy density, there is a non-linear mixing of matter waves. In particular the three overlapping components above can produce a fourth beam. For example, two atoms with momenta zero and $\hbar\mathbf{q}_a$ can be scattered to states with momenta $\hbar\mathbf{q}_b$ and $\hbar(\mathbf{q}_a - \mathbf{q}_b)$, thereby producing a new beam with the latter momentum. This is a four-wave mixing process analogous to that familiar in optics. For the process to occur, the laser frequencies and beam directions must be chosen so that energy and momentum are conserved. Such experiments have been performed on Bose–Einstein condensed clouds, and the fourth beam was estimated to contain up to 11% of the total number of atoms [12]. This yield is impressive compared with what is possible in optics, and it reflects the strongly non-linear nature of Bose–Einstein condensates.

13.4 The atom laser

The phrase 'atom laser' is used to describe sources of coherent matter waves. In the first such sources, a coherent beam of particles was extracted from a Bose–Einstein condensate either as a series of pulses [13, 14] or continuously [15]. This way of producing a coherent beam is referred to as *output coupling*. The optical analogue of these devices is a resonant cavity highly excited in one mode, and from which radiation is allowed to leak out, thereby generating a coherent beam of photons. In optical terminology the latter device would hardly be called a laser, and thus when applied to matter waves the word 'laser' is used in a broader sense than in optics.

Phase-coherent amplification of matter waves is an effect somewhat closer to what occurs in an optical laser, and this has been observed experimentally [16]. A fundamental difference between atom lasers and optical lasers is related to the fact that the number of atoms is conserved, while the number of photons is not. The active medium of a matter-wave amplifier must therefore include a reservoir of atoms. In the experiment described in Ref. [16] phase-coherent amplification was achieved using a Bose–Einstein condensate of sodium atoms. The moving coherent matter wave to be amplified, the input wave, was generated by Bragg diffraction, as described in

Sec. 13.3. This was done by applying a pulse of radiation in two mutually perpendicular laser beams with wave vectors $\mathbf{q}_1$ and $\mathbf{q}_2$, which transferred a small fraction (less than 1%) of the condensate atoms into the input matter wave, whose wave vector is $\mathbf{q}_1 - \mathbf{q}_2$. To ensure that the energy conservation condition (13.48) for Bragg diffraction is satisfied, the frequency of one of the laser beams was detuned slightly with respect to the other.

The input matter wave is amplified by applying another pulse from one of the two laser beams used to produce the input wave. The effect of this second pulse on the part of the condensate at rest is to mix into the condensate wave function a component in which atoms are in the excited state, and have momentum equal to that of the absorbed photon. By emitting a photon, this excited-state component can decay to give ground-state atoms with a range of momenta, provided the Bragg frequency condition is satisfied. Final states in which the momentum of the atom is equal to that of the input beam will be favoured because of stimulated emission, which arises because the matrix element for an atom in the excited state to make a transition to the ground state with momentum $\mathbf{p} = \hbar \mathbf{q}$ is proportional to the matrix element of the creation operator for an atom with momentum $\mathbf{p}$. This matrix element is $\sqrt{N_\mathbf{p} + 1}$, where $N_\mathbf{p}$ denotes the occupation number. Therefore the rate of the process is proportional to $N_\mathbf{p} + 1$. It is thus enhanced as a consequence of the Bose statistics, and the input beam is amplified. Transitions to other momentum states are not correspondingly enhanced.

13.5 The criterion for Bose–Einstein condensation

Since the discovery of Bose–Einstein condensation in dilute gases, renewed attention has been given to the question of how it may be characterized microscopically. For non-interacting particles, the criterion for Bose–Einstein condensation is that the occupation number for one of the single-particle energy levels should be macroscopic. The question is how to generalize this condition to interacting systems.

A criterion for bulk systems was proposed by Penrose [17] and Landau and Lifshitz [18], and subsequently elaborated by Penrose and Onsager [19] and by Yang [20]. Consider the one-particle density matrix[3]

$$\rho(\mathbf{r}, \mathbf{r}') \equiv <\hat{\psi}^\dagger(\mathbf{r}')\hat{\psi}(\mathbf{r})>, \qquad (13.49)$$

which gives the amplitude for removing a particle at $\mathbf{r}$ and creating one at $\mathbf{r}'$. Bose–Einstein condensation is signalled by $\rho(\mathbf{r}, \mathbf{r}')$ tending to a constant

---

[3] With this definition, the density matrix satisfies the normalization condition $\int d\mathbf{r}\rho(\mathbf{r}, \mathbf{r}) = N$. Note that other choices of normalization exist in the literature.

as $|\mathbf{r} - \mathbf{r}'| \to \infty$. For the uniform ideal Bose gas, the eigenstates are plane waves $e^{i\mathbf{p}\cdot\mathbf{r}/\hbar}/V^{1/2}$ with occupancy $N_\mathbf{p}$, and the one-particle density matrix is given by

$$\rho(\mathbf{r},\mathbf{r}') = \frac{1}{V}\sum_\mathbf{p} N_\mathbf{p} e^{i\mathbf{p}\cdot(\mathbf{r}-\mathbf{r}')/\hbar}. \tag{13.50}$$

For large $|\mathbf{r} - \mathbf{r}'|$, the only term that survives is the one for the zero-momentum state, since the contribution from all other states tends to zero because of the interference between different components, and therefore

$$\lim_{|\mathbf{r}-\mathbf{r}'|\to\infty} \rho(\mathbf{r},\mathbf{r}') = \frac{N_0}{V}. \tag{13.51}$$

For an interacting system, the energy eigenstates are not generally eigenstates of the operator for the number of zero-momentum particles, but one can write

$$\lim_{|\mathbf{r}-\mathbf{r}'|\to\infty} \rho(\mathbf{r},\mathbf{r}') = \frac{<N_0>}{V}, \tag{13.52}$$

where $<N_0>$ denotes the expectation value of the occupation number of the zero-momentum state.

In finite systems such as gas clouds in traps, it makes no sense to take the limit of large separations between the two arguments of the single-particle density matrix, so it is customary to adopt a different procedure and expand the density matrix in terms of its eigenfunctions $\chi_j(\mathbf{r})$, which satisfy the equation

$$\int d\mathbf{r}' \rho(\mathbf{r},\mathbf{r}')\chi_j(\mathbf{r}') = \lambda_j \chi_j(\mathbf{r}). \tag{13.53}$$

Since the density matrix is Hermitian and positive definite, its eigenvalues λ_j are real and positive and

$$\rho(\mathbf{r},\mathbf{r}') = \sum_j \lambda_j \chi_j^*(\mathbf{r}')\chi_j(\mathbf{r}). \tag{13.54}$$

For a non-interacting gas in a potential, the χ_j are the single-particle wave functions, and the eigenvalues are the corresponding occupation numbers. At zero temperature, the eigenvalue for the lowest single-particle state is N, and the others vanish. For interacting systems, the natural generalization of the condition for the non-interacting gas that the occupation number of one state be macroscopic is that one of the eigenvalues λ_j be of order N, while the others are finite in the limit $N \to \infty$.

13.5.1 Fragmented condensates

In bulk systems, one usually regards Bose–Einstein condensation as being characterized by the macroscopic occupation of *one* single-particle state. As a simple example of a more complicated situation in which two single-particle states are macroscopically occupied, consider atoms in two potential wells so far apart that the wave functions of particles in the two traps do not overlap. If there are N_1 particles in the ground state of the first well, and N_2 in the ground state of the second one, two single-particle states, one in each well, are macroscopically occupied. The state is given by Eq. (13.20), and the density matrix for it is

$$\rho(\mathbf{r},\mathbf{r}') = N_1 \psi_1^*(\mathbf{r}')\psi_1(\mathbf{r}) + N_2 \psi_2^*(\mathbf{r}')\psi_2(\mathbf{r}), \tag{13.55}$$

which has two large eigenvalues, assuming N_1 and N_2 are both large. Here $\psi_1(\mathbf{r})$ and $\psi_2(\mathbf{r})$ are the ground-state single-particle wave functions for the two wells.

One may ask whether there are more general condensates with properties similar to the two-well system considered above. One example arose in studies of possible Bose–Einstein condensation of excitons [21], but it is of interest more generally. The energy of a uniform system of bosons in the Hartree–Fock approximation, Eq. (8.74), is

$$E = \sum_{\mathbf{p}} \epsilon_p^0 N_\mathbf{p} + \frac{U_0}{2V} N(N-1) + \frac{U_0}{2V} \sum_{\mathbf{p},\mathbf{p}'(\mathbf{p}\neq\mathbf{p}')} N_\mathbf{p} N_{\mathbf{p}'}. \tag{13.56}$$

This shows that for a repulsive interaction the lowest-energy state of the system has all particles in the zero-momentum state, since the Fock term, the last term in (13.56), can only increase the energy. However, for an attractive interaction the Fock term, which is absent if all particles are in the same state, lowers the energy, rather than raises it as it does for repulsive forces. The interaction energy is minimized by distributing the particles over as many states as possible [21].[4] Nozières and Saint James referred to such a state as a *fragmented condensate*. However, they argued that fragmented condensates are not physically relevant for bulk matter because such states would be unstable with respect to collapse.

Another example of a fragmented condensate is the ground state of the dilute spin-1 Bose gas with antiferromagnetic interactions. This problem can be solved exactly, as we have seen in Sec. 12.2.2. The ground state for N particles is $N/2$ pairs of atoms, each pair having zero angular momentum. All particles have zero kinetic energy. The pairing to zero total spin

---

[4] Note that according to Eq. (8.75) the interaction energy is minimized for $U_0 < 0$, when $\sum_\mathbf{p} N_\mathbf{p}^2$ is made as small as possible.

is physically understandable since the energy due to the antiferromagnetic interaction is minimized by aligning the spins of a pair of atoms oppositely. The ground state is

$$|\Psi\rangle \propto (\hat{A}^\dagger)^{N/2}|0\rangle, \qquad (13.57)$$

where $\hat{A}$ is given by (12.57). This state cannot be written in the form $(\hat{a}^\dagger)^N|0\rangle$, where $\hat{a}^\dagger$ is the creation operator for an atom in some state, and therefore it is quite different from the usual Bose–Einstein condensate of atoms in the same internal state. Rather, it is a Bose–Einstein condensate of *pairs* of atoms in the state with zero momentum. The state resembles in this respect the ground state of a BCS superconductor, but whereas the two electrons in a pair in a superconductor are correlated in space, the two bosons in a pair are not. The one-particle density matrix for the original bosons has three eigenvalues equal to $N/3$, and therefore the state is a fragmented condensate (Problem 13.2).

The study of rotating Bose gases in traps has brought to light a number of other examples of fragmented condensates [22]. The simplest is the Bose gas with attractive interactions in a harmonic trap with an axis of symmetry. In the lowest state with a given angular momentum, all the angular momentum is carried by the centre-of-mass motion (the dipole mode), while the internal correlations are the same as in the ground state (see Sec. 9.3.1). The one-particle density matrix for this state generally has a number of large eigenvalues, not just one. One might be tempted to regard it as a fragmented condensate, but since the internal correlations are identical with those in the ground state, this conclusion is surprising. Recently it has been shown that a one-particle density matrix for this state defined using coordinates of particles relative to the centre of mass, rather than the coordinates relative to an origin fixed in space, has a single macroscopic eigenvalue [23]. This indicates that caution is required in formulating a criterion for Bose–Einstein condensation. It is also noteworthy that in a number of cases for which a fragmented condensate is more favourable than one with a single macroscopically-occupied state, the energy difference between the two states is of order $1/N$ compared with the total energy, and consequently the correlations in the fragmented condensate are very delicate.

The examples given above illustrate that Bose–Einstein condensation is a many-faceted phenomenon. The bosons that condense may be single particles, composite bosons made up of a pair of bosonic atoms, as in a gas of spin-1 bosons with antiferromagnetic interaction, or pairs of fermions, as in a superconductor. In addition, the eigenvalues of the one-particle density matrix for particles under rotation in a harmonic trap depend on the coordinates used. The above discussion indicates how the study of dilute Bose gases has deepened our understanding of some fundamental theoretical issues in Bose–Einstein condensation.

Problems

PROBLEM 13.1   Show that in the Hartree–Fock approximation the three-particle correlation function Eq. (13.42) is given by

$$g_3(0) = n_0^3 + 9n_0^2 n_{\text{ex}} + 18 n_0 n_{\text{ex}}^2 + 6 n_{\text{ex}}^3.$$

PROBLEM 13.2   Consider the ground state of an even number of spin-1 bosons with antiferromagnetic interactions for zero total spin. The state is given by (13.57). Show that the eigenvalues of the single-particle density matrix are $N/3$.

PROBLEM 13.3   In the electric field created by two opposed laser beams of the same frequency ω and the same polarization propagating in the $\pm z$ directions, the magnitude of the electric field varies as $\mathcal{E}_0 \cos\omega(t-t_0)\cos qz$. What is the effective potential acting on an atom? [Hint: Recall the results from Secs. 3.3 and 4.2.] Treating this potential as small, calculate the response of a uniform non-interacting Bose–Einstein condensate when the electric field is switched on at $t = t_0$ and switched off at $t = t_0 + \tau$, and show that the magnitude of the density wave induced in the condensate varies periodically with τ. Interpret your result for small τ in terms of phase imprinting and the normal modes of the condensate.

References

[1] M. R. Andrews, C. G. Townsend, H.-J. Miesner, D. S. Durfee, D. M. Kurn, and W. Ketterle, *Science* **275**, 637 (1997).

[2] J. Javanainen and S. M. Yoo, *Phys. Rev. Lett.* **76**, 161 (1996).

[3] Y. Castin and J. Dalibard, *Phys. Rev. A* **55**, 4330 (1997).

[4] R. Hanbury Brown and R. Q. Twiss, *Nature* **177**, 27 (1956).

[5] K. Mølmer, *Phys. Rev. A* **55**, 3195 (1997).

[6] Yu. Kagan, B. V. Svistunov, and G. V. Shlyapnikov, *Pis'ma Zh. Eksp. Teor. Fiz.* **42**, 169 (1985). [*JETP Lett.* **42**, 209 (1985).]

[7] E. A. Burt, R. W. Ghrist, C. J. Myatt, M. J. Holland, E. A. Cornell, and C. E. Wieman, *Phys. Rev. Lett.* **79**, 337 (1997).

[8] J. Denschlag, J. E. Simsarian, D. L. Feder, C. W. Clark, L. A. Collins, J. Cubizolles, L. Deng, E. W. Hagley, K. Helmerson, W. P. Reinhardt, S. L. Rolston, B. I. Schneider, and W. D. Phillips, *Science* **287**, 97 (2000).

[9] S. Burger, K. Bongs, S. Dettmer, W. Ertmer, K. Sengstock, A. Sanpera, G. V. Shlyapnikov, and M. Lewenstein, *Phys. Rev. Lett.* **83**, 5198 (1999).

[10] Yu. B. Ovchinnikov, J. H. Müller, M. R. Doery, E. J. D. Vredenbregt, K. Helmerson, S. L. Rolston, and W. D. Phillips, *Phys. Rev. Lett.* **83**, 284 (1999).

[11] D. M. Stamper-Kurn, A. P. Chikkatur, A. Görlitz, S. Inouye, S. Gupta, and W. Ketterle, *Phys. Rev. Lett.* **83**, 2876 (1999).

[12] L. Deng, E. W. Hagley, J. Wen, M. Trippenbach, Y. Band, P. S. Julienne, D. E. Pritchard, J. E. Simsarian, K. Helmerson, S. L. Rolston, and W. D. Phillips, *Nature* **398**, 218 (1999).

[13] M.-O. Mewes, M. R. Andrews, D. M. Kurn, D. S. Durfee, C. G. Townsend, and W. Ketterle, *Phys. Rev. Lett.* **78**, 582 (1997).

[14] B. P. Anderson and M. A. Kasevich, *Science* **282**, 1686 (1998).

[15] I. Bloch, T. W. Hänsch, and T. Esslinger, *Phys. Rev. Lett.* **82**, 3008 (1999).

[16] S. Inouye, T. Pfau, S. Gupta, A. P. Chikkatur, A. Görlitz, D. E. Pritchard, and W. Ketterle, *Nature* **402**, 641 (1999).

[17] O. Penrose, *Phil. Mag.* **42**, 1373 (1951).

[18] L. D. Landau and E. M. Lifshitz, *Statisticheskaya Fizika* (Fizmatgiz, Moscow, 1951) §133 [English translation: *Statistical Physics* (Pergamon, Oxford, 1958) §133].

[19] O. Penrose and L. Onsager, *Phys. Rev.* **104**, 576 (1956).

[20] C. N. Yang, *Rev. Mod. Phys.* **34**, 4 (1962).

[21] P. Nozières and D. Saint James, *J. Phys. (Paris)* **43**, 1133 (1982). See also P. Nozières, in *Bose–Einstein Condensation*, ed. A. Griffin, D. W. Snoke, and S. Stringari, (Cambridge Univ. Press, Cambridge, 1995), p. 15.

[22] N. K. Wilkin, J. M. F. Gunn, and R. A. Smith, *Phys. Rev. Lett.* **80**, 2265 (1998).

[23] C. J. Pethick and L. P. Pitaevskii, *Phys. Rev. A* **62**, 033609 (2000).

14
Fermions

The laser cooling mechanisms described in Chapter 4 operate irrespective of the statistics of the atom, and they can therefore be used to cool Fermi species. The statistics of a neutral atom is determined by the number of neutrons in the nucleus, which must be odd for a fermionic atom. Since alkali atoms have odd atomic number Z, their fermionic isotopes have even mass number A. Such isotopes are relatively less abundant than those with odd A since they have both an unpaired neutron and an unpaired proton, which increases their energy by virtue of the odd–even effect. To date, ^{40}K [1] and ^{6}Li atoms [2] have been cooled to about one-quarter of the Fermi temperature.

In the classical limit, at low densities and/or high temperatures, clouds of fermions and bosons behave alike. The factor governing the importance of quantum degeneracy is the phase-space density ϖ introduced in Eq. (2.24), and in the classical limit $\varpi \ll 1$. When ϖ becomes comparable with unity, gases become degenerate: bosons condense in the lowest single-particle state, while fermions tend towards a state with a filled Fermi sea. As one would expect on dimensional grounds, the degeneracy temperature for fermions – the Fermi temperature T_F – is given by the same expression as the Bose–Einstein transition temperature for bosons, apart from a numerical factor of order unity.

As we described in Chapter 4, laser cooling alone is insufficient to achieve degeneracy in dilute gases, and it must be followed by evaporative cooling. The elastic collision rate, which governs the effectiveness of evaporative cooling, behaves differently for fermions and bosons when gases become degenerate. For identical fermions, the requirement of antisymmetry of the wave function forces the scattering cross section to vanish at low energy (see Eq. (5.23)), and therefore evaporative cooling with a single species of fermion, with all atoms in the same internal state, cannot work. This difficulty may be overcome by using a mixture of two types of atoms, either two different fermions, which could be different hyperfine states of the same fermionic isotope, or a boson and a fermion. In the experiments on ^{40}K, the two species were the two hyperfine states $|9/2, 9/2\rangle$ and $|9/2, 7/2\rangle$ of the atom [1], and both species were evaporated. In the lithium experiments [2], the second component was the boson ^{7}Li. The ^{6}Li atoms were cooled by collisions with ^{7}Li atoms which were cooled evaporatively, a process referred to as *sympathetic cooling*.

The rate of collisions is influenced not only by the statistics of the atoms, but also by the degree of degeneracy. Due to stimulated emission, degeneracy increases collision rates for bosons. This is expressed by factors $1 + f_{\mathbf{p}'}$ for final states in the expressions for rates of processes. For fermions the sign of the effect is opposite, since the corresponding factors are $1 - f_{\mathbf{p}'}$, which shows that transitions to occupied final states are blocked by the Pauli exclusion principle. Consequently, it becomes increasingly difficult to cool fermions by evaporation or by collisions with a second component when the fermions become degenerate.

In this chapter we describe selected topics in the physics of trapped Fermi gases. We begin with equilibrium properties of a trapped gas of non-interacting fermions in Sec. 14.1. In Sec. 14.2 we consider interactions, and demonstrate that under most conditions they have little effect on either static or dynamic properties of trapped fermions. However, there are conditions under which the effects of interactions can be large. One of the exciting possibilities is that a gaseous mixture of two sorts of fermions with an attractive interaction between the two species could undergo a transition to a superfluid phase similar to that for electrons in metallic superconductors. One promising candidate is ^{6}Li, for which the two-body interaction in the electronic triplet state is large and negative, see Sec. 5.5.1. For other species, Feshbach resonances may make it possible to tune the interaction to a large, negative value. In Sec. 14.3 we calculate the transition temperature to the superfluid phase, and the gap in the excitation spectrum. Section 14.4 deals with mixtures of bosons and fermions, and we discuss the interaction between fermions mediated by excitations in the boson gas. The study of collective modes is a possible experimental probe of superfluidity in Fermi gases, and in Sec. 14.5 we give a brief introduction to collective modes in the superfluid at zero temperature.

14.1 Equilibrium properties

We begin by considering N fermions in the same internal state. The kinetic energy due to the Fermi motion which results from the requirement that no two particles occupy any single-particle state gives a major contribution to the total energy. Interactions are essentially absent at low temperature because there is no s-wave scattering for two fermions in the same internal state. This is in marked contrast to a Bose–Einstein condensate, for which the interaction energy dominates the kinetic energy under most experimental conditions. A very good first approximation to the properties of trapped Fermi gases may be obtained by treating the fermions as non-interacting [3].

In Sec. 2.1.1 we introduced the function $G(\epsilon)$, the number of states with energy less than ϵ. In the ground state of the system all states with energy less than the zero-temperature chemical potential are occupied, while those with higher energy are empty. Since each state can accommodate only one particle we have

$$G(\mu) = N, \tag{14.1}$$

where μ denotes the zero-temperature chemical potential. For a power-law density of states, $g(\epsilon) = C_\alpha \epsilon^{\alpha-1}$, the relation (14.1) yields

$$g(\mu) = \frac{\alpha N}{\mu}. \tag{14.2}$$

For a particle in a box ($\alpha = 3/2$), the total number of states with energy less than ϵ is given by Eq. (2.3), $G(\epsilon) = V(2m\epsilon)^{3/2}/6\pi^2\hbar^3$. According to (14.1) the Fermi energy, which equals the chemical potential at zero temperature, and the particle number are therefore related by the condition

$$V \frac{(2m\epsilon_{\rm F})^{3/2}}{6\pi^2\hbar^3} = N, \tag{14.3}$$

and the Fermi temperature $T_{\rm F} = \epsilon_{\rm F}/k$ is given by

$$kT_{\rm F} = \frac{(6\pi^2)^{2/3}}{2} \frac{\hbar^2}{m} n^{2/3} \approx 7.596 \frac{\hbar^2}{m} n^{2/3}. \tag{14.4}$$

The density of states $g(\epsilon)$ is given by Eq. (2.5). We introduce the density of states per unit volume by $N(\epsilon) = g(\epsilon)/V$, which is

$$N(\epsilon) = \frac{g(\epsilon)}{V} = \frac{m^{3/2}\epsilon^{1/2}}{\sqrt{2}\pi^2\hbar^3} \tag{14.5}$$

and therefore at the Fermi energy it is

$$N(\epsilon_{\rm F}) = \frac{g(\epsilon_{\rm F})}{V} = \frac{3n}{2\epsilon_{\rm F}}. \tag{14.6}$$

in agreement with the general relation (14.2).

For a three-dimensional harmonic trap, the total number of states with energy less than ϵ is given by Eq. (2.9),

$$G(\epsilon) = \frac{1}{6}\left(\frac{\epsilon}{\hbar\bar{\omega}}\right)^3, \qquad (14.7)$$

where $\bar{\omega}^3 = \omega_1\omega_2\omega_3$, and from (14.1) we then obtain the chemical potential μ and the equivalent Fermi temperature T_F as

$$\mu = kT_F = (6N)^{1/3}\hbar\bar{\omega}. \qquad (14.8)$$

A very good approximation for the density distribution of a trapped cloud of fermions in its ground state may be obtained by use of the Thomas–Fermi approximation, which we applied to bosons in Sec. 6.2.2. According to this semi-classical approximation, the properties of the gas at a point $\mathbf{r}$ are assumed to be those of a uniform gas having a density equal to the local density $n(\mathbf{r})$. The Fermi wave number $k_F(\mathbf{r})$ is related to the density by the relation for a homogeneous gas,

$$n(\mathbf{r}) = \frac{k_F^3(\mathbf{r})}{6\pi^2}, \qquad (14.9)$$

and the local Fermi energy $\epsilon_F(\mathbf{r})$ is given by

$$\epsilon_F(\mathbf{r}) = \frac{\hbar^2 k_F^2(\mathbf{r})}{2m}. \qquad (14.10)$$

The condition for equilibrium is that the energy required to add a particle at any point inside the cloud be the same everywhere. This energy is the sum of the local Fermi energy and the potential energy due to the trap, and it is equal to the chemical potential of the system,

$$\frac{\hbar^2 k_F^2(\mathbf{r})}{2m} + V(\mathbf{r}) = \mu. \qquad (14.11)$$

Note that the chemical potential is equal to the value of the local Fermi energy in the centre of the cloud, $\mu = \epsilon_F(0)$. The density profile corresponding to (14.11) is thus

$$n(\mathbf{r}) = \frac{1}{6\pi^2}\left\{\frac{2m}{\hbar^2}[\mu - V(\mathbf{r})]\right\}^{3/2}, \qquad (14.12)$$

if $V(\mathbf{r}) < \mu$ and zero otherwise. In a cloud of fermions the density profile is therefore more concentrated towards the centre than it is for a cloud of bosons, for which the density varies as $\mu - V(\mathbf{r})$ (Eq. (6.31)). The boundary of the cloud is determined by the condition $V(\mathbf{r}) = \mu$. Therefore for an anisotropic harmonic-oscillator potential of the form (2.7) the cloud extends to distances R_1, R_2 and R_3 along the three axes of the trap, where

$$R_i^2 = \frac{2\mu}{m\omega_i^2}, \quad i = 1, 2, 3, \tag{14.13}$$

and in general the cloud is aspherical.

As a measure of the linear dimensions of the cloud we use $\bar{R} = (R_1 R_2 R_3)^{1/3}$, the harmonic mean of the R_i. This may be estimated from (14.8) and (14.13) to be

$$\bar{R} = 48^{1/6} N^{1/6} \bar{a} \approx 1.906 N^{1/6} \bar{a}, \tag{14.14}$$

where $\bar{a} = (\hbar/m\bar{\omega})^{1/2}$. By contrast, the size of a cloud of bosons depends on the strength of the interaction, and ranges from $\bar{a}$ for $Na/\bar{a} \ll 1$ to approximately $15^{1/5}(Na/\bar{a})^{1/5}\bar{a}$ for $Na/\bar{a} \gg 1$. For typical trap parameters and scattering lengths, a cloud of fermions is therefore generally a few times larger than one with the same number of bosons.

The total number of particles is obtained by integrating the density over the volume of the cloud, and one finds

$$N = \frac{\pi^2}{8} n(0) \bar{R}^3, \tag{14.15}$$

or

$$n(0) = \frac{8}{\pi^2} \frac{N}{\bar{R}^3}, \tag{14.16}$$

where $n(0)$ is the density at the centre of the cloud. Thus

$$k_F(0) = \frac{(48N)^{1/3}}{\bar{R}} = 48^{1/6} \frac{N^{1/6}}{\bar{a}} \approx 1.906 \frac{N^{1/6}}{\bar{a}}, \tag{14.17}$$

which shows that the maximum wave number is of order the average inverse interparticle separation, as in a homogeneous gas. The momentum distribution is isotropic as a consequence of the isotropy in momentum space of the single-particle kinetic energy $p^2/2m$ (Problem 14.1).

The Thomas–Fermi approximation for equilibrium properties of clouds of fermions is valid provided the Fermi wavelength is small compared with the dimensions of the cloud, or $k_F(0) R_i \sim N^{1/3} \gg 1$. This condition is generally less restrictive than the corresponding one for bosons, $Na/\bar{a} \gg 1$. The Thomas–Fermi approximation fails at the surface, and it is left as a problem to evaluate the thickness of this region (Problem 14.2).

We turn now to thermodynamic properties. The distribution function is given by the Fermi function[1]

$$f = \frac{1}{e^{(\epsilon-\mu)/kT} + 1}, \qquad (14.18)$$

and the chemical potential μ depends on temperature. The total energy E is therefore given by

$$E(T) = \int_0^\infty d\epsilon\, \epsilon g(\epsilon) f(\epsilon), \qquad (14.19)$$

where $g(\epsilon)$ is the single-particle density of states. We again consider power-law densities of states $g(\epsilon) = C_\alpha \epsilon^{\alpha-1}$. At zero temperature the distribution function reduces to a step function, and one finds $E(0) = [\alpha/(\alpha+1)]N\mu$, where μ is given by (14.8). Therefore for fermions in a three-dimensional harmonic trap

$$E(0) = \frac{3}{4}N\mu, \qquad (14.20)$$

while for particles in a box ($\alpha = 3/2$) the numerical coefficient is $3/5$.

Properties of the system at temperatures less than the Fermi temperature may be estimated by carrying out a low-temperature expansion in the standard way. For the energy one finds the well-known result

$$E \simeq E(0) + \frac{\pi^2}{6} g(\mu)(kT)^2, \qquad (14.21)$$

where μ is the zero-temperature chemical potential, given by Eq. (14.1).

For temperatures high compared with the Fermi temperature one may calculate thermodynamic properties by making high-temperature expansions, as we did for bosons in Sec. 2.4.2. The energy E tends towards its classical value αNkT, as it does for bosons, and the first correction due to degeneracy has the same magnitude as that for bosons, but the opposite sign.

14.2 Effects of interactions

As we have argued above, interactions have essentially no effect on low-temperature properties of dilute Fermi systems if all particles are in the same internal state. However, they can play a role for a mixture of two kinds of fermions, but their effects are generally small. Consider as an example a uniform gas containing equal densities of two kinds of fermions, which we assume to have the same mass. The kinetic energy per particle is of order the Fermi energy $\epsilon_F = (\hbar k_F)^2/2m$. The interaction energy per

---

[1] In this chapter all distribution functions are equilibrium ones, so we shall denote them by f rather than f^0.

particle is of order nU_0, where n is the particle density for one component and $U_0 = 4\pi\hbar^2 a/m$ is the effective interaction between two unlike fermions, a being the corresponding scattering length. Since the density is given by $n = k_F^3/6\pi^2$, the ratio of the interaction energy to the Fermi energy is

$$\frac{nU_0}{\epsilon_F} = \frac{4}{3\pi} k_F a, \tag{14.22}$$

which is of order the scattering length divided by the interparticle spacing. This is typically $\sim 10^{-2}$. For two ^{6}Li atoms in a triplet electronic state the scattering length is negative and exceptionally large in magnitude, and the ratio can be of order 10^{-1}. An equivalent dimensionless measure of the coupling strength is $N(\epsilon_F)U_0 = 3nU_0/2\epsilon_F$. For trapped clouds, the Fermi wave number is of order $N^{1/6}/\bar{a}$, and therefore the ratio of the interaction energy to the Fermi energy is of order $N^{1/6}a/\bar{a}$. This dimensionless quantity is familiar from Sec. 11.1, where it was shown to give the ratio of interaction energy to kinetic energy for a Bose gas at a temperature of order T_c. The fact that it determines the ratio of these two energies also for Fermi systems at low temperatures reflects the similarity between the momentum distribution for a Bose gas near T_c and that for a Fermi gas at or below the Fermi temperature. We therefore expect particle interactions to have little influence on thermodynamic properties of trapped clouds of fermions, except when the magnitude of the scattering length is exceptionally large.

Equilibrium size and collective modes

Let us first consider the equilibrium size of a cloud of fermions in a harmonic trap. For simplicity, we neglect the anisotropy of the potential and take it to be of the form $V(r) = m\omega_0^2 r^2/2$. If the spatial extent of the cloud is $\sim R$, the potential energy per particle is of order $m\omega_0^2 R^2/2$, while the kinetic energy per particle is of order $N^{2/3}\hbar^2/2mR^2$, according to (14.10) and (14.17). Observe that this kinetic energy is a factor $N^{2/3}$ larger than the kinetic energy $\sim \hbar^2/2mR^2$ for a cloud of condensed bosons of the same size. In the absence of interactions the total energy thus varies as $N^{2/3}/R^2$ at small radii and as R^2 at large radii, with a minimum when R is of order $N^{1/6}(\hbar/m\omega_0)^{1/2} = N^{1/6}a_{\text{osc}}$, in agreement with (14.14).

Let us now consider the effect of interactions. If the cloud contains an equal number N of each of two fermion species, the interaction energy per particle is of order $U_0 N/R^3$. As discussed for bosons in Sec. 6.2, interactions shift the equilibrium size of the cloud, tending to increase it for repulsive interactions and reduce it for attractive ones. For fermions the shift is generally small, since the ratio of the interaction energy to the kinetic energy is of order $N^{1/6}a/a_{\rm osc}$, as argued above. From these considerations it is natural to expect that interactions will have little effect on collective modes. This may be demonstrated explicitly by calculating the frequency of the breathing mode of a cloud containing equal numbers of two different fermion species of equal mass in an isotropic harmonic trap. The method of collective coordinates described in Sec. 7.3.3 may be applied to fermions if one works in terms of the density distribution rather than the condensate wave function. We parametrize the density distribution $n(r)$ for a *single* species as

$$n(r) = \frac{AN}{R^3} h(r/R), \qquad (14.23)$$

where A, a pure number, is a normalization constant. The quantity corresponding to the zero-point energy $E_{\rm zp}$ given in (7.97) is the kinetic energy. Due to the Pauli exclusion principle, particles occupy excited single-particle states and, consequently, there is a kinetic contribution to the total energy. We may estimate this energy using the Thomas–Fermi approximation. The energy per particle for a free Fermi gas scales as the Fermi energy, which varies as $n^{2/3}$, and therefore the contribution to the total energy scales as R^{-2}, just as does the kinetic energy of a condensed cloud of bosons. The calculation of the frequency of the breathing mode goes through as before, with the kinetic energy due to the Pauli principle replacing the zero-point energy. The expression (7.113) for the frequency of the monopole mode does not contain the zero-point energy explicitly, and it is therefore also valid for clouds of fermions:

$$\omega^2 = 4\omega_0^2 \left[1 + \frac{3}{8}\frac{E_{\rm int}(R_0)}{E_{\rm osc}(R_0)}\right]. \qquad (14.24)$$

This result was earlier derived from sum rules in Ref. [4].

The ratio $E_{\rm int}/E_{\rm osc}$ is easily calculated within the Thomas–Fermi approximation, for which the density profile is given by (14.12), since

$$E_{\rm int} = U_0 \int d\mathbf{r}\, n^2(r), \qquad (14.25)$$

the density of each species being $n(r)$. The energy due to the trap is

$$E_{\rm osc} = m\omega_0^2 \int d\mathbf{r}\, r^2 n(r). \qquad (14.26)$$

Apart from a numerical constant, the ratio of the two energies is equal to $n(0)U_0/m\omega_0^2 R^2 \approx N^{1/6}a/a_{\rm osc}$. Thus

$$\omega^2 = 4\omega_0^2 \left(1 + c_1 N^{1/6} \frac{a}{a_{\rm osc}}\right), \qquad (14.27)$$

where c_1 is a numerical constant to be determined in Problem 14.5. The result (14.27) shows that interactions have little effect on mode frequencies.

According to Eq. (14.27), frequencies of collective modes in a trap depend linearly on the interaction strength for $N^{1/6}|a| \ll a_{\text{osc}}$. This is to be contrasted with the situation for homogeneous gases, where a weak repulsive interaction between two species of fermions (say, spin-up and spin-down) of the same density gives rise to the zero-sound mode. The velocity of zero sound differs from the Fermi velocity $v_F = p_F/m$ by an amount proportional to $v_F \exp[-1/N(\epsilon_F)U_0]$, which is exponentially small for weak coupling. Here $N(\epsilon_F)$ is the density of states (for one species) per unit volume at the Fermi energy. The reason for this qualitative difference is that the strong degeneracy of the frequencies for single-particle motion in a harmonic trap leads to larger collective effects.

Stability of uniform matter

If interactions are sufficiently strong, a uniform mixture of fermions may be unstable. To investigate the problem quantitatively we use the stability conditions (12.10) and (12.11). The energy density of a uniform mixture of two kinds of fermions, labelled a and b, is given in the Hartree approximation by

$$\mathcal{E} = \frac{3}{5}n_a \epsilon_{Fa} + \frac{3}{5}n_b \epsilon_{Fb} + U_0 n_a n_b, \tag{14.28}$$

where n_i is the density of component $i(=a,b)$, ϵ_{Fi} is the Fermi energy, and U_0 is the effective interaction between two unlike atoms. The stability conditions (12.10) are always satisfied, because $\partial^2 \mathcal{E}/\partial n_i^2 = 1/N_i(\epsilon_{Fi})$, where $N_i(\epsilon)$ is the density of states per unit volume (14.5) for species i at energy ϵ. The stability condition (12.11) is

$$\frac{1}{N_a(\epsilon_{Fa})N_b(\epsilon_{Fb})} > U_0^2. \tag{14.29}$$

For a mixture with equal densities of two species having the same mass, the densities of states occurring in Eq. (14.29) are both equal to $N(\epsilon_F) = mk_F/2\pi^2\hbar^2$. The condition (14.29) then becomes

$$N(\epsilon_F)|U_0| = \frac{2}{\pi}k_F|a| < 1. \tag{14.30}$$

This approximation predicts that for an attractive interaction the mixture becomes unstable with respect to collapse if $a < -\pi/2k_F$, while for repulsive interactions it becomes unstable with respect to phase separation if $a > \pi/2k_F$. However, when $N(\epsilon_F)|U_0|$ becomes comparable with unity the mean-field approach becomes invalid, and the properties of a mixture of fermions with $N(\epsilon_F)|U_0| \sim 1$ is at present an open problem. We shall comment more on this aspect in the next section.

14.3 Superfluidity

While the effects of particle interactions are generally small, they can be dramatic if the effective interaction is attractive. The gas is then predicted to undergo a transition to a superfluid state in which atoms are paired in the same way as electrons are in superconducting metals. The basic theory of the state was developed by Bardeen, Cooper, and Schrieffer (BCS) [5], and it has been applied widely to atomic nuclei and liquid ^{3}He as well as to metallic superconductors. It was used to estimate transition temperatures of dilute atomic vapours in Ref. [6].

The properties of mixtures of dilute Fermi gases with attractive interactions are of interest in a number of contexts other than that of atomic vapours. One of these is in nuclear physics and astrophysics, where dilute mixtures of different sorts of fermions (neutrons and protons with two spin states each) are encountered in the outer parts of atomic nuclei and in the crusts of neutron stars [7, 8]. Another is in addressing the question of how, as the strength of an attractive interaction is increased, the properties of a Fermi system change from those of a BCS superfluid for weak coupling to those of a system of diatomic (bosonic) molecules [9]. The possibility of altering the densities of components and the effective interaction between them indicates that dilute atomic gases may be useful systems with which to explore this question experimentally.

A rough estimate of the transition temperature may be obtained by using a simplified model, in which one assumes the interaction between fermions to be a constant $-|U|$ for states with energies within E_c of the Fermi energy, and zero otherwise. The transition temperature is predicted to be given in order of magnitude by

$$kT_c \sim E_c e^{-1/N(\epsilon_F)|U|}. \tag{14.31}$$

For electrons in metals, the attractive interaction is a consequence of exchange of phonons, and the cut-off energy E_c is comparable with the maximum energy of an acoustic phonon $\hbar\omega_D$, where ω_D is the Debye frequency. In dilute gases, the dominant part of the interaction is the direct interaction between atoms. The effective interaction between particles at the Fermi surface is given by the usual pseudopotential result, and thus for U we take U_0. Since, generally, the interaction between atoms is not strongly dependent on energy, we take the cut-off to be the energy scale over which the density of states varies, namely ϵ_F. This leads to the estimate

$$kT_c \sim \epsilon_F e^{-1/N(\epsilon_F)|U_0|}. \qquad (14.32)$$

Later we shall confirm this result and evaluate the numerical coefficient.

The dimensionless interaction strength $\lambda = N(\epsilon_F)|U_0|$ at the centre of a harmonic trap may be written as

$$\lambda = N(\epsilon_F)|U_0| = \frac{2}{\pi} k_F(0)|a| \approx 1.214 \frac{N^{1/6}|a|}{\bar{a}}. \qquad (14.33)$$

Since $|a|$ is generally much less than $\bar{a}$, λ is small, and estimated transition temperatures are therefore much less than the Fermi temperature.

In the remaining part of this section we describe the quantitative theory of the condensed state. We first calculate the temperature of the transition to the condensed state. For two-body scattering *in vacuo*, the existence of a bound state gives rise to a divergence in the T matrix at the energy of the state. For example, a large, positive scattering length indicates that there is a bound state at a small negative energy. Similarly, for two particles in a medium, the onset of pairing is signalled by a divergence in the two-body scattering, and we calculate the transition temperature by determining when the scattering of two fermions in the medium becomes singular. Following that, we describe how to generalize Bogoliubov's method for bosons to the condensed state of fermions. Subsequently, we demonstrate how induced interactions affect estimates of properties of the condensed state.

14.3.1 Transition temperature

To understand the origin of the pairing phenomenon we investigate the scattering of two fermions in a uniform Fermi gas. In Sec. 5.2.1 we studied scattering of two particles *in vacuo*, and we now extend this treatment to take into account the effects of other fermions. Since there is no scattering at low energies for fermions in the same internal state, we consider a mixture of two kinds of fermions, which we denote by a and b. As described in the introduction to this chapter, the two fermions may be different internal states of the same isotope. For simplicity, we shall assume that the two species have the same mass, and that their densities are equal. We shall denote the common Fermi wave number by k_F.

The medium has a number of effects on the scattering process. One is that the energies of particles are shifted by the mean field of the other particles. In a dilute gas, this effect is independent of the momentum of a particle. Consequently, the equation for the T matrix will be unaltered, provided that

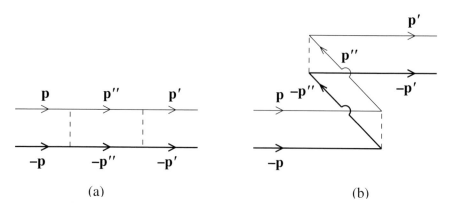

Fig. 14.1. Diagrams representing two-particle (a) and two-hole (b) intermediate states in two-particle scattering. The thin and thick lines correspond to the two kinds of fermions, and the dashed lines to the bare interaction. Time is understood to advance from left to right. Consequently, internal lines directed to the left represent holes, and those directed to the right, particles.

energies are measured relative to the mean-field energy of the two incoming particles. A second effect is that some states are occupied, and therefore, because of the Pauli exclusion principle, these states are unavailable as intermediate states in the scattering process. In the Lippmann–Schwinger equation (5.30) the intermediate state has an a particle with momentum $\mathbf{p}''$ and a b particle with momentum $-\mathbf{p}''$. This process is represented diagrammatically in Fig. 14.1(a). The probabilities that these states are unoccupied are $1 - f_{\mathbf{p}''}$ and $1 - f_{-\mathbf{p}''}$, and therefore the contribution from these intermediate states must include the blocking factors. They are the analogues of the factors $1 + f$ that enhance rates of processes for bosons. Generally the distribution functions for the two species are different, but for the situation we have chosen they are the same, and therefore we omit species labels on the distribution functions. Modifying the Lippmann–Schwinger equation in this way gives

$$T(\mathbf{p}', \mathbf{p}; E) = U(\mathbf{p}', \mathbf{p}) + \frac{1}{V} \sum_{\mathbf{p}''} U(\mathbf{p}', \mathbf{p}'') \frac{(1 - f_{\mathbf{p}''})(1 - f_{-\mathbf{p}''})}{E - 2\epsilon^0_{\mathbf{p}''} + i\delta} T(\mathbf{p}'', \mathbf{p}; E).$$

(14.34)

Here $\epsilon^0_\mathbf{p} = p^2/2m$ is the free-particle energy, and we work in terms of particle momenta, rather than wave vectors as we did in Sec. 5.2.1.

A third effect of the medium is that interactions between particles can initially excite two fermions from states in the Fermi sea to states outside, or, in other words, two particles and two holes are created. The two holes then annihilate with the two incoming particles. This process is represented in

Fig. 14.1(b). The contribution to the Lippmann–Schwinger equation from this process differs from that for two-particle intermediate states in two respects. First, holes can be created only in occupied states, and the probability of this is $f_{\pm \mathbf{p}''}$ for the two states. Thus the thermal factors for each state are $f_{\pm \mathbf{p}''}$ instead of $1 - f_{\pm \mathbf{p}''}$. Second, the energy to create a hole is the negative of that for a particle and, consequently, the energy denominator changes sign. The equation including both particle–particle and hole–hole intermediate states is

$$T(\mathbf{p}', \mathbf{p}; E)$$
$$= U(\mathbf{p}', \mathbf{p}) + \frac{1}{V} \sum_{\mathbf{p}''} U(\mathbf{p}', \mathbf{p}'') \frac{(1 - f_{\mathbf{p}''})(1 - f_{-\mathbf{p}''}) - f_{\mathbf{p}''} f_{-\mathbf{p}''}}{E - 2\epsilon^0_{\mathbf{p}''} + i\delta} T(\mathbf{p}'', \mathbf{p}; E)$$
$$= U(\mathbf{p}', \mathbf{p}) + \frac{1}{V} \sum_{\mathbf{p}''} U(\mathbf{p}', \mathbf{p}'') \frac{1 - f_{\mathbf{p}''} - f_{-\mathbf{p}''}}{E - 2\epsilon^0_{\mathbf{p}''} + i\delta} T(\mathbf{p}'', \mathbf{p}; E). \qquad (14.35)$$

This result may be derived more rigourously by using finite-temperature field-theoretic methods [10].

Equation (14.35) shows that the greatest effect of intermediate states is achieved if $E = 2\mu = p_F^2/m$, since the sign of $1 - f_{\mathbf{p}''} - f_{-\mathbf{p}''}$ is then always opposite that of the energy denominator and, consequently, all contributions to the sum have the same sign. For other choices of the energy there are contributions of both signs, and some cancellation occurs. In future we shall therefore put the energy equal to 2μ.

When the T matrix diverges, the first term on the right hand side of Eq. (14.35) may be neglected, and one finds

$$T(\mathbf{p}', \mathbf{p}; 2\mu) = -\frac{1}{V} \sum_{\mathbf{p}''} U(\mathbf{p}', \mathbf{p}'') \frac{1 - 2f_{\mathbf{p}''}}{2\xi_{\mathbf{p}''}} T(\mathbf{p}'', \mathbf{p}; 2\mu), \qquad (14.36)$$

where $\xi_{\mathbf{p}} = p^2/2m - \mu$. It is no longer necessary to include the infinitesimal imaginary part since the numerator vanishes at the Fermi surface. The right hand side depends on temperature through the Fermi function that occurs there, and this equation determines the temperature at which the scattering diverges.

Eliminating the bare interaction

Atomic potentials are generally strong, and they have appreciable matrix elements for transitions to states at energies much greater than the Fermi energy. If one replaces the bare interaction by a constant, the sum on the right hand side of Eq. (14.36) for momenta less than some cut-off, p_c, diverges as $p_c \to \infty$. To remove this dependence on the high momentum states we eliminate the bare potential in favour of the pseudopotential, as was done in our discussion for bosons. We write Eq. (14.36) formally as

$$T = UG_\mathrm{M}T, \qquad (14.37)$$

where G_M is the propagator for two particles, the subscript indicating that it applies to the medium. The T matrix in free space and for energy $E = 2\mu$, which we here denote by T_0, is given by

$$T_0(\mathbf{p}', \mathbf{p}; 2\mu) = U(\mathbf{p}', \mathbf{p}) - \frac{1}{V}\sum_{\mathbf{p}''} U(\mathbf{p}', \mathbf{p}'') \frac{1}{2\xi_{\mathbf{p}''} - i\delta} T_0(\mathbf{p}'', \mathbf{p}; 2\mu), \qquad (14.38)$$

which we write formally as

$$T_0 = U + UG_0T_0, \qquad (14.39)$$

where G_0 corresponds to propagation of two free particles *in vacuo*. The infinitesimal imaginary part takes into account real scattering to intermediate states having the same energy as the initial state. It gives contributions proportional to the density of states, which varies as p_F. These are small at low densities. Thus we shall neglect them and interpret integrals as principal-value ones. Solving for U, we find

$$U = T_0(1 + G_0T_0)^{-1} = (1 + T_0G_0)^{-1}T_0, \qquad (14.40)$$

where the second form follows from using an identity similar to that used in deriving Eq. (5.110).

Equation (14.37) may therefore be rewritten as

$$T = (1 + T_0G_0)^{-1}T_0G_\mathrm{M}T. \qquad (14.41)$$

Multiplying on the left by $1 + T_0G_0$ one finds

$$T = T_0(G_\mathrm{M} - G_0)T, \qquad (14.42)$$

or

$$T(\mathbf{p}', \mathbf{p}; 2\mu) = \frac{1}{V}\sum_{\mathbf{p}''} T_0(\mathbf{p}', \mathbf{p}''; 2\mu) \frac{f_{\mathbf{p}''}}{\xi_{\mathbf{p}''}} T(\mathbf{p}'', \mathbf{p}; 2\mu). \qquad (14.43)$$

The quantity $f_{\mathbf{p}''}/\xi_{\mathbf{p}''}$ is appreciable only for momenta less than or slightly above the Fermi momentum. This is because contributions from high-momentum states have now been incorporated in the effective interaction. We may therefore replace T_0 by its value for zero energy and zero momentum, $U_0 = 4\pi\hbar^2 a/m$. Consequently T also depends weakly on momentum for momenta of order p_F, and Eq. (14.43) reduces to

$$\frac{U_0}{V}\sum_{\mathbf{p}} \frac{f_{\mathbf{p}}}{\xi_{\mathbf{p}}} = U_0 \int_0^\infty d\epsilon N(\epsilon) \frac{f(\epsilon)}{\epsilon - \mu} = 1, \qquad (14.44)$$

where $f(\epsilon) = \{\exp[(\epsilon - \mu)/kT] + 1\}^{-1}$ and $N(\epsilon)$ is the density of states per unit volume for a single species, which is given by (14.5).

Analytical results

At zero temperature the integral diverges logarithmically as $\epsilon \to \mu$, and at non-zero temperatures it is cut off by the Fermi function at $|\xi_{\mathbf{p}}| \sim kT$. Thus for an attractive interaction ($U_0 < 0$) there is always a temperature T_c at which Eq. (14.44) is satisfied. Introducing the dimensionless variables $x = \epsilon/\mu$ and $y = \mu/kT_c$, we may write Eq. (14.44) as

$$\frac{1}{N(\epsilon_F)|U_0|} = -\int_0^\infty dx \frac{x^{1/2}}{x-1} \frac{1}{e^{(x-1)y}+1}. \qquad (14.45)$$

Since $N(\epsilon_F)|U_0| \ll 1$, transition temperatures are small compared with the Fermi temperature, so we now evaluate the integral in (14.45) at low temperatures, $kT_c \ll \mu$, which implies that $\mu \simeq \epsilon_F$. We rewrite the integrand using the identity

$$\frac{x^{1/2}}{x-1} = \frac{1}{x^{1/2}+1} + \frac{1}{x-1}. \qquad (14.46)$$

The first term in (14.46) has no singularity at the Fermi surface in the limit of zero temperature, and the integral containing it may be replaced by its value at zero temperature ($y \to \infty$), which is

$$\int_0^1 \frac{dx}{x^{1/2}+1} = 2(1 - \ln 2). \qquad (14.47)$$

The integrand involving the second term in Eq. (14.46) has a singularity at the Fermi surface, and the integral must be interpreted as a principal value one. We divide the range of integration into two parts, one from 0 to $1-\delta$, and the second from $1+\delta$ to infinity, where δ is a small quantity which we allow to tend to zero in the end. Integrating by parts and allowing the lower limit of integration to tend to $-\infty$ since $T_c \ll T_F$, we find

$$\int_0^\infty \frac{dx}{x-1} \frac{1}{e^{(x-1)y}+1} = \int_0^\infty dz \ln(z/y) \frac{1}{2\cosh^2 z/2} = -\ln\frac{2y\gamma}{\pi}. \qquad (14.48)$$

Here $\gamma = e^C \approx 1.781$, where C is Euler's constant. By adding the contributions (14.47) and (14.48) one finds that the transition temperature is given by

$$kT_c = \frac{8\gamma}{\pi e^2}\epsilon_F e^{-1/N(\epsilon_F)|U_0|} \approx 0.61\epsilon_F e^{-1/N(\epsilon_F)|U_0|}. \quad (14.49)$$

This result, which has been obtained by a number of groups working in different areas of physics [6–9], confirms the qualitative estimates made at the beginning of this section. However, it is not the final answer because it neglects the influence of the medium on the interaction between atoms. In Sec. 12.1.1 we introduced the concept of induced interactions in discussing the stability of mixtures of bosons, and we now describe induced interactions in mixtures of fermions. They significantly reduce the transition temperature.

14.3.2 Induced interactions

In our description of boson–boson mixtures we saw how the interaction between two species led to an effective interaction between members of the same species. Likewise the interaction between two fermions is affected by the other fermions, and thus the interaction between two fermions in the medium differs from that between two fermions *in vacuo*. Long ago Gorkov and Melik-Barkhudarov calculated the transition temperature of a dilute Fermi gas allowing for this effect [11], and the results are interpreted in simple physical terms in Ref. [12]. Rather surprisingly, the induced interaction changes the prefactor in the expression (14.49). Consider an a fermion with momentum $\mathbf{p}$ and a b fermion with momentum $-\mathbf{p}$ that scatter to states with momenta $\mathbf{p}'$ and $-\mathbf{p}'$ respectively. Up to now we have assumed that this can occur only via the bare two-body interaction between particles. However, another possibility in a medium is that the incoming a fermion interacts with a b fermion in the medium with momentum $\mathbf{p}''$ and scatters to a state with a b fermion with momentum $-\mathbf{p}'$ and an a fermion with momentum $\mathbf{p}+\mathbf{p}'+\mathbf{p}''$. The latter particle then interacts with the other incoming b fermion with momentum $-\mathbf{p}$ to give an a fermion with momentum $\mathbf{p}'$ and a b one with momentum $\mathbf{p}''$. The overall result is the same as for the original process, since the particle from the medium that participated in the process is returned to its original state. A diagram illustrating this process is shown in Fig. 14.2(a), and one for the related process in which the b particle interacts first with an a particle in the Fermi sea is shown in Fig. 14.2(b). For systems of charged particles, processes like these screen the Coulomb interaction.

The change in the effective interaction due to the medium may be calculated by methods analogous to those used to calculate the energy shift

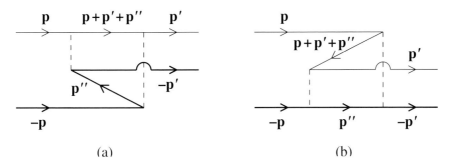

Fig. 14.2. Diagrams representing contributions to the induced interaction (see text). The notation is the same as in Fig. 14.1.

and Landau damping of an excitation in a Bose gas in Sec. 10.5.1. Due to the factor $1/(\epsilon - \mu)$ in the integrand in Eq. (14.44), interactions between particles close to the Fermi surface play an especially important role in pairing, and it is sufficient to consider the interaction between particles with momentum equal to the Fermi momentum and with energies equal to the chemical potential. The change in the effective interaction may be calculated by second-order perturbation theory, including the appropriate thermal factors. It is given by

$$U_{\text{ind}}(\mathbf{p}, \mathbf{p}') = -\frac{1}{V} \sum_{\mathbf{p}''} U_0^2 \left[\frac{f_{\mathbf{p}''}(1 - f_{\mathbf{p}+\mathbf{p}'+\mathbf{p}''})}{\epsilon_{\mathbf{p}''} - \epsilon_{\mathbf{p}+\mathbf{p}'+\mathbf{p}''}} + \frac{f_{\mathbf{p}+\mathbf{p}'+\mathbf{p}''}(1 - f_{\mathbf{p}''})}{\epsilon_{\mathbf{p}+\mathbf{p}'+\mathbf{p}''} - \epsilon_{\mathbf{p}''}} \right]$$

$$= \int \frac{d\mathbf{p}''}{(2\pi\hbar)^3} U_0^2 \frac{f_{\mathbf{p}''} - f_{\mathbf{p}+\mathbf{p}'+\mathbf{p}''}}{\epsilon_{\mathbf{p}+\mathbf{p}'+\mathbf{p}''} - \epsilon_{\mathbf{p}''}} = U_0^2 L(|\mathbf{p}' + \mathbf{p}|). \quad (14.50)$$

The minus sign in the first of Eqs. (14.50) is necessary because in the process considered one of the particles in the final state is created before that in the initial state is destroyed. Expressed in terms of creation and annihilation operators $a^\dagger, a, b^\dagger,$ and b for the two species, the effective interaction would correspond to the combination of operators $a^\dagger_{\mathbf{p}'} b_{-\mathbf{p}} b^\dagger_{-\mathbf{p}'} a_{\mathbf{p}}$ for the diagram in Fig. 14.2(a), and to $b^\dagger_{-\mathbf{p}'} a_{\mathbf{p}} a^\dagger_{\mathbf{p}'} b_{-\mathbf{p}}$ for Fig. 14.2(b). To get the operators in the standard order $a^\dagger_{\mathbf{p}'} b^\dagger_{-\mathbf{p}'} b_{-\mathbf{p}} a_{\mathbf{p}}$ a minus sign is required because fermion creation and annihilation operators anticommute. Expressed more formally, the minus sign is due to the fact that the general rules for evaluating diagrams for fermions require a factor $-f_{\mathbf{p}}$ for every hole line. The quantity $L(q)$ is the static Lindhard screening function, which occurs in the theory of the electron gas [13]. It is given by

$$L(q) = \int \frac{d\mathbf{p}}{(2\pi\hbar)^3} \frac{f_{\mathbf{p}} - f_{\mathbf{p}+\mathbf{q}}}{\epsilon_{\mathbf{p}+\mathbf{q}} - \epsilon_{\mathbf{p}}} \simeq N(\epsilon_F) \left[\frac{1}{2} + \frac{(1-w^2)}{4w} \ln \left| \frac{1+w}{1-w} \right| \right], \quad (14.51)$$

where $w = q/2p_F$. The temperatures of interest are much less than the Fermi temperature, and the second expression is the result for $T = 0$. The Lindhard function is the negative of the density–density (or spin–spin) response function $\chi(q)$ for a one-component Fermi gas, $L(q) = -\chi(q)$. The medium-dependent contribution to the interaction is an example of an induced interaction. Its sign is positive, corresponding to a repulsive interaction, and therefore pairing is suppressed. Solving Eq. (14.36) with the induced interaction included as a perturbation, one finds that T_c is given by Eq. (14.49), but with U_0 replaced by $U_0 + <U_{\text{ind}}>$, where $<\cdots> = \int_{-1}^{1} \ldots d(\cos\theta)/2$ denotes an average over the Fermi surface. Here θ is the angle between $\mathbf{p}$ and $\mathbf{p}'$. Thus from Eq. (14.50) one finds

$$U_0 + <U_{\text{ind}}> \simeq U_0 - U_0^2 <\chi(q)>, \tag{14.52}$$

and therefore

$$\frac{1}{U_0 + <U_{\text{ind}}>} \simeq \frac{1}{U_0} + <\chi(q)>. \tag{14.53}$$

Since $p = p' = p_F$, one has $q^2 = 2p_F^2(1 + \cos\theta)$. The integral of the Lindhard function over angles is proportional to

$$\int_0^1 dw\, 2w \left[\frac{1}{2} + \frac{1}{4w}(1-w^2)\ln\frac{1+w}{1-w}\right] = \frac{2}{3}\ln 2 + \frac{1}{3} = \ln(4e)^{1/3}, \tag{14.54}$$

and therefore the transition temperature is given by

$$kT_c = \left(\frac{2}{e}\right)^{7/3} \frac{\gamma}{\pi} \epsilon_F e^{-1/N(\epsilon_F)|U_0|} \approx 0.28 \epsilon_F e^{-1/N(\epsilon_F)|U_0|}. \tag{14.55}$$

The induced interaction thus reduces the transition temperature by a factor $(4e)^{1/3} \approx 2.22$. An examination of the microscopic processes contributing to U_{ind} shows that the dominant effect is exchange of 'spin' fluctuations, where by 'spin' we mean the variable associated with the species labels a and b [12]. Such processes suppress pairing in an s state, an effect well established for magnetic metals and liquid ^{3}He.

14.3.3 The condensed phase

At the temperature T_c the normal Fermi system becomes unstable with respect to formation of pairs, and below that temperature there is a condensate of pairs in the zero-momentum state. The formation of the pairs in a BCS state is an intrinsically many-body process, and is not simply the formation of molecules made up of two fermions. If the simple molecular picture were correct, molecules would be formed as the temperature is lowered, and when their density became high enough they would form a Bose–Einstein condensate in the zero-momentum state. In the BCS picture, pair formation and condensation of pairs into the state with zero momentum occur at the same temperature.

To make a quantitative theory of the condensed state we again consider a uniform gas with equal densities of two species of fermions. The calculations can be carried out for unequal densities of the two components, but the transition temperatures are lower for a given total density of particles. Interactions between like particles play little role in the pairing of unlike ones, and therefore we neglect these terms. The Hamiltonian for interacting fermions is

$$H = \sum_{\mathbf{p}} \epsilon_p^0 (a_{\mathbf{p}}^\dagger a_{\mathbf{p}} + b_{\mathbf{p}}^\dagger b_{\mathbf{p}}) + \frac{1}{V} \sum_{\mathbf{pp'q}} U(\mathbf{p}, \mathbf{p'}, \mathbf{q}) a_{\mathbf{p+q}}^\dagger b_{\mathbf{p'-q}}^\dagger b_{\mathbf{p'}} a_{\mathbf{p}} , \quad (14.56)$$

where the operators $a^\dagger, a, b^\dagger,$ and b create and destroy particles of the two species. Because the two interacting particles belong to different species, there is no factor of $1/2$ in the interaction term, as there would be for a single species. The Hamiltonian has essentially the same form as for bosons, except that the creation and annihilation operators obey Fermi commutation rules

$$\{a_{\mathbf{p}}, a_{\mathbf{p'}}^\dagger\} = \{b_{\mathbf{p}}, b_{\mathbf{p'}}^\dagger\} = \delta_{\mathbf{p},\mathbf{p'}}$$

and

$$\{a_{\mathbf{p}}, b_{\mathbf{p'}}\} = \{a_{\mathbf{p}}^\dagger, b_{\mathbf{p'}}\} = \{a_{\mathbf{p}}, b_{\mathbf{p'}}^\dagger\} = \{a_{\mathbf{p}}^\dagger, b_{\mathbf{p'}}^\dagger\} = 0, \quad (14.57)$$

where $\{A, B\}$ denotes the anticommutator. As in Sec. 8.2 for bosons, it is convenient to work with the operator $K = H - \mu \hat{N}$, where μ is the chemical potential, which is chosen to keep the average number of particles fixed. This is given by

$$K = \sum_{\mathbf{p}} (\epsilon_{\mathbf{p}}^0 - \mu)(a_{\mathbf{p}}^\dagger a_{\mathbf{p}} + b_{\mathbf{p}}^\dagger b_{\mathbf{p}}) + \frac{1}{V} \sum_{\mathbf{pp'q}} U(\mathbf{p}, \mathbf{p'}, \mathbf{q}) a_{\mathbf{p+q}}^\dagger b_{\mathbf{p'-q}}^\dagger b_{\mathbf{p'}} a_{\mathbf{p}} . \quad (14.58)$$

In the dispersion relation for elementary excitations of a Bose system, the Hartree and Fock terms played an important role. For a dilute Fermi system, they change the total energy of the system but they have little influence on pairing. The Hartree and Fock contributions to the energy of an excitation are independent of momentum. However, the chemical potential is changed by the same amount, and therefore the energy of an excitation measured relative to the chemical potential is unaltered. In the Hartree–Fock approximation, K is shifted by a constant amount which we shall neglect.

Elementary excitations

In Chapters 7 and 8 we described the Bogoliubov method for calculating properties of a gas of bosons, and we now generalize it to describe a condensate of pairs of fermions. For a Bose gas, the Bogoliubov approach amounts to assuming that the creation and annihilation operators may be written as a classical part, which is a c number, and a fluctuation term. We now make an analogous approximation for fermions. However, since the condensate consists of pairs of fermions, the quantities we assume to have a c-number part are operators that create or destroy pairs of particles. For a condensate with total momentum zero, the two fermions that make up a pair must have equal and opposite momenta. Also they must be in different internal states, otherwise there is no interaction between them. We therefore write

$$b_{-\mathbf{p}}a_{\mathbf{p}} = C_{\mathbf{p}} + (b_{-\mathbf{p}}a_{\mathbf{p}} - C_{\mathbf{p}}), \tag{14.59}$$

where $C_{\mathbf{p}}$ is a c number. Since the relative phases of states whose particle numbers differ by two is arbitrary, we shall choose it so that $C_{\mathbf{p}}$ is real. As in the analogous calculations for bosons, we substitute this expression into Eq. (14.58) and retain only terms with two or fewer creation and annihilation operators. This leads to

$$K = \sum_{\mathbf{p}} (\epsilon_{\mathbf{p}}^0 - \mu)(a_{\mathbf{p}}^\dagger a_{\mathbf{p}} + b_{-\mathbf{p}}^\dagger b_{-\mathbf{p}}) + \sum_{\mathbf{p}} \Delta_{\mathbf{p}} (a_{\mathbf{p}}^\dagger b_{-\mathbf{p}}^\dagger + b_{-\mathbf{p}} a_{\mathbf{p}})$$
$$- \frac{1}{V} \sum_{\mathbf{p}\mathbf{p}'} U(\mathbf{p},\mathbf{p}') C_{\mathbf{p}} C_{\mathbf{p}'}, \tag{14.60}$$

where we have omitted the Hartree–Fock contribution. Here

$$\Delta_{\mathbf{p}} = \frac{1}{V} \sum_{\mathbf{p}'} U(\mathbf{p},\mathbf{p}') C_{\mathbf{p}'} \tag{14.61}$$

and, for simplicity, we omit the final argument in the interaction and write

$$U(\mathbf{p},-\mathbf{p},\mathbf{p}'-\mathbf{p}) = U(\mathbf{p},\mathbf{p}'). \tag{14.62}$$

The quantity $C_{\mathbf{p}}$ must be determined self-consistently, just as the mean particle distribution function is in Hartree–Fock theory. Thus

$$C_{\mathbf{p}} = <b_{-\mathbf{p}}a_{\mathbf{p}}> = <a_{\mathbf{p}}^\dagger b_{-\mathbf{p}}^\dagger>, \tag{14.63}$$

where $<\cdots>$ denotes an expectation value.

The Hamiltonian now has the same form as that for bosons in the Bogoliubov approximation in Sec. 8.1, and it is a sum of independent terms of the type,

$$H = \epsilon_0 (a^\dagger a + b^\dagger b) + \epsilon_1 (a^\dagger b^\dagger + ba), \tag{14.64}$$

where $a = a_{\mathbf{p}}$ and $b = b_{-\mathbf{p}}$ satisfy Fermi commutation rules. As in Sec. 8.1, we introduce new operators α and β defined by

$$\alpha = ua + vb^\dagger \quad \text{and} \quad \beta = ub - va^\dagger, \tag{14.65}$$

where u and v are real, and demand that they too satisfy Fermi commutation rules,

$$\{\alpha, \alpha^\dagger\} = \{\beta, \beta^\dagger\} = 1, \quad \{\alpha, \beta^\dagger\} = \{\beta, \alpha^\dagger\} = 0. \tag{14.66}$$

When (14.65) is inserted into (14.66) one obtains the condition

$$u^2 + v^2 = 1, \tag{14.67}$$

which differs from the analogous result (8.17) for bosons only by the sign of the v^2 term. Inverting the transformation (14.65), one finds

$$a = u\alpha - v\beta^\dagger, \quad \text{and} \quad b = u\beta + v\alpha^\dagger. \tag{14.68}$$

The subsequent manipulations follow closely those for bosons. We insert (14.68) into (14.64), and the result is

$$H = 2v^2 \epsilon_0 - 2uv\epsilon_1 + [\epsilon_0(u^2 - v^2) + 2uv\epsilon_1](\alpha^\dagger\alpha + \beta^\dagger\beta) \\ - [\epsilon_1(u^2 - v^2) - 2uv\epsilon_0](\alpha\beta + \beta^\dagger\alpha^\dagger). \tag{14.69}$$

To eliminate the term proportional to $\alpha\beta + \beta^\dagger\alpha^\dagger$, we choose u and v so that

$$\epsilon_1(u^2 - v^2) - 2uv\epsilon_0 = 0. \tag{14.70}$$

The condition (14.67) is satisfied by writing

$$u = \cos t, \quad v = \sin t, \tag{14.71}$$

where t is a parameter to be determined. Equation (14.70) then becomes

$$\epsilon_1(\cos^2 t - \sin^2 t) - 2\epsilon_0 \sin t \cos t = 0 \tag{14.72}$$

or

$$\tan 2t = \frac{\epsilon_1}{\epsilon_0}. \tag{14.73}$$

Solving for $u^2 - v^2 = \cos 2t$ and $2uv = \sin 2t$, one finds

$$u^2 - v^2 = \frac{\epsilon_0}{\sqrt{\epsilon_0^2 + \epsilon_1^2}} \quad \text{and} \quad 2uv = \frac{\epsilon_1}{\sqrt{\epsilon_0^2 + \epsilon_1^2}}, \tag{14.74}$$

which on insertion into Eq. (14.69) give the result

$$H = \epsilon(\alpha^\dagger\alpha + \beta^\dagger\beta) + \epsilon_0 - \epsilon, \tag{14.75}$$

where

$$\epsilon = \sqrt{\epsilon_0^2 + \epsilon_1^2}. \tag{14.76}$$

We choose the positive sign for the square root to ensure that $\alpha^\dagger$ and $\beta^\dagger$ create excitations with positive energy. The excitation energy (14.76) has the same form as for bosons, Eq. (8.25), except that the ϵ_1^2 term has the opposite sign.

To diagonalize the Hamiltonian (14.60) we therefore introduce the operators

$$\alpha_{\mathbf{p}} = u_{\mathbf{p}} a_{\mathbf{p}} + v_{\mathbf{p}} b^\dagger_{-\mathbf{p}} \quad \text{and} \quad \beta_{-\mathbf{p}} = u_{\mathbf{p}} b_{-\mathbf{p}} - v_{\mathbf{p}} a^\dagger_{\mathbf{p}}. \tag{14.77}$$

The normalization condition is

$$u_{\mathbf{p}}^2 + v_{\mathbf{p}}^2 = 1, \tag{14.78}$$

and terms of the type $\alpha\beta$ or $\beta^\dagger\alpha^\dagger$ in the Hamiltonian vanish if we choose

$$u_{\mathbf{p}} v_{\mathbf{p}} = \frac{\Delta_{\mathbf{p}}}{2\epsilon_{\mathbf{p}}}. \tag{14.79}$$

Here the excitation energy is given by

$$\epsilon_{\mathbf{p}}^2 = \Delta_{\mathbf{p}}^2 + \xi_{\mathbf{p}}^2, \tag{14.80}$$

where

$$\xi_{\mathbf{p}} = \epsilon_{\mathbf{p}}^0 - \mu. \tag{14.81}$$

Close to the Fermi surface, $\xi_{\mathbf{p}}$ is approximately $(p-p_F)v_F$, where $v_F = p_F/m$ is the Fermi velocity, and therefore the spectrum exhibits a gap Δ equal to $\Delta_{\mathbf{p}}$ for $p = p_F$. Excitations behave as free particles for $p - p_F \gg \Delta/v_F$. The momentum Δ/v_F corresponds in Fermi systems to the momentum ms at which the excitation spectrum for Bose systems changes from being sound-like to free-particle-like. The length scale $\hbar v_F/\Delta$, which depends on temperature, is the characteristic healing length for disturbances in the Fermi superfluid, and it is the analogue of the length ξ for bosons given by Eq. (6.62).

The expressions for the coefficients u and v are

$$u_{\mathbf{p}}^2 = \frac{1}{2}\left(1 + \frac{\xi_{\mathbf{p}}}{\epsilon_{\mathbf{p}}}\right) \quad \text{and} \quad v_{\mathbf{p}}^2 = \frac{1}{2}\left(1 - \frac{\xi_{\mathbf{p}}}{\epsilon_{\mathbf{p}}}\right). \tag{14.82}$$

The Hamiltonian (14.60) then assumes the form

$$K = \sum_{\mathbf{p}} \epsilon_{\mathbf{p}}(\alpha^\dagger_{\mathbf{p}}\alpha_{\mathbf{p}} + \beta^\dagger_{\mathbf{p}}\beta_{\mathbf{p}}) - \sum_{\mathbf{p}}(\epsilon_{\mathbf{p}} - \epsilon_{\mathbf{p}}^0 + \mu) - \frac{1}{V}\sum_{\mathbf{p}\mathbf{p}'} U(\mathbf{p},\mathbf{p}') C_{\mathbf{p}} C_{\mathbf{p}'}, \tag{14.83}$$

which describes non-interacting excitations with energy $\epsilon_{\mathbf{p}}$.

The gap equation

The gap parameter $\Delta_\mathbf{p}$ is determined by inserting the expression for $C_\mathbf{p}$ into Eq. (14.61). Calculating the average in Eq. (14.63) one finds

$$\begin{aligned} C_\mathbf{p} &= <(u_\mathbf{p}\beta_{-\mathbf{p}} + v_\mathbf{p}\alpha_\mathbf{p}^\dagger)(u_\mathbf{p}\alpha_\mathbf{p} - v_\mathbf{p}\beta^\dagger_{-\mathbf{p}})> \\ &= -[1 - 2f(\epsilon_\mathbf{p})]\frac{\Delta_\mathbf{p}}{2\epsilon_\mathbf{p}}, \end{aligned} \quad (14.84)$$

and therefore the equation (14.61) for the gap becomes

$$\Delta_\mathbf{p} = -\frac{1}{V}\sum_{\mathbf{p}'} U(\mathbf{p},\mathbf{p}')\frac{1-2f(\epsilon_{\mathbf{p}'})}{2\epsilon_{\mathbf{p}'}}\Delta_{\mathbf{p}'}, \quad (14.85)$$

since the thermal averages of the operators for the numbers of excitations are given by

$$<\alpha_\mathbf{p}^\dagger\alpha_\mathbf{p}> = <\beta_\mathbf{p}^\dagger\beta_\mathbf{p}> = f(\epsilon_\mathbf{p}) = \frac{1}{\exp(\epsilon_\mathbf{p}/kT)+1}. \quad (14.86)$$

At the transition temperature the gap vanishes, and therefore the excitation energy in the denominator in Eq. (14.85) may be replaced by the result for $\Delta_\mathbf{p} = 0$, that is $\epsilon_\mathbf{p} = |\epsilon_\mathbf{p}^0 - \mu|$. The equation then becomes

$$\begin{aligned} \Delta_\mathbf{p} &= -\frac{1}{V}\sum_{\mathbf{p}'} U(\mathbf{p},\mathbf{p}')\frac{1-2f(|\epsilon_{\mathbf{p}'}^0 - \mu|)}{2|\epsilon_{\mathbf{p}'}^0 - \mu|}\Delta_{\mathbf{p}'} \\ &= -\frac{1}{V}\sum_{\mathbf{p}'} U(\mathbf{p},\mathbf{p}')\frac{1-2f(\epsilon_{\mathbf{p}'}^0 - \mu)}{2(\epsilon_{\mathbf{p}'}^0 - \mu)}\Delta_{\mathbf{p}'}, \end{aligned} \quad (14.87)$$

where the latter form follows because $[1 - 2f(\epsilon)]/\epsilon$ is an even function of ϵ. The equation is identical with Eq. (14.36), which gives the temperature at which two-particle scattering in the normal state becomes singular. The formalism for the condensed state appears somewhat different from that for the normal one because in the condensed state it is conventional to work with excitations with positive energy, irrespective of whether the momentum of the excitation is above or below the Fermi surface. For the normal state, one usually works with particle-like excitations. These have positive energy relative to the Fermi energy above the Fermi surface, but negative energy below. The positive-energy elementary excitations of a normal Fermi system for momenta below the Fermi surface correspond to creation of a hole, that is, removal of a particle. As T_c is approached in the condensed state, the description naturally goes over to one in which excitations in the normal state are particles for $p > p_F$ and holes for $p < p_F$.

The gap at zero temperature

As an application of the formalism, we determine the gap at $T = 0$. Again it is convenient to express the bare interaction in terms of the T matrix for two-particle scattering in free space. Forgetting the effects of the induced interaction for a moment, one finds

$$\Delta_{\mathbf{p}} = -\frac{U_0}{V} \sum_{\mathbf{p}'} \left[\frac{1}{2\epsilon_{\mathbf{p}'}} - \frac{1}{2(\epsilon_{\mathbf{p}'}^0 - \mu)} \right] \Delta_{\mathbf{p}'}. \qquad (14.88)$$

First, we see that $\Delta_{\mathbf{p}}$ is independent of the direction of $\mathbf{p}$, and therefore it corresponds to pairing in an s-wave state. Second, since the main contributions to the integral now come from momenta of order p_F, we may replace the gap by its value at the Fermi surface, which we have denoted by Δ. The result is

$$\begin{aligned}
1 &= -\frac{U_0}{2V} \sum_{\mathbf{p}} \left[\frac{1}{(\xi_{\mathbf{p}}^2 + \Delta^2)^{1/2}} - \frac{1}{\xi_{\mathbf{p}}} \right] \\
&= -\frac{U_0 N(\epsilon_F)}{2} \int_0^\infty dx\, x^{1/2} \left[\frac{1}{[(x-1)^2 + (\Delta/\epsilon_F)^2]^{1/2}} - \frac{1}{x-1} \right].
\end{aligned} \qquad (14.89)$$

As we did when evaluating the integrals in the expression for T_c, we split the integral into two by writing $x^{1/2} = (x^{1/2} - 1) + 1$. The integral involving the first term is well behaved for small Δ and may be evaluated putting $\Delta = 0$. The second part of the integral may be evaluated directly, and since $\Delta/\epsilon_F \ll 1$, one finds

$$\Delta = \frac{8}{e^2} \epsilon_F e^{-1/N(\epsilon_F)|U_0|}. \qquad (14.90)$$

When the induced interaction is included, the gap is reduced by a factor $(4e)^{-1/3}$, as is the transition temperature, and thus

$$\Delta = \left(\frac{2}{e}\right)^{7/3} \epsilon_F e^{-1/N(\epsilon_F)|U_0|}. \qquad (14.91)$$

The ratio between the zero-temperature gap and the transition temperature is given by

$$\frac{\Delta(T=0)}{kT_c} = \frac{\pi}{\gamma} \approx 1.76 . \qquad (14.92)$$

The pairing interaction in dilute gases is different from the phonon-exchange interaction in metals. The latter extends only over an energy interval of order the Debye energy $\hbar\omega_D$ about the Fermi surface, whereas in dilute gases the interaction is of importance over a range of energies in excess of the Fermi energy. Often, when treating superconductivity in metals one adopts the BCS schematic model, in which there is an attractive interaction of constant strength for momenta such that $\epsilon_F - \hbar\omega_0 < \epsilon_\mathbf{p}^0 < \epsilon_F + \hbar\omega_0$, and one neglects the momentum dependence of the density of states. This yields an energy gap at zero temperature equal to $\Delta = 2\hbar\omega_0 \exp(-1/N(\epsilon_F)|U_0|)$. To obtain the same results in the schematic model with an effective interaction $\tilde{U}$ as from the full treatment given above, the cut-off energy must, according to (14.90), be chosen to be $\hbar\omega_0 = (4/e^2)\epsilon_F \approx 0.541\epsilon_F$. In the weak-coupling limit the relation (14.92) is a general result, which is also obtained within the BCS schematic model.

The reduction of the energy due to formation of the paired state may be calculated from the results above, and at zero temperature it is of order Δ^2/ϵ_F per particle, which is small compared with the energy per particle in the normal state, which is of order ϵ_F. Consequently, the total energy and pressure are relatively unaffected by the transition, since Δ/ϵ_F is expected to be small.

When the dimensionless coupling parameter $N(\epsilon_F)|U_0|$ becomes comparable with unity it is necessary to take into account higher order effects than those considered here. Such calculations predict that the maximum transition temperature will be significantly lower, and generally no more than about $0.02T_F$ [14]. However, we stress that at present no reliable theory exists for $N(\epsilon_F)|U_0| \sim 1$.

14.4 Boson–fermion mixtures

The experimental techniques that have been developed also open up the possibility of exploring properties of mixtures of bosons and fermions. These systems are the dilute analogues of liquid mixtures of ^{3}He and ^{4}He, and just as the interaction between two ^{3}He atoms is modified by the ^{4}He, the interaction between two fermions in a dilute gas is modified by the bosons. The calculation of T_c for two fermion species illustrated how induced interactions can have a marked effect. However, the modification of the effective interaction there was modest, because for fermions the interaction energy is generally small compared with the Fermi energy. For mixtures of fermions and bosons the effects can be much larger, because the boson gas is very compressible.

14.4.1 Induced interactions in mixtures

Consider a uniform mixture of a single fermion species of density n_F and bosons of density n_B at zero temperature. The energy per unit volume is given by

$$\mathcal{E} = \frac{1}{2}n_B^2 U_{BB} + n_B n_F U_{BF} + \frac{3}{5}\epsilon_F n_F, \quad (14.93)$$

where U_{BB} is the effective interaction between two bosons and U_{BF} is that between a boson and a fermion. Let us now calculate the change in energy when a long-wavelength, static density fluctuation is imposed on the densities of the two components, just as we did in deriving the stability condition for binary boson mixtures in Sec. 12.1.1. The contribution of second order in the density fluctuations is

$$\delta^2 E = \frac{1}{2}\int d\mathbf{r} \left[\frac{2\epsilon_F}{3n_F}(\delta n_F)^2 + U_{BB}(\delta n_B)^2 + 2U_{BF}\delta n_F \delta n_B\right]. \quad (14.94)$$

We wish to calculate the effect of the response of the bosons to the presence of the fluctuation in the fermion density. At long wavelengths the Thomas–Fermi approximation is valid, so in equilibrium the chemical potential μ_B of the bosons is constant in space. The chemical potential is given by

$$\mu_B = \frac{\partial \mathcal{E}}{\partial n_B} = U_{BB}n_B + U_{BF}n_F, \quad (14.95)$$

and therefore in equilibrium the boson and fermion density fluctuations are related by the expression

$$\delta n_B = -\frac{1}{U_{BB}}\delta V_B, \quad (14.96)$$

where

$$\delta V_B = U_{BF}\delta n_F \quad (14.97)$$

is the change in boson energy induced by the change in fermion density. Substituting this result into Eq. (14.94) one finds

$$\delta^2 E = \frac{1}{2}\int d\mathbf{r}\left(\frac{2\epsilon_F}{3n_F} - \frac{U_{BF}^2}{U_{BB}}\right)(\delta n_F)^2. \quad (14.98)$$

The second term shows that the response of the bosons leads to an extra contribution to the effective interaction between fermions which is of exactly the same form as we found in Sec. 12.1.1 for binary boson mixtures.

The origin of the induced interaction is that a fermion density fluctuation gives rise to a potential $U_{BF}\delta n_F$ acting on a boson. This creates a boson density fluctuation, which in turn leads to an extra potential $U_{BF}\delta n_B$ acting on a fermion. To generalize the above calculations to non-zero frequencies and wave numbers, we use the fact that the response of the bosons is given in general by

$$\delta n_B = \chi_B(\mathbf{q}, \omega) \delta V_B, \tag{14.99}$$

where $\chi_B(\mathbf{q}, \omega)$ is the density–density response function for the bosons. Thus the induced interaction is

$$U_{\text{ind}}(q, \omega) = U_{\text{BF}}^2 \chi_B(q, \omega). \tag{14.100}$$

This result is similar to those discussed in Sec. 14.3.2 for the induced interaction between two species of fermions. This interaction is analogous to the phonon-induced attraction between electrons in metals, and it is attractive at low frequencies, irrespective of the sign of the boson–fermion interaction. In a quantum-mechanical treatment, the wave number and frequency of the density fluctuation are related to the momentum transfer $\hbar \mathbf{q}$ and energy change $\hbar \omega$ of the fermion in a scattering process.

In the Bogoliubov approximation, the density–density response function of the Bose gas is given by (7.38)

$$\chi_B(q, \omega) = \frac{n_B q^2}{m_B (\omega^2 - \omega_q^2)}, \tag{14.101}$$

where the excitation frequencies are the Bogoliubov ones

$$\hbar \omega_q = \left[\epsilon_q^0 (\epsilon_q^0 + 2 n_B U_{\text{BB}}) \right]^{1/2}. \tag{14.102}$$

In the static limit, and for $q \to 0$ the response function is $\chi_B(q \to 0, 0) = -1/U_{\text{BB}}$, and therefore the effective interaction reduces to the result

$$U_{\text{ind}}(q \to 0, 0) = -\frac{U_{\text{BF}}^2}{U_{\text{BB}}}. \tag{14.103}$$

The static induced interaction for general wave numbers is

$$U_{\text{ind}}(q, 0) = -U_{\text{BF}}^2 \frac{n_B}{n_B U_{\text{BB}} + \hbar^2 q^2 / 4 m_B}. \tag{14.104}$$

In coordinate space this is a Yukawa, or screened Coulomb, interaction

$$U_{\text{ind}}(r) = -\frac{m_B n_B U_{\text{BF}}^2}{\pi \hbar^2} \frac{e^{-\sqrt{2} r / \xi}}{r}, \tag{14.105}$$

where ξ is the coherence (healing) length for the bosons, given by (6.62),

$$\xi^2 = \frac{\hbar^2}{2 m_B n_B U_{\text{BB}}}. \tag{14.106}$$

A noteworthy feature of the induced interaction is that at long wavelengths it is independent of the density of bosons. In addition, its value $-U_{\rm BF}^2/U_{\rm BB}$ is of the same order of magnitude as a typical bare interaction if the boson–boson and boson–fermion interactions are of comparable size. The reason for this is that even though the induced interaction involves two boson–fermion interactions, the response function for the bosons at long wavelengths is large, since it is inversely proportional to the boson–boson interaction. At wave numbers greater than the inverse of the coherence length for the bosons, the magnitude of the induced interaction is reduced, since the boson density–density response function for $q \gtrsim 1/\xi$ has a magnitude $\sim 2n_{\rm B}/\epsilon_q^0$ where $\epsilon_q^0 = (\hbar q)^2/2m_{\rm B}$ is the free boson energy. The induced interaction is thus strongest for momentum transfers less than $m_{\rm B} s_{\rm B}$, where $s_{\rm B} = (n_{\rm B} U_{\rm BB}/m_{\rm B})^{1/2}$ is the sound speed in the boson gas. For momentum transfers of order the Fermi momentum, the induced interaction is of order the 'diluteness parameter' $k_{\rm F} a$ times the direct interaction if bosons and fermions have comparable masses and densities, and the scattering lengths are comparable.

We have calculated the induced interaction between two identical fermions, but the mechanism also operates between two fermions of different species (for example two different hyperfine states) when mixed with bosons. When bosons are added to a mixture of two species of fermions, the induced interaction increases the transition temperature to a BCS superfluid state [12, 15]. The effect can be appreciable because of the strong induced interaction for small momentum transfers.

14.5 Collective modes of Fermi superfluids

One issue under current investigation is how to detect superfluidity of a Fermi gas experimentally. The density and momentum distributions of a trapped Fermi gas are very similar in the normal and superfluid states, because the energy difference between the two states is small. Consequently, the methods used in early experiments to provide evidence of condensation in dilute Bose gases cannot be employed.

One possibility is to measure low-lying oscillatory modes of the gas. In the normal state, properties of modes may be determined by expressions similar to those for bosons, apart from the difference of statistics. If interactions are unimportant, mode frequencies for a gas in a harmonic trap are sums of multiples of the oscillator frequencies, Eq. (11.28). When interactions are taken into account, modes will be damped and their frequencies shifted, as we have described for bosons in Secs. 11.2–11.3 and for fermions in Sec.

14.2.1. In a superfluid Fermi system, there are two sorts of excitations. One is the elementary fermionic excitations whose energies we calculated from microscopic theory in Sec. 14.3.3. Another class is collective modes of the condensate, which are bosonic degrees of freedom. These were not allowed for in the microscopic theory above because we assumed that the gas was spatially uniform. We now consider the nature of these modes at zero temperature, first from general considerations based on conservation laws, and then in terms of microscopic theory.

At zero temperature, a low-frequency collective mode cannot decay by formation of fermionic excitations because of the gap in their spectrum. For a boson to be able to decay into fermions, there must be an even number of the latter in the final state, and therefore the process is forbidden if the energy $\hbar\omega$ of the collective mode is less than twice the gap Δ. In addition, since no thermal excitations are present, the only relevant degrees of freedom are those of a perfect fluid, which are the local particle density n and the local velocity, which we denote by $\mathbf{v}_s$, since it corresponds to the velocity of the superfluid component. The equations of motion for these variables are the equation of continuity

$$\frac{\partial n}{\partial t} + \boldsymbol{\nabla}\cdot(n\mathbf{v}_s) = 0, \quad (14.107)$$

and the Euler equation (7.24), which when linearized is

$$\frac{\partial \mathbf{v}_s}{\partial t} = -\frac{1}{mn}\boldsymbol{\nabla}p - \frac{1}{m}\boldsymbol{\nabla}V = -\frac{1}{m}\boldsymbol{\nabla}\mu - \frac{1}{m}\boldsymbol{\nabla}V. \quad (14.108)$$

In deriving the second form we have used the Gibbs–Duhem relation at zero temperature, $dp = nd\mu$, to express small changes in the pressure p in terms of those in the chemical potential μ. Since this relation is valid also for a normal Fermi system in the hydrodynamic limit at temperatures low compared with the Fermi temperature, the equations for that case are precisely the same [16, 17].

We linearize Eq. (14.107), take its time derivative, and eliminate $\mathbf{v}_s$ by using Eq. (14.108), assuming the time dependence to be given by $\exp(-i\omega t)$. The result is

$$-m\omega^2 \delta n = \boldsymbol{\nabla}\cdot\left[n\boldsymbol{\nabla}\left(\frac{d\mu}{dn}\delta n\right)\right]. \quad (14.109)$$

This equation is the same as Eq. (7.59) for a Bose–Einstein condensate in the Thomas–Fermi approximation, the only difference being due to the specific form of the equation of state. For a dilute Bose gas, the energy density is given by $\mathcal{E} = n^2 U_0/2$ and $d\mu/dn = U_0$, while for a Fermi gas the effects of interactions on the total energy are small for $k_F|a| \ll 1$ and therefore we may use the results for an ideal gas, $\mathcal{E} = (3/5)n\epsilon_F \propto n^{5/3}$ and $d\mu/dn = 2\epsilon_F/3n$. The density profile is given by Eq. (14.12), and for an isotropic harmonic trap Eq. (14.109) may therefore be written as

$$-\omega^2 \delta n = \frac{\omega_0^2 R^2}{3} \nabla \cdot \left\{ (1 - r^2/R^2)^{3/2} \nabla \left[(1 - r^2/R^2)^{-1/2} \delta n \right] \right\}$$

$$= \omega_0^2 \left[1 - \frac{r}{3} \frac{\partial}{\partial r} + \frac{1}{3}(R^2 - r^2)\nabla^2 \right] \delta n. \quad (14.110)$$

The equation may be solved by the methods used in Sec. 7.3.1, and the result is [16, 17]

$$\delta n = Cr^l (1 - r^2/R^2)^{1/2} F(-n, l+n+2, l+3/2, r^2/R^2) Y_{lm}(\theta, \varphi), \quad (14.111)$$

where C is a constant and $R = 48^{1/6} N^{1/6} a_{\text{osc}}$ is the radius of the cloud for an isotropic trap (see Eq. (14.14)). Here F is the hypergeometric function (7.73). The mode frequencies are given by

$$\omega^2 = \omega_0^2 \left[l + \frac{4}{3} n(2 + l + n) \right], \quad (14.112)$$

where $n = 0, 1, \ldots$. Thus the frequencies of collective oscillations of a superfluid Fermi gas differ from those of a normal Fermi gas except in the hydrodynamic limit. Consequently, measurements of collective mode frequencies provide a way of detecting the superfluid transition, except when the hydrodynamic limit applies to the normal state. In addition, the damping of collective modes will be different for the normal and superfluid states.

We return now to microscopic theory. In Chapters 6 and 7 we saw that the wave function of the condensed state is a key quantity in the theory of condensed Bose systems. In microscopic theory, this is introduced as the c-number part or expectation value of the boson annihilation operator.[2] The analogous quantity for Fermi systems is the expectation value of the operator $\hat{\psi}_b(\mathbf{r} - \boldsymbol{\rho}/2)\hat{\psi}_a(\mathbf{r} + \boldsymbol{\rho}/2)$ that destroys two fermions, one of each species, at the points $\mathbf{r} \pm \boldsymbol{\rho}/2$. In the equilibrium state of the uniform system, the average of the Fourier transform of this quantity with respect to the relative coordinate corresponds to $C_\mathbf{p}$ in Eq. (14.59). If a Galilean transformation to a frame moving with velocity $-\mathbf{v}_s$ is performed on the system, the wave function is multiplied by a factor $\exp(im\mathbf{v}_s \cdot \sum_j \mathbf{r}_j/\hbar)$, where the sum is over

---

[2] By assuming that the expectation value is non-zero, we work implicitly with states that are not eigenstates of the particle number operator. However, we showed at the end of Sec. 8.1 how for Bose systems it is possible to work with states having a definite particle number, and similar arguments may also be made for fermions. For simplicity we shall when discussing fermions work with states in which the operator that destroys pairs of fermions has a non-zero expectation value, as we did in describing the microscopic theory.

all particles. Thus the momentum of each particle is boosted by an amount $m\mathbf{v}_s$ and $<\hat{\psi}_b(\mathbf{r}-\boldsymbol{\rho}/2)\hat{\psi}_a(\mathbf{r}+\boldsymbol{\rho}/2)>$ is multiplied by a factor $\exp i2\phi$, where $\phi = m\mathbf{v}_s \cdot \mathbf{r}/\hbar$. The velocity of the system is thus given by $\mathbf{v}_s = \hbar\boldsymbol{\nabla}\phi/m$, which has the same form as for a condensate of bosons, Eq. (7.14). An equivalent expression for the velocity is $\mathbf{v}_s = \hbar\boldsymbol{\nabla}\Phi/2m$, where $\Phi = 2\phi$ is the phase of $<\hat{\psi}_b(\mathbf{r}-\boldsymbol{\rho}/2)\hat{\psi}_a(\mathbf{r}+\boldsymbol{\rho}/2)>$. The change in the phase of the quantity $<\hat{\psi}_b(\mathbf{r}-\boldsymbol{\rho}/2)\hat{\psi}_a(\mathbf{r}+\boldsymbol{\rho}/2)>$ is independent of the relative coordinate $\boldsymbol{\rho}$, and henceforth we shall put $\boldsymbol{\rho}$ equal to zero.

We turn now to non-uniform systems, and we shall assume that the spatial inhomogeneities are on length scales greater than the coherence length $\hbar v_F/\Delta$. Under these conditions the system may be treated as being uniform locally, and therefore the natural generalization of the result above for the superfluid velocity is

$$\mathbf{v}_s(\mathbf{r}) = \frac{\hbar}{m}\boldsymbol{\nabla}\phi(\mathbf{r}), \qquad (14.113)$$

where $2\phi(\mathbf{r})$ is the phase of $<\hat{\psi}_b(\mathbf{r})\hat{\psi}_a(\mathbf{r})>$, and also the phase of the local value of the gap, given by Eq. (14.61), but with Δ and $C_\mathbf{p}$ both dependent on the centre-of-mass coordinate $\mathbf{r}$. For long-wavelength disturbances, the particle current density may be determined from Galilean invariance, and it is given by $\mathbf{j} = n(\mathbf{r})\mathbf{v}_s(\mathbf{r})$. Unlike dilute Bose systems at zero temperature, where the density is the squared modulus of the condensate wave function, the density of a Fermi system is not simply related to the average of the annihilation operator for pairs.

At non-zero temperature, thermal Fermi excitations are present, and the motion of the condensate is coupled to that of the excitations. In the hydrodynamic regime this gives rise to first-sound and second-sound modes, as in the case of Bose systems. The basic formalism describing the modes is the same as for Bose systems, but the expressions for the thermodynamic quantities entering are different.

Problems

PROBLEM 14.1 Determine the momentum distribution for a cloud of fermions in an anisotropic harmonic-oscillator potential at zero temperature and compare the result with that for a homogeneous Fermi gas.

PROBLEM 14.2 Consider a cloud of fermions in a harmonic trap at zero temperature. Determine the thickness of the region at the surface where the Thomas–Fermi approximation fails.

PROBLEM 14.3 By making a low-temperature expansion, show that the chemical potential of a single species of non-interacting fermions in a harmonic trap at low temperatures is given by

$$\mu \simeq \epsilon_\text{F} \left(1 - \frac{\pi^2}{3}\frac{T^2}{T_\text{F}^2}\right).$$

Determine the temperature dependence of the chemical potential in the classical limit, $T \gg T_\text{F}$. Plot the two limiting forms as functions of T/T_F and compare their values at $T/T_\text{F} = 1/2$.

PROBLEM 14.4 Verify the expression (14.21) for the temperature dependence of the energy at low temperatures. Carry out a high-temperature expansion, as was done for bosons in Sec. 2.4.2, and sketch the dependence of the energy and the specific heat as functions of temperature for all values of T/T_F.

PROBLEM 14.5 Consider a cloud containing equal numbers of two different spin states of the same atom in an isotropic harmonic-oscillator potential. Use the method of collective coordinates (Sec. 7.3.3) to show that the shift in the equilibrium radius due to interactions is given by

$$\Delta R \approx \frac{3}{8}\frac{E_\text{int}}{E_\text{osc}}R,$$

and evaluate this for the Thomas–Fermi density profile. Prove that the frequency of the breathing mode can be written in the form (14.27), and determine the value of the coefficient c_1.

References

[1] B. DeMarco and D. S. Jin, *Science* **285**, 1703 (1999); B. DeMarco, S. B. Papp, and D. S. Jin, *Phys. Rev. Lett.* **86**, 5409 (2001).

[2] A. G. Truscott, K. E. Strecker, W. I. McAlexander, G. B. Partridge, and R. G. Hulet, *Science* **291**, 2570 (2000); F. Schreck, G. Ferrari, K. L. Corwin, J. Cubizolles, L. Khaykovich, M.-O. Mewes, and C. Salomon, *Phys. Rev. A* **64**, 011402 (2001).

[3] D. A. Butts and D. S. Rokhsar, *Phys. Rev. A* **55**, 4346 (1997).

[4] L. Vichi and S. Stringari, *Phys. Rev. A* **60**, 4734 (1999).

[5] J. Bardeen, L. N. Cooper, and J. R. Schrieffer, *Phys. Rev.* **108**, 1175 (1957).

[6] H. T. C. Stoof, M. Houbiers, C. A. Sackett, and R. G. Hulet, *Phys. Rev. Lett.* **76**, 10 (1996).

[7] V. A. Khodel, V. V. Khodel, and J. W. Clark, *Nucl. Phys. A* **598**, 390 (1996).

[8] T. Papenbrock and G. F. Bertsch, *Phys. Rev. C* **59**, 2052 (1999).

[9] C. A. R. Sá de Melo, M. Randeira, and J. R. Engelbrecht, *Phys. Rev. Lett.* **71**, 3202 (1993); J. R. Engelbrecht, M. Randeira, and C. A. R. Sá de Melo, *Phys. Rev. B* **55**, 15 153 (1997).

[10] See, e.g., G. D. Mahan, *Many-Particle Physics*, (Plenum, New York, 1981), p. 778.

[11] L. P. Gorkov and T. K. Melik-Barkhudarov, *Zh. Eksp. Teor. Fiz.* **40**, 1452 (1961) [*Sov. Phys.-JETP* **13,** 1018 (1961)].

[12] H. Heiselberg, C. J. Pethick, H. Smith, and L. Viverit, *Phys. Rev. Lett.* **85**, 2418 (2000).

[13] See, e.g., N. W. Ashcroft and N. D. Mermin, *Solid State Physics*, (Holt, Rinehart, and Winston, New York, 1976), p. 343.

[14] R. Combescot, *Phys. Rev. Lett.* **83**, 3766 (1999).

[15] M. Bijlsma, B. A. Heringa, and H. T. C. Stoof, *Phys. Rev. A* **61**, 053601 (2000).

[16] G. M. Bruun and C. W. Clark, *Phys. Rev. Lett.* **83**, 5415 (1999).

[17] M. A. Baranov and D. S. Petrov, *Phys. Rev. A* **62**, 041601 (2000).

Appendix. Fundamental constants and conversion factors

Based on CODATA 1998 recommended values. (P. J. Mohr and B. N. Taylor, *Rev. Mod. Phys.* **72**, 351 (2000).) The digits in parentheses are the numerical value of the standard uncertainty of the quantity referred to the last figures of the quoted value. For example, the relative standard uncertainty in $\hbar$ is thus $82/1\,054\,571\,596 = 7.8 \times 10^{-8}$.

Quantity	Symbol	Numerical value	Units
Speed of light	c	$2.997\,924\,58 \times 10^{8}$	$\mathrm{m\,s^{-1}}$
		$2.997\,924\,58 \times 10^{10}$	$\mathrm{cm\,s^{-1}}$
Permeability of vacuum	μ_0	$4\pi \times 10^{-7}$	$\mathrm{N\,A^{-2}}$
Permittivity of vacuum	$\epsilon_0 = 1/\mu_0 c^2$	$8.854\,187\,817\ldots \times 10^{-12}$	$\mathrm{F\,m^{-1}}$
Planck constant	h	$6.626\,068\,76(52) \times 10^{-34}$	$\mathrm{J\,s}$
		$6.626\,068\,76(52) \times 10^{-27}$	$\mathrm{erg\,s}$
	hc	$1.239\,841\,857(49) \times 10^{-6}$	$\mathrm{eV\,m}$
(Planck constant)$/2\pi$	$\hbar$	$1.054\,571\,596(82) \times 10^{-34}$	$\mathrm{J\,s}$
		$1.054\,571\,596(82) \times 10^{-27}$	$\mathrm{erg\,s}$
Inverse Planck constant	h^{-1}	$2.417\,989\,491(95) \times 10^{14}$	$\mathrm{Hz\,eV^{-1}}$
Elementary charge	e	$1.602\,176\,462(63) \times 10^{-19}$	C
Electron mass	m_e	$9.109\,381\,88(72) \times 10^{-31}$	kg
		$9.109\,381\,88(72) \times 10^{-28}$	g
	$m_\mathrm{e} c^2$	$0.510\,998\,902(21)$	MeV
Proton mass	m_p	$1.672\,621\,58(13) \times 10^{-27}$	kg
		$1.672\,621\,58(13) \times 10^{-24}$	g
	$m_\mathrm{p} c^2$	$938.271\,998(38)$	MeV
Atomic mass unit	$m_\mathrm{u} = m(^{12}\mathrm{C})/12$	$1.660\,538\,73(13) \times 10^{-27}$	kg
	$m_\mathrm{u} c^2$	$931.494\,013(37)$	MeV

Appendix. Fundamental constants and conversion factors

Quantity	Symbol	Numerical value	Units
Boltzmann constant	k	$1.3806503(24) \times 10^{-23}$	J K^{-1}
		$1.3806503(24) \times 10^{-16}$	erg K^{-1}
		$8.617342(15) \times 10^{-5}$	eV K^{-1}
	k/h	$2.0836644(36) \times 10^{10}$	Hz K^{-1}
		$20.836644(36)$	Hz nK^{-1}
Inverse Boltzmann constant	k^{-1}	$11\,604.506(20)$	K eV^{-1}
Inverse fine structure constant	α_{fs}^{-1}	$137.03599976(50)$	
Bohr radius	a_0	$0.5291772083(19) \times 10^{-10}$	m
		$0.5291772083(19) \times 10^{-8}$	cm
Classical electron radius	$e^2/4\pi\epsilon_0 m_e c^2$	$2.817940285(31) \times 10^{-15}$	m
Atomic unit of energy	$e^2/4\pi\epsilon_0 a_0$	$27.2113834(11)$	eV
Bohr magneton	μ_{B}	$9.27400899(37) \times 10^{-24}$	J T^{-1}
	μ_{B}/h	$13.99624624(56) \times 10^{9}$	Hz T^{-1}
	μ_{B}/k	$0.6717131(12)$	K T^{-1}
Nuclear magneton	μ_{N}	$5.05078317(20) \times 10^{-27}$	J T^{-1}
	μ_{N}/h	$7.62259396(31) \times 10^{6}$	Hz T^{-1}
	μ_{N}/k	$3.6582638(64) \times 10^{-4}$	K T^{-1}

Index

absorption process 73–79
action principle 166
alkali atoms 1, 40–44, 51–53
angular momentum
    in a rotating trap 251
    of excited states 258
    of vortex state 243, 246–248
    operators 331, 336
    orbital 41
    spin 40–48
    total 41
angular velocity 251–253
    critical 251
anisotropy parameter 35, 182–185, 305, 306
annihilation operator 166, 204, 205, 213, 214
atomic number 40, 361
atomic structure 40–42
atomic units 43, 142
atom laser 354
attractive interaction 103, 119, 153, 177, 178, 199, 250, 260, 261

BCS theory 175, 370–385
Bernoulli equation 256
Bogoliubov dispersion relation 172, 175, 197, 387
Bogoliubov equations 174, 175, 215, 216
Bogoliubov transformation 207–209, 215, 380–382
Bohr magneton 42, 395
Bohr radius 43, 395
Boltzmann constant 3, 395
Boltzmann distribution 17, 29, 31
Boltzmann equation 86–88, 307–316
Born approximation 113, 114, 128
Bose distribution 8, 16–18, 27, 218, 268, 308
Bose–Einstein condensation 1
    criterion for 5, 17, 355–358
    in lower dimensions 23, 36
    theoretical prediction of 2
Bose–Einstein transition temperature 4, 5, 17, 21–23
    effect of finite particle number 35

effect of interactions 224, 292
Bose gas
    interacting 204–214
    non-interacting 16–38, 276
boson 40
    composite 335, 358
bound state 114, 116, 119, 132, 371
Bragg diffraction 352
breathing mode 182, 184, 186–193

canonical variables 271
central interaction 122
centre-of-mass motion 180
    in rotating gas 250, 259
centrifugal barrier 129, 241
cesium 41, 42, 52, 105, 139, 142–144, 236
channel 102, 120, 121, 131–139
    definition of 102
chemical potential 17, 18, 27–29, 32–36, 149, 166, 256, 271, 290, 363–366, 389
Cherenkov radiation 266
chirping 58
circulation 239, 244
    average 253
    conservation of 254
    quantum of 7, 239
classical electron radius 128, 395
Clebsch–Gordan coefficient 82, 83
clock shift 230, 236
closed shells 41, 43
clover-leaf trap 66, 67
coherence 338, 342, 348–353
coherence length 162, 170, 197, 226, 241, 243, 245, 387, 391
    definition of 162
coherent state 213
collapse 153, 369
collective coordinates 186–193
collective modes
    frequency shifts of 287
    in traps 178–195, 298–306
    of Fermi superfluids 388–390
    of mixtures 326–328

397

of uniform superfluids 273–279
collisional shift 230–236
collision integral 308
collision invariants 309
collisionless regime 306–317
collisions
    elastic 95, 107–120, 131, 132
    inelastic 93, 122–131, 139
    with foreign gas atoms 93
composite bosons 335, 358
condensate fraction 23, 24, 210, 295, 297
conservation law 171, 272
    of energy 195, 200, 273
    of entropy 274
    of momentum 168, 273
    of particle number 167, 171, 272, 302, 389
contact interaction 146, 206
continuity equation 168
cooling time 86, 88, 89
covalent bonding 103
creation operator
    for elementary excitations 209, 210, 382
    for particles 204
cross section 94, 108–110, 125, 308
    for identical bosons 110, 140
    for identical fermions 110
cryogenic cooling 97

d'Alembert's paradox 257
damping 280–287, 315–318
    Beliaev 281
    collisional 315–318
    Landau 281–287
de Broglie wavelength 4
decay 56, 70–73
    of excited state 56, 70
    of mode amplitude 281–283
    of oscillating cloud 307, 315–318
    of temperature anisotropies 307, 310–314
Debye energy 384
Debye temperature 1
degeneracy 218, 258, 316
density correlations 174, 348–350
density depression in soliton, 197–201
density distribution
    for fermions 364
    for trapped bosons 24–26
    in mixtures 324–326
density of particles 1
density of states 18–20, 36
    for particle in a box 19, 20, 363
    for particle in a harmonic trap 20, 364
    per unit volume 363
density profile, *see* density distribution
depletion of the condensate 2, 147, 210–212, 333
    thermal 219, 227, 229, 230, 294–298
detuning 70–72, 74, 77
    blue 70, 77
    red 70, 74, 77, 79
diffusion coefficient 85, 89

dipolar losses 93
dipole approximation 49, 56, 67, 98
dipole–dipole interaction
    electric 106, 107, 140, 141
    magnetic 41, 93, 123, 126–130
dipole force 72
dipole moment, electric 49–51, 67–72
dispersion relation 7, 177, 194
dissipation 201, 274
distribution function 17, 27, 86–88, 94, 308, 366
Doppler effect 58, 74–78
doubly polarized state 47, 63, 93, 127
drag force 257
dressed atom picture 71

effective interaction, *see* interaction, effective
Einstein summation convention 255
electric field
    oscillating 53–54, 68–73
    static 49–53
electron mass 394
electron pairs 8, 370
elementary charge 43, 394
elementary excitations 7, 8, 71, 171–178, 382
    in a rotating gas 259
    in a trapped gas 174–175, 214–218
    in a uniform gas 171–178, 205–214
energy density 166, 213, 271, 273, 294, 322, 323, 386, 389
energy flux density 273
energy gap 7, 9, 177, 225, 382–385
energy scales 55–57, 290
entangled state 233
entropy 30, 31, 219, 223–225, 274
    per unit mass 274–278
equipartition theorem 31
Euler equation 169, 257, 302, 389
evaporative cooling 3, 90–96
exchange 220–222
exchange interaction 126, 329
excitons 9

f sum rule 51
Fermi energy 363
Fermi function 366
fermions 4, 40, 41, 361–392
Fermi's Golden Rule 125, 281
Fermi temperature 1, 361
    definition of 363
Fermi wave number 364
Feshbach resonance 73, 102, 103, 105, 131–139, 142–144, 212, 362
fine structure constant 43, 128, 129, 395
finite particle number 35
fluctuations 36, 205, 294
Fock state 343–348
Fourier transformation 111
four-wave mixing 354
fragmented condensates 357, 358
Fraunhofer lines 51

free expansion 195, 196, 340
friction coefficient 75, 85
fugacity 17, 18, 28
functional derivative 169

g factor 43, 47, 57
Galilean invariance 268, 271
Galilean transformation 265, 268, 352, 390
gamma function 21, 22, 30
gap equation 383–385
Gaussian density distribution 151
Gibbs–Duhem relation 169, 274, 389
grating 352–354
gravity 63, 64, 194
gravity wave 194
Green function 135, 374
Gross–Pitaevskii equation 146
    for single vortex 237
    time-dependent 165–171, 326, 332
    time-independent 148, 149, 228, 322
ground state
    for harmonic oscillator 25
    for spinor condensate 334
    for uniform Bose gas 210
ground-state energy
    for non-interacting fermions 361–365
    for trapped bosons 148–156
    for uniform Bose gas 148, 210–213

harmonic-oscillator potential 19
    energy levels of 20
harmonic trap 19
Hartree approximation 146, 147, 206
Hartree–Fock theory 206, 219–225
    and thermodynamic properties 295
healing length, see coherence length
Heisenberg uncertainty principle 89, 196
helium
    liquid ^{3}He 6, 9, 283, 333, 378, 385
    liquid ^{4}He 2, 6–8, 36, 173, 218 239, 243, 266, 280, 283, 287, 385
    metastable ^{4}He atoms 4, 41
Helmholtz coils 60, 64
high-field seeker 60, 61, 97
hole 372
hydrodynamic equations 167–171
    for perfect fluid 169
    of superfluids 273–280
    validity of 301, 302
hydrogen 3, 36, 40–43, 48–50, 52, 53, 55–57, 93, 96–99, 105, 107, 126, 129–131, 142, 144, 204, 230, 236
    atomic properties 41, 52
    hyperfine interaction 40–49
    line shift in 98
    spin-polarized 3, 37, 96–99
hydrostatic equilibrium 185
hyperfine interaction 41–49
hyperfine states 41–49, 80, 320, 328
hypergeometric function 181, 390

imaging 24
    absorptive 24
    phase-contrast 24
inelastic processes, see collisions, inelastic
instability 153, 154, 177, 178, 209
    in mixtures 322–324, 328, 369
interaction 2
    antiferromagnetic 332
    between excitations 280–287
    between fermions 362, 366–369
    effective 111–114, 132–139, 212, 213, 222, 279
    ferromagnetic 332
    induced 324, 376–378, 385–388
    interatomic 103–107
interaction energy
    for weak coupling 217
    in Hartree–Fock theory 221
interference
    of electromagnetic waves 339, 340
    of two condensates 338–348
internal energy 30–33
internal states 120–122
Ioffe bars 65, 66
Ioffe–Pritchard trap 64–67, 97
irrotational flow 168, 238

Josephson relation 166, 169

Kelvin's theorem 254
kinetic theory 86–88, 272, 307–317
kink 198
Kosterlitz–Thouless transition 36

Lagrange equation 191
Lagrange multiplier 148
Lagrangian 166, 191
lambda point 4, 6, 280
Landau criterion 265–267
Landau damping 281–287
Laplace equation 64, 180, 194
laser beams 78–79
    circularly polarized 78
laser cooling 1, 55, 74–78, 71–90, 96, 97
Legendre polynomials 64, 108
length scales
    of anisotropic cloud 153, 156
    of spatial variation 162
    of spherical cloud 150
    of surface structure 158, 159
lifetime
    of cold atomic clouds 24
    of excited state 67, 70, 71
lift 257
linear ramp potential 158, 193
line shift 98, 230–236
linewidth 55–57, 70, 71, 76–78, 81
Lippmann–Schwinger equation 112, 121, 372
liquid helium, see helium
lithium 1, 6, 40–42, 52, 53, 105, 131, 139, 142, 143, 151, 361, 362
Lorentzian 71, 74

low-field seeker 60, 61, 97
Lyman-α line 52, 56, 97, 98

macroscopic occupation 18
magnetic bottle 65
magnetic field 44–49, 55, 58–67, 73
  critical 252
magnetic moment 40, 41, 43, 44, 59, 60, 138
  nuclear 40, 41
magnetic trap 38, 59–67, 73, 97
magneto-optical trap (MOT) 58, 78–80
  dark-spot MOT 80
Magnus force 257
mass number 25, 40, 81, 361
matter waves 350–355
maximally stretched state 47, 93, 127, 142, 143
Maxwellian distribution 17
Maxwell relation 276
mean field 2, 11, 146, 307, 330, 331
mean free path 182, 301
mean occupation number 17, 27
mixtures 320–328, 361, 385–388
momentum density 167, 267–270, 278
momentum distribution 25, 26
momentum flux density 255, 302
MOT, see magneto-optical trap

neutron number 41, 361
neutron scattering 2
normal component 267
normal density 7, 269
normalization conditions 107, 111, 134, 147, 148, 176, 208, 210, 355
normal modes, see collective modes
nuclear magnetic moment 40, 41
nuclear magneton 41, 395
nuclear spin 40, 41

occupation number 17, 220–222, 270
one-particle density matrix 355–358
optical lattice 84, 90
optical path length 24
optical pumping 80
optical trap 73, 74, 320, 328
oscillator strength 51–54, 56, 73, 106

pair distribution function 349
pairing 8, 9
  in atomic gases 370–385
  in liquid ^{3}He 9
  in neutron stars 9
  in nuclei 9
  in superconductors 8, 370
parity 50
particle number
  operator for 207, 210
  states with definite 213, 390
Pauli exclusion principle 362, 372
permeability of vacuum 394
permittivity of vacuum 394
phase imprinting 351, 352

phase shift 109
phase space 19, 27
phase-space density 22, 96, 361
phase state 341–348
phase velocity 266
phonons 7, 201, 279–281, 287
  in solids 1, 370
photoassociative spectroscopy 103, 140–143
Planck constant 4, 394
plane wave 17
  expansion in spherical waves 128
plasmas 60, 65, 281
polarizability 49
  dynamic 53–54, 68–73
  static 49–53, 56, 107
Popov approximation 225–230, 298
potassium 41–43, 52, 105, 143, 361, 362
potential flow 168, 170, 238
pressure 31, 169, 273
  in ideal fluid 255
  quantum 170, 271
projection operators 122, 133, 329
propagator, see Green function
proton mass 394
proton number 41
pseudopotential 114, 235, 370
pumping time 84–88

quadrupole trap 38, 60–64, 72
quark–antiquark pairs 9
quasi-classical approximation 43
quasiparticle 71

Rabi frequency 71
radiation pressure 73, 78
random phase approximation 235
recoil energy 81, 90
reduced mass 107
resonance line 51, 52, 55, 107
retardation 107
Riemann zeta function 21, 22
rotating traps 251–254
roton 7, 8, 173, 266
rubidium 1, 5, 6, 40–42, 44, 52, 104, 105, 127, 131, 139, 142, 143, 212, 236, 253, 254, 291, 298, 320

s-wave scattering 108
scattering
  as a multi-channel problem 118
  basic theory of 107–114
  real 230
  virtual 230
scattering amplitude 106–110, 121
  Breit–Wigner form of 139
  for identical particles 110
scattering length
  definition of 106
  for alkali atoms 102, 142–144
  for a r^{-6}-potential 114–120
  for hydrogen 98, 142
  scale of 105

scattering theory 107–140
Schrödinger equation 133, 149, 166
    for relative motion 108, 115
scissors mode 184–186, 305
second quantization 220, 221
semi-classical approximation 16, 27–29, 43, 120, 228–230
Sisyphus 85
Sisyphus cooling 81–90, 97
singlet
    ground state 334
    potential 104, 122
sodium 1, 6, 40–42, 51, 52, 55, 57, 58, 74, 77, 105, 107, 131, 139, 143, 254, 328, 354
solitary wave 197
solitons 196–201, 351, 352
    bright 200
    dark 199
    energy of 200
    velocity of 199
sound 7, 172, 173, 200, 273–280
    first 226, 277, 278, 391
    second 278, 391
    velocity of 172, 210, 219, 226, 275, 304
specific heat 30–34, 38, 219
speed of light 394
spherical harmonics 64, 123
spherical tensor 123
spin 4
    electronic 40
    nuclear 40–48
    operators 123, 331
    total 40–42
spin-exchange collisions 97, 122, 126–127, 129
spinor 331
    condensates 328–335
spin–orbit interaction 51, 55, 130
spin waves 332
spontaneous emission 56
stability 150, 322, 369
    of mixtures 322–324
Stern–Gerlach experiment 347
stimulated emission 56
strong-coupling limit 150
superconductors 8, 9, 252, 370, 385
superfluid component 267
superfluid density 7, 269
superfluid helium, see helium
superfluidity 264–287, 370–385, 388–391
surface structure 158–161
surface tension 195
surface waves 180, 183, 193–195, 249, 250
sympathetic cooling 362

T matrix 112–114, 134–138, 371–374
Tartarus 85
temperature wave 276
thermal de Broglie wavelength 5
thermodynamic equilibrium 170
thermodynamic properties
    of interacting gas 218, 219, 294–298

of non-interacting gas 29–34
Thomas–Fermi approximation 155–157, 228, 229, 245, 246, 342, 364, 365
three-body processes 93, 130, 131, 350
threshold energy 94, 121
TOP trap 62–64
transition temperature 4
    Bose–Einstein, see Bose–Einstein transition temperature
    for pairing of fermions 370–379
trap frequency 25
trap loss 61, 92–96, 141
traps
    Ioffe–Pritchard 64–67
    magnetic 59–67, 320
    optical 73, 74, 320
    TOP 62–64
triplet potential 104, 122
triplet state 4
two-component condensates 253, 321–328
two-fluid model 7, 267–270
two-photon absorption 98, 99, 230, 233

uncertainty principle 26, 89, 196

vacuum permeability 394
vacuum permittivity 43, 394
van der Waals coefficients, table of 105
van der Waals interaction 103–107
variational calculation 148, 151–154, 191
variational principle 166, 190
velocity
    critical 266
    mean relative 95
    mean thermal 95, 313
    of condensate 167, 268–271, 274, 391
    of expanding cloud 196
    of normal component 268, 274, 302
    of sound, see sound velocity
velocity distribution 24
virial theorem 100, 157, 163
viscosity 311–313
viscous relaxation time 312
vortex 168, 239
    angular momentum of 243, 246, 247
    energy of 240–243, 246, 247
    force on a 255–257
    in trapped cloud 245–247
    lattice 252
    multiply-quantized 244, 245
    off-axis 247, 248
    quantized 7, 239

water waves 194, 197
wave function
    as product of single-particle states 147, 321
    in Hartree–Fock theory 220
    of condensed state 148, 205
weak coupling 152, 216–218, 257–261
WKB approximation 43, 120

Yukawa interaction 387

Zeeman effect 44–49, 55
Zeeman energy 55
Zeeman slower 58, 78

zero-momentum state 2, 205
  occupancy of 2
zero-point kinetic energy 187
zero-point motion 20, 35
zeta function 21, 22